Enzymatic Bioelectrocatalysis

Enzymatic Bioelectrocatalysis

Editors

Elisabeth Lojou
Xinxin Xiao

MDPI • Basel • Beijing • Wuhan • Barcelona • Belgrade • Manchester • Tokyo • Cluj • Tianjin

Editors
Elisabeth Lojou
BIP-CNRS-GLM
Aix Marseille University
Marseille
France

Xinxin Xiao
Department of Chemistry
Technical University of Denmark
Lyngby
Denmark

Editorial Office
MDPI
St. Alban-Anlage 66
4052 Basel, Switzerland

This is a reprint of articles from the Special Issue published online in the open access journal *Catalysts* (ISSN 2073-4344) (available at: www.mdpi.com/journal/catalysts/special_issues/enzy_bioelect).

For citation purposes, cite each article independently as indicated on the article page online and as indicated below:

LastName, A.A.; LastName, B.B.; LastName, C.C. Article Title. *Journal Name* **Year**, *Volume Number*, Page Range.

ISBN 978-3-0365-3460-2 (Hbk)
ISBN 978-3-0365-3459-6 (PDF)

© 2022 by the authors. Articles in this book are Open Access and distributed under the Creative Commons Attribution (CC BY) license, which allows users to download, copy and build upon published articles, as long as the author and publisher are properly credited, which ensures maximum dissemination and a wider impact of our publications.

The book as a whole is distributed by MDPI under the terms and conditions of the Creative Commons license CC BY-NC-ND.

Contents

About the Editors vii

Elisabeth Lojou and Xinxin Xiao
Enzymatic Bioelectrocatalysis
Reprinted from: *Catalysts* **2021**, *11*, 1373, doi:10.3390/catal11111373 1

Charlène Beaufils, Hiu-Mun Man, Anne de Poulpiquet, Ievgen Mazurenko and Elisabeth Lojou
From Enzyme Stability to Enzymatic Bioelectrode Stabilization Processes
Reprinted from: *Catalysts* **2021**, *11*, 497, doi:10.3390/catal11040497 3

Hongqi Xia and Jiwu Zeng
Rational Surface Modification of Carbon Nanomaterials for Improved Direct Electron Transfer-Type Bioelectrocatalysis of Redox Enzymes
Reprinted from: *Catalysts* **2020**, *10*, 1447, doi:10.3390/catal10121447 49

Xiaomei Yan, Jing Tang, David Tanner, Jens Ulstrup and Xinxin Xiao
Direct Electrochemical Enzyme Electron Transfer on Electrodes Modified by Self-Assembled Molecular Monolayers
Reprinted from: *Catalysts* **2020**, *10*, 1458, doi:10.3390/catal10121458 65

Huijie Zhang, Rosa Catania and Lars J. C. Jeuken
Membrane Protein Modified Electrodes in Bioelectrocatalysis
Reprinted from: *Catalysts* **2020**, *10*, 1427, doi:10.3390/catal10121427 91

Taiki Adachi, Yuki Kitazumi, Osamu Shirai and Kenji Kano
Recent Progress in Applications of Enzymatic Bioelectrocatalysis
Reprinted from: *Catalysts* **2020**, *10*, 1413, doi:10.3390/catal10121413 121

Simin Arshi, Mehran Nozari-Asbemarz and Edmond Magner
Enzymatic Bioreactors: An Electrochemical Perspective
Reprinted from: *Catalysts* **2020**, *10*, 1232, doi:10.3390/catal10111232 141

About the Editors

Elisabeth Lojou

Elisabeth Lojou is a senior researcher in the Bioenergetic and Protein Engineering Laboratory (Aix Marseille University, CNRS) in France. Her group conducts research in the domain of fundamental bioelectrochemistry, with a special interest in electron transfers between functionalized nanostructured electrodes and redox enzymes, such as hydrogenases and multicopper oxidases. For 8 years, the group has been engaged in the design of H_2/O_2 enzymatic fuel cells.

Xinxin Xiao

Xinxin Xiao achieved his PhD in 2019 at the University of Limerick, Ireland. Currently, he is a researcher at the Technical University of Denmark. His research focuses on bioelectrochemistry, and especially electrochemical sensors and electrochemical energy harvesters.

Editorial

Enzymatic Bioelectrocatalysis

Elisabeth Lojou [1,*] and Xinxin Xiao [2,*]

1. National Center for Scientific Research (CNRS), Aix Marseille University, BIP, UMR 7281, 31 Chemin Aiguier, 13009 Marseille, France
2. Department of Chemistry, Technical University of Denmark, 2800 Lyngby, Denmark
* Correspondence: lojou@imm.cnrs.fr (E.L.); xixiao@kemi.dtu.dk (X.X.)

Citation: Lojou, E.; Xiao, X. Enzymatic Bioelectrocatalysis. *Catalysts* **2021**, *11*, 1373. https://doi.org/10.3390/catal11111373

Received: 9 November 2021
Accepted: 12 November 2021
Published: 14 November 2021

Publisher's Note: MDPI stays neutral with regard to jurisdictional claims in published maps and institutional affiliations.

Copyright: © 2021 by the authors. Licensee MDPI, Basel, Switzerland. This article is an open access article distributed under the terms and conditions of the Creative Commons Attribution (CC BY) license (https://creativecommons.org/licenses/by/4.0/).

Enzymatic bioelectrocatalysis relies on immobilizing oxidoreductases on electrode surfaces, leading to different applications, such as biosensors [1], biofuel cells [2], and bioelectrosynthesis [3]. Based on their intrinsic properties, i.e., high specificity and high affinity for the substrate, enzymes may provide sustainable alternatives to currently used chemical catalysts for human health monitoring, biopower generation or high-value product synthesis. However, enzyme bioelectrocatalysis suffers from low catalytic efficiency, imposing fundamental investigations on mechanisms of enzyme immobilization, including molecular basis knowledge of the efficient electronic communication between enzymes and the electrode. Bioelectrocatalysis is also limited by long-term stability requiring groundbreaking strategies for enzyme protection and bioelectrode survival.

The "Enzymatic Bioelectrocatalysis" Special Issue comprises six reviews contributed by research groups from different countries, covering fundamentals and applications, as well as the recent research progress in this field.

Enzyme immobilization and electron transfer mechanisms are two crucial and closely interrelated aspects, which ultimately determine the stability and efficiency of the bioelectrode. On the basis of understanding the parameters governing protein stability, Beaufils et al. discuss the major strategies to improve redox enzyme stability, with a focus on the immobilization as an important route [4]. The authors further discuss additional factors specific for bioelectrocatalysis, such as enzyme reorientation, effect of the electric field and protection against reactive oxygen species (ROS) production, that should be considered to achieve highly stable bioelectrodes undergoing electron transfer. For future directions, they emphasize the need to screen the diversity for the discovering of new outstanding enzymes with enhanced stability, as well as the requirement of in situ and in operando methodologies to get new insights on enzyme behavior in the immobilized state.

Gold and carbon electrodes represent two key materials for enzyme immobilization, which are respectively reviewed by Xia et al. [5] and Yan et al. [6]. Strategies used to rationalize surface modification for high-performance direct electron transfer (DET) of redox enzymes are especially discussed. Yan et al. emphasizes the employment of self-assembled monolayers (SAM) as tunable bridges between redox enzymes and gold electrode surfaces permitting DET [6]. The authors overview the characterization methods of SAMs and structural properties of common enzymes, highlighting the strategic selection of a specific SAM to control proper enzyme orientation. Xia et al. highlight the key parameters allowing enzymatic DET on carbon nanomaterials, and review the various methods for the oriented immobilization. Interestingly, they also present the tools currently developed to probe redox enzymes [5].

Membrane proteins, constituting 20–30% of all proteins secreted by living organisms, are a major subject of bioelectrocatalysis. In the review by Zhang et al., various and eventually coupled techniques, i.e., electrochemistry, spectroscopy, microscopy, and quartz crystal microbalance [7] are discussed toward the understanding and use of membrane enzymes active in bioenergy conversion. Electrode designs with a special focus on the specificity required for membrane proteins are highlighted. Two emerging directions: (i) the membrane protein based hybrid vesicles for improved lifetime and (ii) microorganisms for microbial electrosynthesis and semi-artificial photosynthesis, are emphasized for future research.

Regarding to the applications of enzymatic bioelectrocatalysis, Adachi et al. describe the recent progress in emerging bioelectrochemical fields such as biosupercapacitor, bioelectrosynthesis and photo-bioelectrocatalysis [8]. The authors claim how crucial further research on protein-engineering, rational selection of electrode materials and mediators, immobilization of enzymes, and layout of electrodes to improve bioelectrocatalysis will be. The review by Arshi et al. focuses on the development of electrochemically based enzymatic reactors [9]. Together with the discussion on the mechanisms of electron transfer involving immobilized enzyme, the authors review the usage of electrochemically based batch and flow bio-reactors. They highlight the importance of high surface area electrodes, enzyme engineering and enzyme cascades, joining the general opinion formulated in the review by Adachi et al. [8]. Finally, both reviews provide some examples of enzyme-based electrosynthesis especially relevant in a sustainable world.

The common message conveyed from all the contributions of this special issue is that enzymatic electrochemistry is expected to play an increasingly role towards electrocatalysis in mild conditions. We look forward to further new developments in this exciting field.

Conflicts of Interest: The authors declare no conflict of interest.

References

1. Schachinger, F.; Chang, H.; Scheiblbrandner, S.; Ludwig, R. Amperometric Biosensors Based on Direct Electron Transfer Enzymes. *Molecules* **2021**, *26*, 4525. [CrossRef] [PubMed]
2. Xiao, X.; Xia, H.-q.; Wu, R.; Bai, L.; Yan, L.; Magner, E.; Cosnier, S.; Lojou, E.; Zhu, Z.; Liu, A. Tackling the Challenges of Enzymatic (Bio)Fuel Cells. *Chem. Rev.* **2019**, *119*, 9509–9558. [CrossRef] [PubMed]
3. Chen, H.; Simoska, O.; Lim, K.; Grattieri, M.; Yuan, M.; Dong, F.; Lee, Y.S.; Beaver, K.; Weliwatte, S.; Gaffney, E.M.; et al. Fundamentals, Applications, and Future Directions of Bioelectrocatalysis. *Chem. Rev.* **2020**, *120*, 12903–12993. [CrossRef] [PubMed]
4. Beaufils, C.; Man, H.-M.; de Poulpiquet, A.; Mazurenko, I.; Lojou, E. From Enzyme Stability to Enzymatic Bioelectrode Stabilization Processes. *Catalysts* **2021**, *11*, 497. [CrossRef]
5. Xia, H.; Zeng, J. Rational Surface Modification of Carbon Nanomaterials for Improved Direct Electron Transfer-Type Bioelectrocatalysis of Redox Enzymes. *Catalysts* **2020**, *10*, 1447. [CrossRef]
6. Yan, X.; Tang, J.; Tanner, D.; Ulstrup, J.; Xiao, X. Direct Electrochemical Enzyme Electron Transfer on Electrodes Modified by Self-Assembled Molecular Monolayers. *Catalysts* **2020**, *10*, 1458. [CrossRef]
7. Zhang, H.; Catania, R.; Jeuken, L.J.C. Membrane Protein Modified Electrodes in Bioelectrocatalysis. *Catalysts* **2020**, *10*, 1427. [CrossRef]
8. Adachi, T.; Kitazumi, Y.; Shirai, O.; Kano, K. Recent Progress in Applications of Enzymatic Bioelectrocatalysis. *Catalysts* **2020**, *10*, 1413. [CrossRef]
9. Arshi, S.; Nozari-Asbemarz, M.; Magner, E. Enzymatic Bioreactors: An Electrochemical Perspective. *Catalysts* **2020**, *10*, 1232. [CrossRef]

Review

From Enzyme Stability to Enzymatic Bioelectrode Stabilization Processes

Charlène Beaufils, Hiu-Mun Man, Anne de Poulpiquet, Ievgen Mazurenko and Elisabeth Lojou *

Aix Marseille University, CNRS, BIP, Bioénergétique et Ingénierie des Protéines, UMR 7281, 31, Chemin Joseph Aiguier, CS 70071 13402 Marseille CEDEX 09, France; cbeaufils@imm.cnrs.fr (C.B.); hman@imm.cnrs.fr (H.-M.M.); adepoulpiquet@imm.cnrs.fr (A.d.P.); imazurenko@imm.cnrs.fr (I.M.)
* Correspondence: lojou@imm.cnrs.fr; Tel.: +33-(0)491-164-144

Abstract: Bioelectrocatalysis using redox enzymes appears as a sustainable way for biosensing, electricity production, or biosynthesis of fine products. Despite advances in the knowledge of parameters that drive the efficiency of enzymatic electrocatalysis, the weak stability of bioelectrodes prevents large scale development of bioelectrocatalysis. In this review, starting from the understanding of the parameters that drive protein instability, we will discuss the main strategies available to improve all enzyme stability, including use of chemicals, protein engineering and immobilization. Considering in a second step the additional requirements for use of redox enzymes, we will evaluate how far these general strategies can be applied to bioelectrocatalysis.

Keywords: enzyme; metalloenzyme; catalysis; stability; electrochemistry; bioelectrochemistry

1. Introduction

In the search for a more sustainable way of life, our society must adapt nowadays to reduce the use of harmful products or high carbon energy sources deleterious for our planet, hence for our lives. Many industrial sectors are concerned by this revolution, from fine product synthesis to environmental monitoring, clinical diagnosis or energy production. Catalysis of the involved chemical reactions is a key process towards sustainability by accelerating kinetics, or increasing selectivity. However, chemical catalysts themselves are very often not produced and used in eco-friendly manner, requiring organic solvents for their synthesis or being based on metals rare on earth. In that sense, enzymes can be regarded as sustainable alternatives with great advantages directly linked to their intrinsic properties required to sustain life [1] In particular, many different enzymes are involved in a variety of reactions in microorganisms where they operate in mild conditions, developing high catalytic activity and specificity for their substrate. Their amazing functional diversity directly originates from the chemistry and polarity of amino acids that fold in a diversity of structures. In addition, they are produced in mild aqueous conditions and are totally biodegradable. However, their weak stability and ability to maintain biological activity during storage and operation, freeze-thaw steps, and in non-physiological environments lower their attractiveness. Development of strategies to enhance enzyme shelf life while maintaining catalytic activity have been the subject of many researches during the last decades [2–4].

Redox enzymes belong to a class of enzymes containing redox cofactors that are necessary for their activity. These oxidoreductases transform their substrate by exchanging electrons with their physiological partners in the metabolic chain. Intra-enzyme redox components can be organic cofactors such as flavin adenine dinucleotide (FAD) for sugar oxidation [5]. They can be metal centers, including copper for oxygen reduction [6], iron and nickel for hydrogen evolution and uptake or CO_2 reduction [7–9], manganese for water oxidation in photosynthesis [10], etc. Some others are dependent on nicotinamide adenine dinucleotide phosphate (NADPH) to realize enzymatic transformations [11]. Such enzymes

have been envisioned as biocatalysts in biosensors [12–14], biosynthesis reactors [15,16], or biofuel cells [17,18]. This domain is referred as Bioelectrochemistry. As for non-redox enzymes, one of the main limitations in the large-scale development of these biotechnologies is the weak stability of bioelectrodes based on redox enzymes. In particular, in the different biotechnological devices listed just before, enzymes may face high temperatures or high salt concentrations. They may have to operate at gas-liquid interfaces or even at the three phase boundaries. In addition, to exploit the catalytic properties of these redox enzymes in biotechnology, a further step is to ensure electron exchange between the protein and a conductive surface [19–21]. This prerequisite imposes the enzyme adopts certain configurations once immobilized, so that not only the structural conformation and dynamic remains for substrate accessibility and activity, but also enzyme positioning and orientation on the solid interface allows electron transfer.

The purpose of this review is to discuss whether general strategies available for all enzyme stabilization can be applied for the special case of redox enzymes used in bioelectrochemistry. To reach this objective, the fundamental bases for all enzyme stabilization by general reported strategies will be first detailed, with a focus on the immobilization procedure as an important way for stabilization. In a second step, we will emphasize the particular features of redox enzymes that must be considered before applying any available stabilization strategies. Ultimately, we will discuss which additional specific strategies must be set up for redox protein, and by extension for bioelectrode, stabilization. Because the topic is very large, this review will not provide an exhaustive list of references but will instead aim to highlight major issues related to enzyme and redox enzyme stability with illustrations taken from papers published during the last few years.

2. Intrinsic (in)Stability of Enzymes

The metabolisms essential for life are catalyzed by enzymes. Over time, these proteins may have undergone different changes resulting from mutations, which cause enzymes to evolve to function in a particular cellular environment and will be very sensitive to exogenous conditions. In addition, in the cell, enzymes can be subjected to various stresses, which can lead to their denaturation or the formation of aggregates. Thanks to a perfectly designed proteolysis machine, new fresh proteins are produced to replace the deactivated proteins. These two fundamental concepts mean that using enzymes outside a cell is expected to induce some loss of activity and eventually progressive denaturation, which can be irreversible. The will to exploit enzymes for other works than predicted by evolution, although highly attractive, imposes finding out strategies to preserve activity within durations compatible with fundamental laboratory studies and with applications [3]. Before reaching that objective, the prerequisite is to know which factors affect enzyme activity and stability. A subtle balance between stability and flexibility is required to maintain activity, so that stability can be the price to pay against activity. Each enzyme is composed of a linear chain of amino acids that fold to produce a unique three-dimensional structure which confers specific and unique enzymatic properties. The overall structure of an enzyme is described according to fours levels which contribute to its specific function and stability (Figure 1). The secondary structure refers to the arrangement of the polypeptide chain (the primary structure) as random coil, α-helix, β-sheet and turn stabilized by hydrogen bonds between amino acid residues. While the primary structure is stable thanks to the strength of the peptide bonds which are unlikely to be broken upon changes in the environmental conditions, the secondary structure can be altered even during protein storage. The tertiary structure is the three-dimensional arrangement of the amino acid residues, giving the overall conformation of the polypeptidic chain. Hydrophobic interactions between nonpolar chains as well as disulfide bonds and ionic interactions between charged residues are involved in the stabilization of the tertiary structure. To fulfill the requirement of enzyme localization in the cell, hydrophobic and hydrophilic groups are arranged within the protein. For soluble proteins, hydrophobic side groups tend to be hidden in the protein core while hydrophilic groups are exposed to the surrounding environment. These amino acid residues on the surface of the enzyme are more likely to be sensitive to the external environment than the core amino acids.

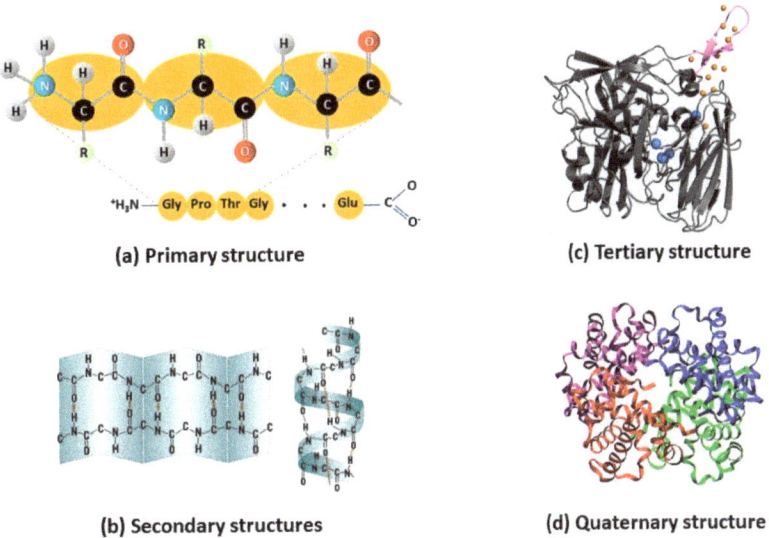

Figure 1. The four levels of protein structures determine the enzyme function.

Denaturation occurs upon unfolding of the tertiary structure to a disordered polypeptide where residues are no longer arranged for functional or structural stabilizing interactions [22]. This process may be reversible. Irreversible loss of activity may also occur upon external stresses. This leads to two different concepts to evaluate in vitro stability of a protein. The thermodynamic stability refers to the ability for a protein to unfold/refold after being subjected to stresses (elevated temperature, extreme pH, or high organic solvent concentration). Methods to evaluate this stability range from scanning calorimetry to circular dichroism (CD) and fluorimetry. Kinetic stability represents the duration of enzyme activity before irreversible denaturation and is measured by activity assays. Actually, the native active state of an enzyme is in equilibrium with the partially denatured, enzymatically inactive state, and the folded/unfolded transition involves mainly intramolecular hydrogen bonds and hydrophobic interactions. This scheme is most certainly more complicated since it was shown that a protein can oscillate between many different folded/unfolded configurations. Each oscillation is governed by the second thermodynamic law, the entropy decreasing as the protein folds, and hydrogen bond formation contributing to the enthalpy. Each strategy that tends to increase rigidity in the enzyme architecture decreases entropy and enhances stability [23]. Thermodynamic instability of enzymes can be partly attributed to the lack of rigidity within the tertiary structure, caused by flexible random coils in contrast to stable β-sheets [24]. In addition to higher rigidity and compact packing, the presence of α-helices with antiparallel arrangement contributes to stability. Water is the natural environment in which protein molecules do exist and operate, and water molecules play a key role in maintaining the entropic stability of the enzyme structure.

Any engineering towards enzyme structural rigidification must maintain the hydration shell and flexibility required for activity. Based on enzyme structural features, many different agents are potential enzyme "killers". When temperature increases, the enzyme tends to unfold in a cooperative process. Disulfide bridges can be broken by reducing agents, extreme pHs are going to affect hydrogen bonding and salt bridges, high salt concentration may aggregate the protein, while charged components may form bonds with charged amino acids modifying the tertiary structure. Organic solvents or detergents will alter hydrophobic interactions, and stability of enzymes in organic solvents will be dependent on the ability of the solvent to strip the essential hydration layer from the protein surface [25,26]. Last but not least, mechanical stress or radiation may disrupt the delicate balance of forces that maintain protein structure [27]. Stabilizing the enzyme means preventing these changes.

3. Strategies for All Enzyme Stabilization

Enzyme stability involves a balance between intramolecular interactions of functional groups and their interaction with the protein environment. Water, as many protein natural solvent, plays a key role in enzyme stability by controlling the hydration shell structure and dynamics and by providing required plasticity for activity [28]. As a direct consequence, many stabilization strategies target water activity in the environment of the enzyme or inside the protein moiety, with the main objective of decreasing the enzyme motion (Figure 2). However, this enhanced stability should not be achieved at the expense of activity, so a delicate balance must be maintained between rigidification and sufficient plasticity controlled by water. As another mere general rule, stabilization over one stress tends to enhance stability against another stress.

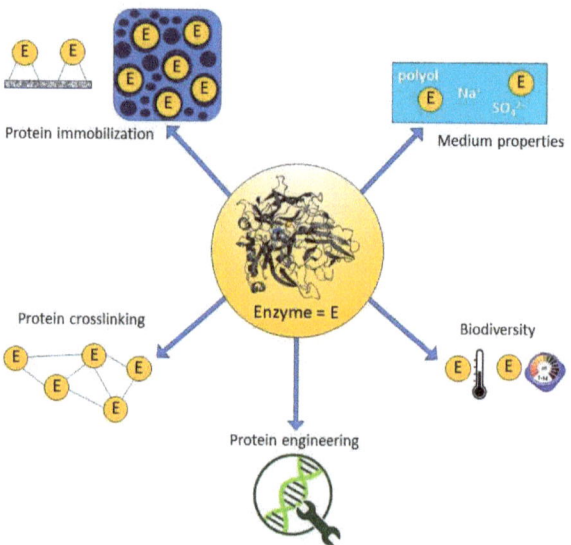

Figure 2. Strategies for all enzyme stabilization.

3.1. Addition of Stabilizers in the Medium
3.1.1. Chemical Stabilizer Addition

It has been known for long time that addition of polyols and polysaccharides such as trehalose, glycerol, dextran, etc. helps stabilizing enzymes [29–32]. Protective effect of polyols was reported in many recent studies in different experimental situations and exposition to stresses (high pressure or high temperature, presence of non-aqueous solvents) [33–37].

Generally speaking, polyols stabilize proteins either in the dried state by serving as water substitutes and preventing dehydration through hydrogen bonding, or in solution by altering protein-solvent interactions [38]. Several mechanisms are discussed to account for the effect of such chemical additives on the protein stabilization [23]. Hydrophilic polyols tend to strengthen hydrophobic interactions among nonpolar amino acid residues resulting in a more compact and spherical enzyme form with smaller surface area, preventing protein dynamics and enhancing stability. Compressibility of proteins, a crucial thermodynamic parameter that determines flexibility, is affected by polyol addition [39]. One accepted mechanism is based on the different sizes between water and stabilizing molecules that preferentially exclude the latter from the protein surface. The preferential hydration of proteins causes an unfavorable free energy change that the proteins tend to minimize by favoring the more compact state over the structurally expanded state. Protective effect is also explained in terms of influence of additives on water activity that results in an increase of structural organization of water molecules contributing to the conservation of low energy interactions favoring native protein conformation [25].

Surfactants, maltodextrine, sodium azide, or special buffers have also been used as additives to maintain the native structure of proteins through purification steps [25]. Dimethyl sulfoxide (DMSO) acts as a stabilizer because it is preferentially excluded from the protein surface [22,40,41]. Addition of DMSO but also of glycerol or ethylene glycol serve as cryoprotectants preventing protein solutions from freezing at −20 °C and allowing multiple use of a unique sample without freeze-thaw cycles [38,42]. Adding polymers in solution such as poly(ethylene glycol) (PEG), alginate, or chitosan maintains a hydration shell around the protein according to the exclusion mechanism [38,43–45]. Addition of polymers also prevents protein aggregation by modifying protein-protein interactions and increasing medium viscosity, thus decreasing enzyme motion. Macromolecular crowding, which is the natural environment of enzymes in vivo, can explain the stabilizing effect of some additives [46–49]. It can be mimicked in vitro through the addition of high concentrations of macromolecules such as dextran or PEG [50] and would favor the folded state of proteins and compact conformation through the excluded volume effect [51–55]. However, the stabilizing effect of in vitro crowding would not be universal as the properties of all macromolecules in vivo are finely regulated under different physiological conditions [56]. Aggregation under crowded conditions can be enhanced because the activity of water is decreased, and the refolding rate of proteins as a consequence of increased viscosity is also decreased.

3.1.2. Salt Effect

Salt addition affects protein stability according to a combination of binding effect, screening of protein surface electrostatic potential, and effect on protein/water interface. Salts may contribute to hydrophobic interaction strengthening in proteins, hence to stabilization, but salt binding to amino acids on the protein surface decreases the repulsion between proteins and induces aggregation. The salting out process, or salt-induced precipitation, depends on the structure of the protein and in particular on the population of hydrophilic amino acid residues. It also largely depends on the salt concentration. The effect of salts on protein stabilization has been tentatively explained on the basis of the Frank Hofmeister series [4,57–59] (Figure 3A). This series early ordered anions and cations composing salts according at first to their ability to precipitate lysozyme, and later to their ability to stabilize protein secondary and tertiary structure. They were respectively classified at that time as water-structure formers (kosmotropes) or water-structure breakers (chaotropes). Kosmotropes and chaotropes would be nowadays more understood as a characterization of the degree of hydration, and different models are developed to consider the various effects observed experimentally [60]. Ionic liquids (IL) would follow the same tendency as salts where kosmotropic anions and chaotropic cations stabilize the enzyme, while chaotropic anions and kosmotropic cations destabilize it [61]. While they can be considered as eco-friendly solvents, many proteins are inherently inactive in ILs, requiring the addition of water for activity recovery. This suggests ILs could affect the internal water shell [61,62].

Recent examples in the literature illustrate the effect of salts on enzyme stability, and most of them agree with the Hofmeister series. Changes in the secondary structure of two proteins with helical and beta structural arrangement, respectively, were followed by Fourier Transform Infra-Red spectroscopy (FTIR) in the presence of various salts. It was shown that the stabilization effect of the salt follows the Hofmeister series of ions, although some exceptions were observed with formation of intermolecular β-sheets typical of amorphous aggregates [62]. Glucose oxidase (GOx) stability was studied using microcalorimetry in the presence of various salts [63]. At high salt concentrations (over 1 M), it was shown that the Hofmeister effect on the temperature of inactivation was determined by the ion-specific effect on the protein/water interface. Correlation between stability and activity of lysozyme in the presence of various salts from the Hofmeister series suggested a role of local stability/flexibility in enzyme activity [64]. The thermal stability of *Aspergillus terreus* glucose dehydrogenase (GDH) was substantially improved by kosmotropic anions, retaining more than 90% activity after 60 min of heat treatment at 60 °C. The stabilizing effect followed the Hofmeister series and was anion concentration-dependent and strongly related to the structural stabilization of the enzyme, which involved enzyme compaction [65] (Figure 3B). It was further shown that salts can stabilize proteins not only in vitro but also in vivo or intracellularly, the stabilization level correlating with the Hofmeister series of ions [66].

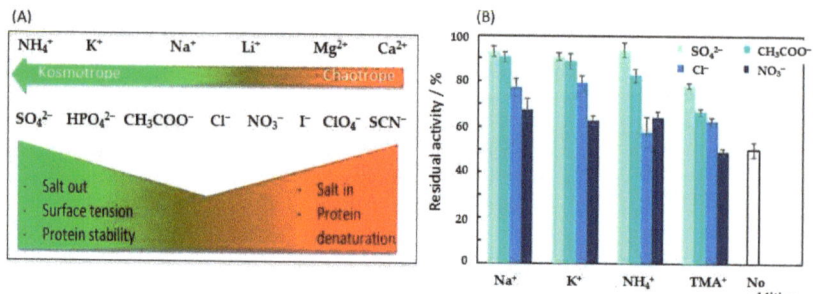

Figure 3. Salt effect on enzyme stability. (**A**) Hofmeister series and effect on protein properties. (**B**) Stabilization effect of kosmotropic anions and cations on *Aspergillus terreus* GDH. Residual activity after 1 h incubation at 50 °C as a function of added anions and cations. Reproduced with permission from [65].

3.2. Chemical Modification of Enzymes, Directed Evolution

More than 95% of all charged components are located on the surface of proteins, consisting mostly of hydrophilic moieties, while most of the hydrophobic ones are buried deep inside the core. As already discussed above, the physico-chemical microenvironment of the protein will be sensed intrinsically through those moieties, demonstrating the protein surface to be an attractive target for protein engineering to enhance protein stability. The efforts toward engineering enzyme catalysts with increased stability can be divided into two groups: (i) chemical protein modification and (ii) protein engineering, such as site-directed mutagenesis and directed evolution. The combination of these two approaches is appealing for improving the catalytic properties of enzymes.

3.2.1. Chemical Modification of Enzymes

Chemical protein modification can be achieved either by modification of one sort of amino acid on the protein surface or by polyfunctional modification using reticulating agents or by conjugation to water-soluble polymers. One widely used method consists in reticulation through cross linking via glutaraldehyde (GA) that increases the protein rigidity and mimics disulfide physiological bonds or salt bridges [25]. Activity is however very often affected. Stabilization of proteins through cross-linked-enzyme-aggregates (CLEAs) will be further discussed below in Section 4 [67,68]. Conjugation through water-soluble polymers such as PEG, chitosan, alginate, dextran, etc. has also been widely reported [38,69,70]. Those polymers present either one functionality or are bifunctional to enable reaction with N- or C-terminal or with one individual amino-acid residue on the protein. Increase in the half-life of proteins is frequently observed [71,72]. Other polymers responding to temperature stimuli can tailor the temperature dependence of enzyme stability [73]. Two different methods can be carried out to synthesize the bioconjugate. In the "grafting to" strategy, the polymer is first synthetized; then, an end-group functionality is attached to one amino-acid on the protein surface. As an example, a mono-PEGylated arginase was constructed by linkage of PEG-maleimide to a cysteine residue on the enzyme surface [74]. The protein was protected against proteases thanks to the shielding effect of PEG, allowing the modified arginase to operate as a promising anticancer drug candidate. In the "grafting from" method, radical polymerization occurs at sites attached to the protein. A recent work describes the ligation of Atom Transfer Radical Polymerization (ATRP) initiators to lysine residues on the surface of a laccase [75]. The polymer-enzyme hybrid obtained by this "grafting-from" procedure showed enhanced solvent and thermal stability, as well as a clear enhancement of activity in a much wider pH range than the free enzyme. Combination of PEGylation and chemical modification by GA can also be efficient [76]. GOx was first PEGylated to provide steric protection; then, it was cross-linked with GA which stabilized the tertiary and quaternary structures. Intermolecular crosslinking-induced aggregation was prevented by the PEG shield. The GA-modified PEGylated GOx retained 73% of activity after 4 weeks at 37 °C against 8.2% for the control (Figure 4).

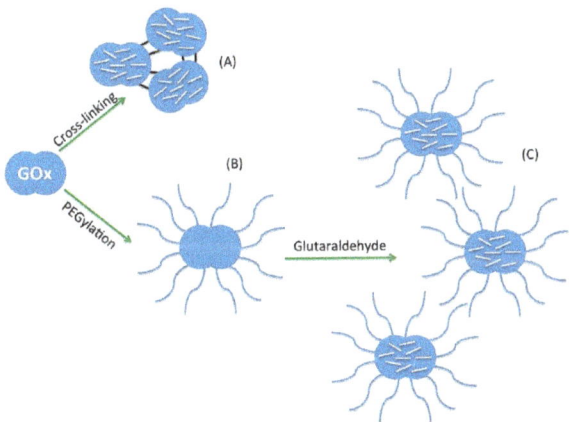

Figure 4. PEGylation of GOx (**B**) followed by chemical modification (**C**) enhances enzyme stability, while preventing intermolecular crosslinking contrary to direct cross-linking (**A**). Reprinted with authorization from [76].

3.2.2. Enzyme Engineering

Protein engineering is another relevant strategy to enhance stability [4,77,78] (Figure 5). Rational design takes benefit of structures and sequences of already known stable proteins. Through identification of amino acids which are assumed to participate in (de)stabilization, new variants of proteins of interest are created through site-directed mutagenesis [79]. As an example, asparagine is a thermolabile residue prone to deamination which can be mutated into threonine or isoleucine, with similar geometry but more thermostable. Replacement of lysines (or histidines) by arginine residues increases thermostability by increasing intramolecular or inter-subunit salt bridges. Otherwise, we will discuss further in this review the interest in the screening of biodiversity to search for more stable enzymes such as in extremophilic organisms. However, because the production of extremophile enzymes may be delicate, genes from hyperthermophiles can be implemented into suitable mesophilic hosts, coupling advantages of high productivity and high thermostability [22,80]. Use of directed evolution to design more stable enzymes is now common. With this method, mutant libraries are created by random changes, and the most promising variants are subjected to further rounds of evolution [81–85]. Rational approach based on molecular dynamic (MD) and QM/MM simulations may help in identifying and redesigning variants with increased stability [86].

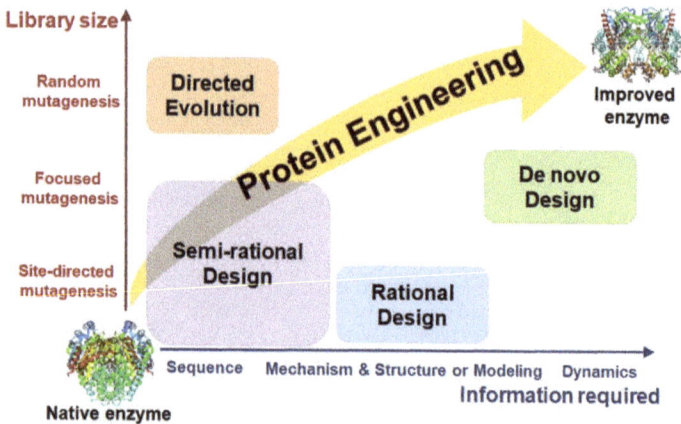

Figure 5. Protein engineering methods to improve enzyme stability. Reprinted with permission from [87].

3.3. Screening of the Biodiversity and Outstanding Properties of Extremophiles

Screening in the biodiversity for "exotic" enzymes such as extremophiles presenting unusual and/or outstanding properties is an attractive strategy to be considered. Although this research topic is increasingly growing, there is no doubt that Nature still retains many secrets that would allow new opportunities for stable biocatalysis. However, the harsh conditions required for the growth of extremophiles very often limit their laboratory studies. Extremophiles have evolved to survive in ecological niches presenting extreme temperatures (thermophiles and psychrophiles living at high (>80 °C) and low (<20 °C) temperatures, respectively), extreme pH (acidophiles and alkalophiles), high pressures (barophiles), high salt concentrations (halophiles), or in the presence of heavy metals (metallophiles)). These extreme environments are found in deep-sea hydrothermal vents, hot springs, volcanic areas, or mine drainage [88–90]. Ancestral microorganisms are another source of stable enzymes because distant ancestors of current organisms were thermophiles and would be composed of proteins that are more thermostable than their current homologues [91–93].

Extremophilic enzymes present specific structural characteristics that afford them to resist in extreme conditions. Interestingly, the same structural features often induce enhanced stability of such enzymes at normal conditions or in the presence of non-aqueous solvents. Moreover, "extremostable" enzymes, thermostable, halostable, acidostable, etc. isolated from extreme environments or obtained by protein engineering can support large number of mutations due to their robustness, leading to eventually even more stable enzymes.

3.3.1. Thermophiles

Thermostability refers to two different concepts: thermotolerance, which is the transient ability to maintain activity at high T°, and thermostability, which is the ability to resist irreversible inactivation at high T° [94]. In addition to the ability to resist to high temperatures, thermostable enzymes usually also present higher resistance to chemical denaturants and extreme pHs. Such cross-adaptations are highly interesting to get stable enzymes in many different extreme conditions. Furthermore, due to their stability at elevated temperatures, enzymatic reactions are faster and less susceptible to microbial contaminations. Many different structural characteristics have been demonstrated to be involved in thermostability. Compared to mesophiles, protein packing, a high number of hydrophobic residues, increased helical fold content, a high number of disulfide bonds, density of internal hydrogen bonds and salt bridges, and distribution of charged residues on the surface are some of the features often shared by thermostable enzymes [95–97]. Proportion of certain amino acids is also significantly different between extremophiles and mesophiles. For instance, in thermostable enzymes, lysines are replaced by arginines; asparagine/glutamine content is lower while proline content is higher [22,98–100]. The most widespread explanation is that these structural features contribute to reducing the flexibility of the enzyme and to allowing optimal conformation at higher temperatures than mesophiles, with no denaturation [101]. MD simulations confirmed that hydrophobic packing and electrostatic interaction network provided by salt bridges explain enzyme thermostability [102]. Dimerization can also be a key factor for enhanced thermostability [103].

3.3.2. Halophiles

Halophiles grow in the presence of high concentrations of salts. In order to avoid osmotic shock, they accumulate inorganic salts or small organic molecules in the cytoplasm until the intracellular osmolarity equals the extracellular ion concentration [101]. They also require their proteins to operate under extreme ionic conditions. Actually, halophilic enzymes require high salt concentrations for activity and stability (1–4 M range) [104]. Stabilization of proteins in high salinity environments is linked to the interaction of hydrated ions with negatively charged surface residues. A highly ordered shell of water molecules is formed that protects the protein and prevents denaturation [105]. Compared to non-halophilic proteins, halophilic enzymes present less non-polar residues and more charged residues on the protein surface, a higher frequency of acidic (Asp and Glu) over basic residues (Lys), and low hydrophobicity [105] (Figure 6). This high surface charge is neutralized mainly by tightly bound water dipoles [22]. Such excess of acidic over basic amino acid residues makes them more flexible at high salt concentrations in conditions where non-halophilic proteins tend to aggregate [106,107]. Halophilic proteins seem to be specially adapted to have multiple crossover adaptations, especially to pH and temperature.

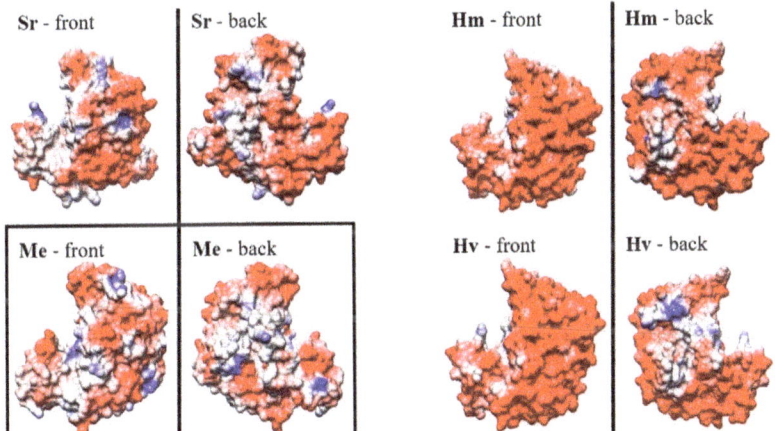

Figure 6. Coulombic surface maps for both sides of the *Salinibacter ruber* (Sr), *Haloarcula morimortui* (Hm), *Haloferax volcanii* (Hv) and *Methylobacterium extorquens* (Me) malate dehydrogenases. The halophilic structures display negative surface areas (in red), a common feature of halophilic proteins. Reprinted with permission from [101].

3.3.3. Cross-Stabilities of Extremophiles

Many examples are reported in the recent literature that highlight the enhanced and cross-stabilities of extremophilic enzymes. Thermostability can be accompanied with pH, salts, metals, or solvent resistance [108,109]. Key enzymatic reaction may benefit from these cross-stabilities. Transformation of cellulose into biofuels is a promising eco-friendly process. However, it requires the enzymes involved in the process to be thermostable, halostable and organic solvent-stable. A cellulase presenting all these characteristics was identified through metagenomic from a deep sea foil reservoir which could be a valuable candidate [110]. Hydrogen is considered as an energy vector for a low carbon economy. Its production and conversion by enzymes should be promoted through operation on a wide range of temperatures. The hyperthermophilic hydrogenase (HASE) from *Aquifex aeolicus* was demonstrated not only to be able to operate at high temperatures for H_2 oxidation but was also much more stable at room temperature than the HASE from *Ralstonia eutropha*, a mesophilic homologue [111] (Figure 7). The structure of this hyperthermostable enzyme reveals more salt-bridges compared to mesophilic HASE, that contribute to its thermostability [112].

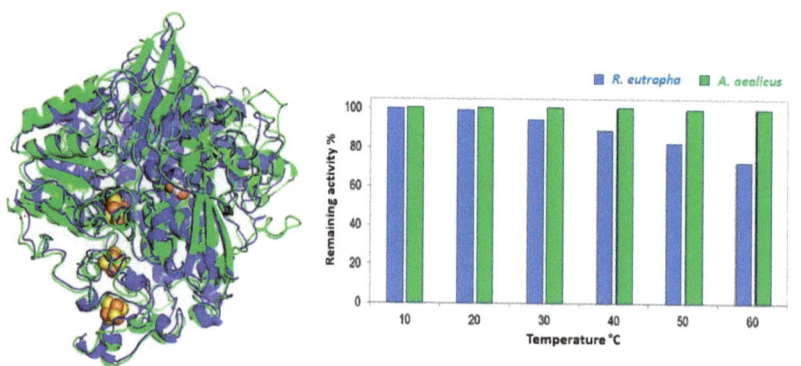

Figure 7. *A. aeolicus* and *R. eutropha* are two homologous NiFe HASEs with different thermostability. (**Left**) Structural alignment of mesophilic *R. eutropha* (blue) and hyperthermophilic *A. aeolicus* (green) HASEs, (**right**) remaining activity of the two HASEs after 360 s incubation at increasing temperatures. Adapted with permission from [111].

4. Strategies for Enzyme Stabilization in the Immobilized State

The discussion above has emphasized that the less flexible enzymes are, the most stable they are. In line with this assessment, it is quite straightforward that the immobilization of an enzyme on a support will decrease the movement of the protein, tightening the structure by single or multipoint binding and hence will increase its stability. The extent of the stabilization will depend on the enzyme, the nature of the support and the mode of enzyme attachment to the support (simple adsorption, covalent attachment, entrapment in pores, etc.) This means that the support must be amenable to surface modification for further enzyme attachment. The material acting as a support must also be biocompatible, stable, and able to host high enzyme loadings. A delicate balance between stability and activity of the enzyme will be engaged, with an additional advantage of immobilization being that the support itself can help in the stabilization by consumption of inhibitors or by providing buffer properties as examples [47]. In the following, we will especially discuss immobilization strategies that can enhance enzyme stability. We will not report an exhaustive list of the coatings that can simply prevent enzyme leaching, but we will emphasize the role of the support structuration on the enzyme conformation, hence its stability, although both concepts are closely linked. We will extend the discussion to modes of immobilization that do not require any carrier, like cross-linking of enzyme aggregates (Figure 8).

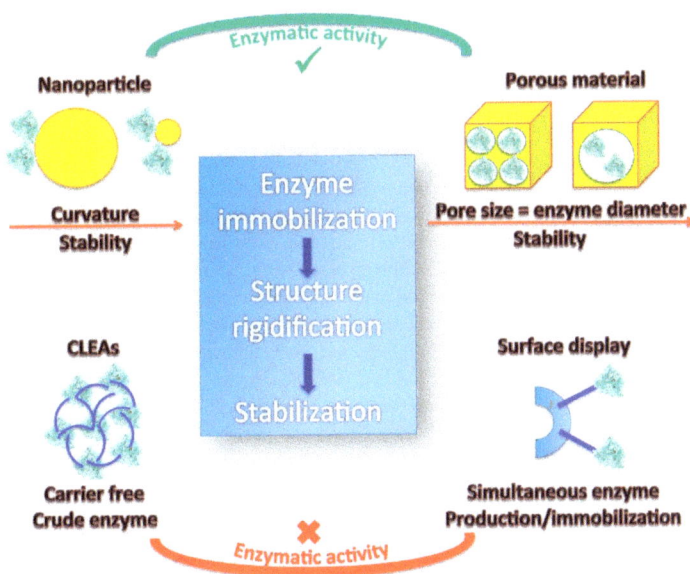

Figure 8. Strategies for enzyme stabilization in the immobilized state.

4.1. Some Fundamentals on Enzyme Immobilization on a Solid Support

4.1.1. Various Types of Interactions

The key point to develop immobilization strategies able to enhance protein stability is to understand how the interactions between the support and the immobilized enzyme may modulate the enzyme internal motion and conformation. The specific structural features of enzymes make their immobilization process very different from rigid particles that simply attach to or detach from a support with certain adsorption and desorption probabilities. The interactions that drive protein adsorption range from high energy covalent bonds (disulfide bridges 320 kJ/mol) to electrostatic interactions (35–90 kJ/mol), hydrogen bonds (8–40 kJ/mol), van der Waals interactions and hydrophobic interactions (4–12 kJ/mol) [113]. Using the concept of "hard" and "soft" proteins based on structural rigidity scale, it can be categorized that adsorption of "hard" proteins will be driven by electrostatic interactions, while "soft" proteins will be able to adsorb either on hydrophobic or hydrophilic surfaces [114].

4.1.2. Enzyme Orientation on the Surface

Upon immobilization of an enzyme onto a support, several processes are going to take place: change in the state of hydration of the enzyme surface, enzyme structural rearrangements, redistribution of charge groups, or surface aggregation [115]. These are slow processes since they involve a whole cascade of rotations. While proteins can rotate freely in solution, they will adopt on a surface one preferential orientation that exposes one part of the enzyme to the surface and the other part to the bulk solution. However, the final orientation can be very different from requirements for optimal activity. Favored orientation is linked to the minimum free energy resulting from attractive coulomb and van der Waals interactions, hydrogen bonds, and the entropy gain of solvent molecules or counter ion release [115]. In addition, rotation of the enzyme can occur even in the adsorption state when local conditions change [116]. In the case of immobilized enzyme, it was expected that only the amino acids on the surface of the enzyme will sense the support. However, by looking at the rigidity profile obtained by Brownian dynamics simulations, it was very recently shown that even amino acids in the core of the enzyme can be affected depending on the surface charge of the support [117].

4.1.3. Conformation Changes upon Immobilization

While electrostatic interactions between the enzyme and the support may drive the adsorption process and prevent enzyme leaching, strong interactions between two charged entities can affect the structure of the enzymes. Flattening of the structure was for example observed when proteins were immobilized on gold metal [118], while no change in the secondary structure was detected when immobilization occurred on a gold surface modified by a self-assembled-monolayer (SAM) of thiol [116]. In the course of adsorption, protein density increases on the surface, and enzyme/enzyme interactions may surpass enzyme/support interactions. Low surface coverage can induce modification of the conformation because the enzyme molecules maximize their contact surface. At higher enzyme loading, it can be expected that protein-protein interactions decrease the interaction with the surface and preserve enzyme conformation. Protein–protein interactions may otherwise induce repulsion between the neighbors which increases with the decrease of the distances between the adsorbed proteins. A variety of methods exists to gain information on the conformation of the enzyme in the immobilized state including CD, surface plasmon resonance (SPR), ellipsometry, fluorescence, and Raman and FTIR spectroscopies. However, characterization methods that can separate effects due to complex coupled mechanisms upon enzyme immobilization are lacking, and data often result in spatial and temporal averages.

4.1.4. Adsorption Versus Multipoint Attachment

Adsorption of proteins on surfaces is a straightforward protocol that is susceptible to maintain high enzyme activity, but it may induce enzyme leaching. Covalent attachment fixes the enzyme on the surface, thus preventing enzyme leaching, but must be carefully considered because the mode of attachment can greatly affect the protein structure, hence both stability and activity. Different protocols can be used to attach an enzyme covalently to a support [20]. The most widely used methods are maleimide and carbodiimide coupling, amine aldehyde condensation and various click chemistry reactions. They allow covalent coupling between one available functional group naturally available or engineered on the surface of the enzyme and a complementary function created on the support where the enzyme is immobilized [20]. A spacer can be added to induce some flexibility required for the activity [119]. Multipoint attachment has been widely suggested to be more suitable than one-point attachment for increased stability because of reduced dynamics of the protein [120–122]. On the other hand, a multipoint attachment is challenging, and both the reactive groups on the support and on the enzyme must be carefully chosen to prevent steric hindrance. Furthermore, the decrease in the essential motion of the folded enzyme state required for catalysis often affects the activity. Stabilization factors more than 1000 were for example reported for formate dehydrogenase or alcohol dehydrogenase after immobilization on activated glyoxyl-agarose, while activity decreased by 50 and 10%, respectively [123].

4.2. Nanomaterials for Stabilizing Enzyme in the Immobilized State

Interest in nanomaterials is increasingly growing thanks to their intrinsic physicochemical properties that allow to envision many applications when combined with enzymes for energy conversion, biosensing, drug delivery, etc. The most attractive feature of nanomaterials is their large surface-to-volume ratio that enables enhanced catalysis thanks to an increased loading of enzymes. Nanomaterials are easily functionalized for further enzyme attachment via classical covalent coupling or click chemistry. However, they also present heterogeneity in terms of size of particles [124], number of sites on the surface available for enzyme attachment, or local curvature that will greatly influence enzyme immobilization. Particle agglomeration, caused by low colloidal stability of the particles in buffer, can also hamper the storage stability of the immobilized biosystems [125].

Nanoparticles (NP) are widely used in medicine, cosmetics, or food industries. Hence, the effect of enzyme immobilization on NPs on both activity and stability has been largely studied over the last ten years. Enhanced thermal or storage stability are reported, the reasons being often related to the mode of attachment of the enzyme [126,127]. As examples, enhanced thermal and long term stabilities, resistance to urea or to acidic conditions were shown for GOx on ferritin [128] or on Fe_3O_4-based NPs [129], for lipase on Fe_3O_4-based NPs [130], for galactosidase on ZnO-NPs [131], or for cellulase on magnetic NPs [132]. The linkage to the NP was assumed to prevent unfolding of the enzyme no matter of the NP intrinsic property.

Other aspects that account for enzyme stability enhancement on NPs need to be discussed. One of the challenges is the control of the number of grafted enzymes, which may have a direct impact on enzyme flexibility, hence stability. Beyond conjugation of the enzyme to the NP that can affect aggregation, role of the charge, size, and morphology of the NP, local pH and ionic strength may be key parameters involved in the stabilization process [127]. For instance, NP size was shown to have a direct effect on the stability of the enzyme [133]. Varying the size of NPs induces the variation of the NP surface curvature, that will offer distinguished surface of contact for the enzyme. High NP surface curvature (diameters of NP less than 20 nm) was shown to preserve enzyme native conformation [134]. Smaller area for protein contact as well as suppression of unfavorable protein–protein lateral interactions are most probably the reasons that can account for stabilization effect of nanomaterials [135,136]. Curvature-based stabilization of enzymes can be extended to other nanomaterials such as carbon nanotubes (CNT) [4,135]. Morphology of the NP was also reported to tune the enzyme stability. For example, switching from nanospheres to nanorods decreases enzyme stability, most probably because of the impact of the flat cylindrical axial surface of nanorods [137].

4.3. Encapsulation in Porous Materials

4.3.1. Key Role of Pore Size

Porous materials are attractive and versatile supports for enzyme stabilization. Once encapsulated, the enzyme should be protected from the environment, while being able to be attached via suitable chemical modification of the walls of the pores. The chemical conditions used for the porous material synthesis, as well as pore size and interconnectivity in the matrix for diffusion of substrate and product in and out, while ensuring that the protein cannot diffuse out are main issues to be considered. Depending on the materials and the synthetic procedure, the diameter of pores can be tuned and adapted to a specific enzyme. We should distinguish here micro, meso and macroporosity that are described by the diameter of pores being, respectively, less than 2 nm, between 2 and 50 nm and more than 50 nm. While mesopores are required for enzyme encapsulation, macroporosity will enable substrate diffusion. Hence, getting a material with hierarchical porosity is highly desirable in most cases. In general, pore size close to enzyme size favors stabilization. Multipoint interactions may be one explanation for such a result [138]. It was also suggested that modification of the water structure in nano-containers can induce higher enzyme stability [23]. Given the average size of enzymes in the range of about 3–10 nm, some microporous materials with controlled pore size and geometry are unfortunately not suitable for enzyme encapsulation. This is the case of silica films presenting vertically aligned pores [139], or of the well-known silica SBA15 porous matrix [140]. However, porous materials, with pore size compatible with enzyme diameter, can be obtained by different strategies [141]. Carbonization of MgO-templated precursors yields a pore size-tunable material (from around 30 to 150 nm) with interconnected mesopores [142]. Increased half-life stability was shown when the pore size was the closest to the hydrodynamic enzyme diameter [143].

4.3.2. Metal-Based Porous Matrix

Nanoporous gold (NPG) can be obtained by acidic treatment of alloys containing 20–50% gold, or by electrochemical treatment (dealloying process). NPG structure presents interconnected ligaments and pores of width between 10 and several hundred nm. NPG displays various surface curvatures that offer a different environment for enzyme immobilization. Different enzymes have been immobilized in NPG for various purposes showing a pore size-dependent enhanced stability as compared to enzyme in solution [144]. Alternatively, enzymes can be encapsulated in gold nanocages within a 3D gold network obtained by in situ reduction of gold salts in the presence of the enzyme [145]. Complexes formed between cations such as Cu(II) and protein molecules can serve as nucleation sites for micrometer-size particles with unique flower petal-like morphology. Enzymes confined in such structures exhibit enhanced stability [146]. Other hierarchical materials presenting mesoporous structures can be obtained by using networks of CNT or nanofibers [19].

Polymeric metal-organic-frameworks (MOFs) are matrices currently under great consideration. They consist of metal ions or clusters coordinated to organic ligands to form two- or three-dimensional structures. Cavities of mesoporous MOFs are greatly suitable for enzyme immobilization. In addition to enhanced enzyme loading and reduced leaching, the strictly controlled pore size can provide selectivity for substrates. Enzymes can be immobilized by infiltration process, requiring MOF pore size larger than protein size. Enzymes can alternatively be encapsulated in the lattices of the MOF structure by mixing the enzyme with the metal and the ligand. In this case, encapsulation proceeds through a nucleation mechanism where the enzyme acts as a nucleus for MOF growth, leading to enzymes protected in pores with size close to the radius of gyration of the protein [147]. However, in the latter case, synthesis conditions must be mild enough to avoid protein denaturation. It is admitted that the framework around the enzyme maintains the conformation of the active enzyme species [148]. Many studies report increased stability of various enzymes when embedded in the pores of MOFs. Enhanced thermal and pH stability was observed for different enzymes, as well as protection against inhibitors, or maintaining of activity in non-aqueous solvents [149–154]. MOF can also be used to encapsulate cascade of enzymes with enhanced stability and spatial control of the proteins inside the lattices [155]. However, care should be taken with solubility of MOFs in the presence of various compounds such as amino acids, some organic acids or buffers that present high affinity for the MOF-metal, inducing leaching of the enzyme via MOF dissolution [156,157]. Covalent organic frameworks, only composed of light elements (H, C, B, N, and O) and covalent bonds, with higher stability than MOFs, can be alternatives as efficient matrices for enzymes [158].

In MOFs, enzymes are statistically entrapped. It should be even more elegant to encage each individual enzyme with well-defined protecting shell surrounding it [159]. This strategy, known as single enzyme nanoparticles (SEN), is based on the controlled formation of a shell of polymer around one enzyme by in situ polymerization from the enzyme surface. The shell is thin and permeable so that substrates can freely diffuse to the enzyme core. SEN strategy is reported to enhance thermal stability, organic solvent, and acid/base tolerance even in aggressive environments [160,161]. Both protection by biocompatible polymer shell and multipoint covalent attachments within the nanocapsule were suggested to explain the enhanced enzyme stability. This last strategy recalls less recent examples reported in Section 2 of this review and based on polymer protection around the enzyme.

4.3.3. Other Polymeric Matrices

Hydrogels are polymeric networks that are of great interest for protein immobilization because they present the double advantage of high water content and a tunable porous structure [162]. Natural polysaccharides such as alginate can produce beads when cross-linked with certain metals. Alginate concentration can be tuned to control the porosity of alginate beads (in the range 5 to 200 nm diameter) and to find the best compromise between substrate diffusion and enzyme conformation preservation and leaching [163]. Enzymes embedded in DNA hydrogels often show enhanced thermal stability and improved stability during freeze-thaw cycles or in the presence of denaturants such as organic solvents. Conformation of the enzyme is proposed to be maintained thanks to extensive inter- and intra-strand weaving of long DNA building blocks. The hydrophilicity of the hydrogel further protects the enzyme against organic solvent [164]. Gel nanofiber made of Zn^{2+} and adenosine monophosphate was also used to encapsulate an enzyme cascade for glucose detection. The biomaterial exhibited enhanced stability against temperature variation, protease attack, extreme pH, and organic solvents. Porosity and water content of the gel allowed to maintain 70% of the activity after 15 days storage while free enzymes lost activity [165].

4.4. Cross-Linked Enzyme Aggregates (CLEAs)

The addition of salts, water miscible organic solvents, or non-ionic polymers to protein solution induces their precipitation as physical protein aggregates, held together by non-covalent bonding. Enzyme tertiary structure can be maintained in the aggregates. Note this is also the traditional procedure for protein purification (in particular see in Section 3 the discussion about salting-out processes) [166]. Subsequent cross-linking of these physical aggregates renders them permanently insoluble while maintaining their pre-organized structure (Figure 9). Covalent binding or cross-linking of enzyme to enzyme to form aggregates is commonly referred to as CLEAs. CLEAs constitute a simple way of enzyme immobilization which is carrier-free and can be used in any reaction medium [22,167,168]. One more advantage of CLEAs is that they can be formed starting from crude enzyme preparations and do not require highly purified enzymes. Typical cross-linking agents are GA, ethylene glycol diglycidyl ether (EGDGE), and dextran, but also polyethyleneimine (PEI) that can generate covalent bonds between two or more biomolecules in the system through the enzyme amino and/or carboxyl surface groups [169]. FTIR analysis of CLEAs highlighted changes in the secondary structure of enzymes, with an increase in β-sheet and α-helix components and a decrease in β-turns compared to free enzymes, showing the ability of CLEAs to stabilize enzymes [169–171].

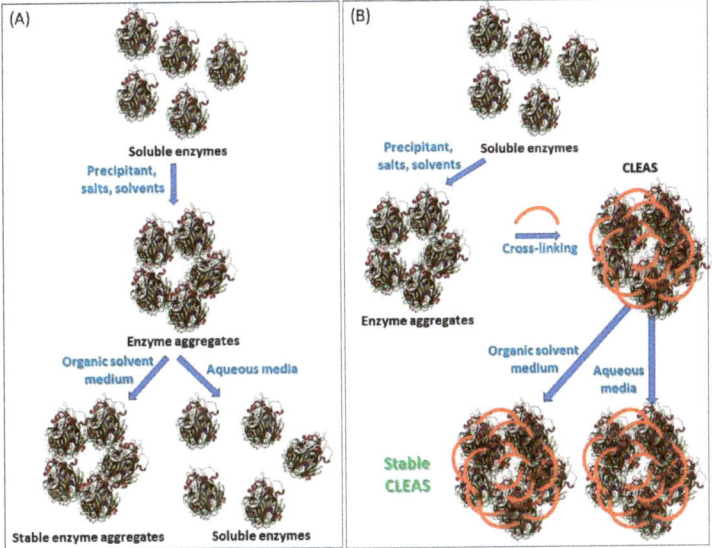

Figure 9. (**A**) formation and redissolution of enzyme aggregates, (**B**) stable CLEA preparation.

A wide variety of enzymes, including hydrolases, oxidoreductases, lyases, and transferases have been used as CLEAs especially for organic synthesis [172,173], displaying enhanced stability under various conditions: protection against organic solvents [174], enhanced thermostability [175], reduced substrate inhibition [176,177], enhanced stability in acidic or alkaline conditions [178,179] or under storage [180]. Magnetic CLEAs combined the advantages of CLEAs and covalent immobilization onto Fe_3O_4 magnetic particles and may show increased stability to changes of pH and temperature [181,182]. Enzyme cascade can also be combined as CLEAs showing enhanced thermal and pH stability [183,184].

Despite a high enzyme concentration per unit volume, decrease in activity is often observed using CLEAs. Diffusional limitation of mass transfer is one of the main issues. Alteration of the enzyme tertiary structure by one of the components required for CLEA preparation or steric hindrance within the aggregate of high particle size hampering inner enzyme activity can also explain the decrease in activity. Hence, the choice of precipitant, cross-linker, and ratio of cross-linker protein is a critical step for maintaining required flexibility of the structure for high activity [185]. These parameters strongly depend on the enzyme, making the CLEA technology delicate [174].

4.5. Bio-Surface Display of Enzymes

Bacteriophages, spores, yeast, and bacterial cells can be used as enzyme carriers in so-called surface protein display systems [186,187]. *E. coli* bacterium is one of the most studied cells for enzyme surface display. Surface display of the targeted enzyme (the passenger) is realized by genetically fusing it to a carrier protein, which facilitates export across the cell envelope (inner membrane and outer membrane separated by the periplasm). The protein of interest is ultimately secreted and immobilized on the cell wall. Interestingly, enzymes are thus simultaneously produced and immobilized on the bio-component. The anchoring motif can be fused to the N- or C- end of the protein, or the protein can be inserted within the sequence of the anchor protein. The anchor protein must be able to ensure transport of the heterologous protein through the secretory pathway, proper folding of the fusion protein, and strong binding at the cell wall surface. Contrary to purified immobilized proteins, expressing enzymes on cell surfaces is considered advantageous as enzymes can be used repeated for extended periods with no loss of enzyme activity. Stability of immobilized proteins by such a natural procedure against organic solvents, extreme pH, or high temperature has been reported [188,189]. Yeast and spores are also currently explored as enzyme display surfaces allowing enzyme enhanced thermal stability [190,191]. However, the efficiency of the display systems in terms of immobilized enzyme amount and specific activity needs to be improved for envisioning any biotechnological application [192].

5. Specificity of the Stabilization of Redox Enzymes on Electrodes for Bioelectrocatalysis

5.1. Bioelectrocatalysis: When Enzymes Meet a Foreign Conductive Surface

5.1.1. Electron Transfer in Metabolic Energy Chains

Most cellular functions, such as cell motility, key molecule biosynthesis or transport across membranes, require energy. Life energy is generated by redox transformations of nutrients present in the environment or by sunlight to create the proton gradient across the membrane. This gradient is later used to produce ATP, the universal molecular "coin" driving metabolic reactions. In these processes, complex but highly specific molecular chains are involved that couple electron transfer (ET) to the diffusion of protons across cell membranes. Focusing on bacteria, a wide range of energy substrates can be metabolized: lactate, glucose, H_2, H_2S, O_2, sulfate, fumarate, etc. thanks to the presence of a large variety of redox enzymes interacting with membrane lipids or with redox proteins acting as electron shuttles. To enable metabolism achievement, ET must proceed fast through highly specific protein–protein interactions. According to Marcus theory, the distance between the donor and the acceptor will drive the rate of ET [193]. It is admitted that electrostatic interactions serve for pre-orientation of interacting partners; then, hydrophobic interactions drive a rearrangement of the transitory complex allowing ET [194]. Pre-orientation is guided by the heterogeneity of charge distribution on the protein surface which induces dipole moments that can be as high as thousands of debyes [112]. As such, one or more amino acids defining a charged patch on the surface of one protein partner are often involved for the specific interaction with the other partner.

5.1.2. Electron Transfer Mechanisms Involving Redox Enzymes at Electrochemical Interfaces

Redox enzymes can operate very efficiently in bioelectrocatalytic processes for the redox transformation of a variety of substrates. As stated in the introduction of this review, various applicative domains are concerned from bioelectrosensing, bioenergy to bioelectrosynthesis. One prerequisite is the use of a conductive support, the electrode, to provide or accept the electrons for the redox transformation to take place. The question directly arises whether the electrode would gently replace the physiological redox partner? Actually, many materials used as electrodes (carbon, gold, nano- or porous structures, etc.) can be chemically functionalized to provide an enzyme environment mimicking the physiological one. It was clearly demonstrated that the oriented approach of the enzyme to the electrode surface was guided by electrostatic interactions driven by the enzyme dipole moment. The example of multicopper oxidases (MCO), the key enzymes for O_2 reduction into water, is highly relevant in that sense. Four copper sites are involved in the catalytic process, CuT1 being the first electron acceptor, and electrons generated travelling intramolecularly to the trinuclear Cu site where O_2 binds and is reduced. Bilirubin oxidase (BOD) from the fungus *Myrothecium verrucaria* has a dipole moment of 910 D at pH 5 pointing positive towards the CuT1. At the same pH, BOD from the bacterium *B. pumilus* and laccase (Lac) from *Thermus thermophilus* have dipoles moments of 1830 and 860 D, respectively, but pointing positive opposite to the CuT1. In accordance with this structural feature, ET was obtained on negatively or positively charged electrodes in the case of *M. verrucaria* BOD or *B. pumilus* BOD and *T. thermophilus* LAC, respectively [116,195,196]. One specificity of bioelectrocatalysis stems from the substrate and product coming in and out to/from the active site that may be hampered by the required enzyme orientation for ET. Not only activity but also stability in case of substrate or product accumulation may be affected [197]. This point is however understudied in bioelectrochemistry. When direct ET (DET) is not possible, mainly because there is no surface electron relay identified, either because the enzyme 3D structure is not known or because the active site is electrically isolated in the protein moiety, mediated ET (MET) can be used instead. Typically, MCOs, enzymes with hemic cofactor such as cellulose dehydrogenase, and enzymes with surface FeS clusters such as HASEs are reported to undergo DET, while GOx is only electrochemically addressed through MET. In this latter case, a small molecule acting as a fast and reversible electrochemical system will make the electron shuttle between the electrode and the FAD cofactor in the enzyme. Redox potential of the redox mediator as well as its affinity for the enzyme will drive the electrocatalysis. One important question is whether the enzyme will be more stable under DET or MET processes (Figure 10).

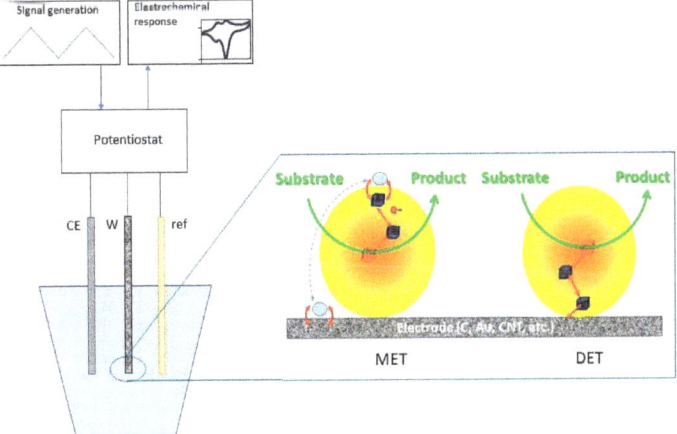

Figure 10. Scheme of a typical bioelectrochemical experiment showing the set up and the adsorption of a protein on the electrode surface either in a MET or a DET mode. The active site of the enzyme, the electronic relays and the redox mediator are represented as red squares, dark blue squares, and light blue spheres, respectively. CE, W, and ref represent counter, working, and reference electrodes, respectively.

5.2. Stabilization Strategies for Bioelectrocatalysis

5.2.1. How Far Can General Enzyme Stabilization Strategies Be Extended to Bioelectrocatalysis?

Table 1 reports bioelectrode stability for two main enzymatic reactions, i.e., O_2 reduction and H_2 oxidation. These two enzymatic reactions have been chosen as relevant for the development of biofuel cells, devices currently developed as alternatives to platinum-based fuel cells. As can be seen, many of the strategies detailed in the previous chapters of this review are involved in bioelectrode stabilization. It should be noted that in most cases stabilization refers to operational duration. Thus, the stability parameter is mainly evaluated through the percentage of preserved catalytic current. Although crucial for improving bioelectrocatalysis, only rare studies report enzyme conformational changes upon redox immobilization of enzymes on electrode.

From the survey of the literature reported in Table 1, general features of enzyme stabilization inducing bioelectrode stability can be recognized. This is the case for redox enzymes extracted from extremophiles yielding efficient bioelectrocatalysis while expected to be more stable than their mesophilic homologues even at room temperature. We already provided in Section 3 one relevant example showing that the NiFe HASE from the hyperthermophilic bacterium *A. aeolicus* was much more stable than its mesophilic homologue from *R. eutropha* [111]. This enhanced stability of the hyperthermophilic enzyme is fruitfully exploited to design an H_2/O_2 fuel cell displaying at the same time high catalytic currents reported to the mass of enzymes (in the order 1 A/mg enzyme), delivering mWs of power in a range of temperature 20–60 °C, and a half-life for the bioanode of one month, a quite encouraging stability compared to other reported devices [198]. It should be also mentioned that at the cathodic side, the thermostable BOD from *B. pumilus* is used which also shows enhanced stability compared to the widely used *M. verrucaria* BOD. Other recent works proved the efficiency of extremophilic enzymes in bioelectrochemical systems, as illustrated in the two following examples. *Bacillus* sp. FNT thermophilic LAC showed 60–80% of remaining activity after two weeks of storage at room temperature when the activity of mesophilic *T. versicolor* LAC was zero in the same conditions [199]. The activity of cellobiose dehydrogenase from *Corynascus thermophilus* retained more than 50% activity after 5 days of multicycle voltammetry mode and about 30% after 9 days [200].

As for non-redox enzymatic catalysis, stability of enzymatic electrodes requires to avoid enzyme leaching from the electrode. Entrapment in porous matrices and covalent immobilization through suitable functionalization of the electrode are ways to improve redox enzyme-based bioelectrode stabilization. However, it must be considered that the redox enzyme will sense an electric field during a bioelectrocatalytic process and that some species can be generated through the electrochemical potential application, thus requiring stabilization strategies more specific to bioelectrochemistry. In addition, the electron exchange between the electrode and the enzyme which is the basis of bioelectrocatalysis imposes specific constraints that may prevent the use of the strategies developed above. Among these constraints, the host matrix must be conductive, cross-linking that yields heterogeneities in enzyme orientation can make a population of enzymes incapable of electron transferring; addition of polyols may electrically isolate the enzyme preventing any electron transfer, or use of high concentrations of salts in the electrolyte solution can favor the leaching of the enzymes from the electrode surface. Hence, any stabilization protocols described above must be evaluated in light of their adaptability to bioelectrochemistry with redox enzymes.

Table 1. Bioelectrode stability: cases of multicopper oxidases (MCOs) and hyperthermophilic hydrogenase (HASEs).

Enzyme	Bioelectrode Design	Electrochemical Conditions	Reaction Followed/Purpose	Stability	
Cerrena unicolor C-139 LAC	Enzyme polymerized at a glassy carbon electrode through cold plasma	Acetate buffer pH 5, 25 °C	Rutin biosensing	42% of current is recovered after 8 days at pH 5.	[201]
Streptomyces coelicor LAC	Enzyme adsorbed on zinc oxide NPs capped with p-amino thiophenol and attached to graphene oxide	0.1 M phosphate buffer pH 5	Sucralose sensing	83% and 73% of the initial activity after 5 and 10 days, respectively.	[202]
Trametes hirsuta LAC GreeDo mutant obtained by combining computational design	Enzyme covalently attached to AuNPs grafted on graphite electrode	0.1 M acetate buffer pH 4	O_2 reduction	80% of initial activity after 96 h incubation at pH 4. 60% of initial activity after 2 h of continuous operation	[203]
LAC	Enzyme adsorbed on AuNPs electrodeposited on a screen-prinzed carbon electrode (SPCE) modified with polypyrrole	0.1 M acetate buffer pH 3.5 25 °C	Polyphenol sensing	85% of the initial activity after 1 month (storage at 4 °C)	[204]
Trametes versicolor LAC	Enzyme and Cu^{2+} co-adsorbed on pyrene-terminated block polymer on pyrolytic graphite surface and graphene papers	0.1 M acetic acid buffer pH 5	Pyrocatechol sensing	70% of original activity upon 90 days of the freeze-dried powder. 80% of activity after 3 freeze-thaw cycles. 31% activity at 70 °C. 96% of the initial activity after 30 days storage at 4 °C.	[205]
Trametes versicolor LAC	Enzyme entrapped within graphene-cellulose microfiber composite modified SPCE.	0.1 M sodium phosphate buffer pH 5	Catechol sensing	97% of the initial activity after 130 h storage.	[206]
Trametes versicolor LAC	Enzyme with Cu-nanoflowers mixed with CNTs	0.01 M phosphate buffer pH 7.4	O_2 reduction in a H_2/O_2 hybrid biofuel cell	85% of biofuel cell initial power density for 15 days at room temperature.	[207]
LAC	Enzyme absorbed on AuNPs-MoS₂ composite glued on a glassy carbon electrode (GCE) by Nafion	0.2 M acetate buffer pH 4	Catechol sensing	95% initial activity after 15 days of storage stability.	[208]
Trametes versicolor LAC	Enzyme entrapped in chitosan-CNT matrix	0.1 M phosphate buffer pH 7.4, 20 °C	O_2 reduction	82% of the initial activity within 10 days.	[209]

Table 1. *Cont.*

Enzyme	Bioelectrode Design	Electrochemical Conditions	Reaction Followed/Purpose	Stability	
Trametes versicolor LAC	Enzyme entrapped with Fe$_3$O$_4$ magnetic nanoparticles in chitosan matrix	0.2 M phosphate buffer	O$_2$ reduction	50% of the initial activity after 20 days.	[210]
Trametes versicolor LAC	Enzyme crosslinked with GA in a matrix made of room temperature IL-CNTs	50 mM acetate buffer pH 5	Polyphenol biosensing	33% of the initial activity after 20 days.	[211]
Bacillus subtilis LAC	Enzyme adsorbed on the electrode modified with thiol graphene-AuNP nanocomposite film	0.1 M HAc-NaAc buffer solution pH 4.6, 25 °C	Hydroquinone sensing	87% of the initial current response after 100 days	[212]
Trametes versicolor LAC	Enzyme and nanocopper incorporated in polyacrylonitrile/polyvinylidene electrospun fibrous membranes		Trichlorophenol removal	68% of the initial activity after 30 days.	[213]
Bacillus subtilis LAC	Enzyme covalently bound on functionalized CNTs		O$_2$ reduction	70% of initial activity after two weeks of storage at room temperature.	[199]
LAC	Enzyme cross-linked with electropolymerized L-lysine molecules on a graphene modified GCE	0.1 M sodium phosphate buffer pH 6	17β-estradiol sensing	68% of initial activity after 30 days storage at 4 °C.	[214]
Aspergillus sp. LAC	Enzyme adsorbed on polypyrrole-modified nanotubes and SrCuO$_2$ microseeds composite on graphite electrode	0.1 M phosphate buffer pH 7	2,4-dichlorophenol sensing	80% of initial activity after 14 days of storage at 4 °C.	[215]
Trametes versicolor LAC	Enzyme deposited by electrospray on carbon black modified screen-printed electrodes	0.1 M citric acid/sodium citrate buffer pH 4.5	Catechol detection	Valuable working and storage stability of the biosensor that remains 100% of its performances for up to 25 measurements and can be preserved at room temperature for at least 90 days.	[216]
LAC	Enzyme mixed with polyvinylacetate and AuNPs electrospun on platinum electrode	Phosphate buffer	Ascorbic acid sensing	Stability for 76 days.	[217]
LAC	Enzyme adsorbed on polyaniline/magnetic graphene composite	pH 5	Hydroquinone sensing	90% initial response after 30 days.	[218]
Trametes versicolor LAC	Enzyme embedded in polydopamine NPs with anchored AuNPs	0.1 M ABS pH 5	Norepinephrine sensing	91% initial response after 30 days.	[219]

Table 1. Cont.

Enzyme	Bioelectrode Design	Electrochemical Conditions	Reaction Followed/Purpose	Stability	
LAC	Enzyme covalently bound to a composite made of acrylate microspheres and AuNPs coated on a carbon paste screen-printed electrode	0.1 M phosphate buffer pH 5	Tartrazine sensing	51% initial response after 90 days.	[220]
Botryosphaeria rhodina LAC	Enzyme entrapment into nanostructured carbon black paste electrode	0.1 M phosphate buffer pH 6	Epinephrine sensing	93% initial response after 7 days.	[221]
Myrothecium. Sp BOD	Enzyme crosslinked into a biogel matrix made of Nafion deposited on carbon NPs	0.1 M phosphate buffer pH 7.2, 25 °C	O_2 reduction	Electrode potential decrease from 0.37 to 0.31 V after 24 h of operation at 1 mA·cm^{-2}	[222]
Magnaporthe orizae BOD	Enzyme entrapped in osmium-based hydrogels deposited on CNTs modified by electrografting of aryldiazonium salts	0.1 M phosphate buffer pH 5	O_2 reduction	After 1 day loss of 27% of the initial activity then stable for several days.	[223]
Myrothecium verrucaria BOD	Enzyme adsorbed on carbon cloth modified with aminobenzoic acid. Gas diffusion electrode.	0.1 M phosphate buffer pH 7.4	O_2 reduction	80% of initial activity after 1.3 h of operation of the H_2/O_2 biofuel cell.	[224]
Myrothecium verrucaria BOD	Enzyme covalently linked to carbon cloth modified by aminobenzoic acid and coated with a microporous hydrophobic Nafion layer. Gas diffusion electrode.	0.1 M phosphate buffer pH 7, 25 °C	O_2 reduction	80% of initial current after 6 h of continuous operation at 0 V vs. Ag/AgCl.	[225]
Myrothecium verrucaria BOD	Enzyme adsorbed on CNTs and protected by a carbon-coated magnetic nanoparticle layer	0.1 M HEPES buffer pH 7,	O_2 reduction	80% of initial activity after 48 h of continuous operation at 0 V vs. Ag/AgCl in the presence of 1 nM proteinase. 70% initial activity after 30 days of storage at room temperature.	[226]
Magnaporthe orizae BOD	Wild type enzyme adsorbed and mutants covalently bound to a maleimide CNT modified electrode	200 mM phosphate buffer 100 mM citrate buffer pH 7	O_2 reduction	40% and 10% initial activity after 3 days and 6 days, respectively, in the adsorbed mode. 95% and 52% initial activity after 3 days and 6 days, respectively, in the covalent attachment mode	[227]

Table 1. *Cont.*

Enzyme	Bioelectrode Design	Electrochemical Conditions	Reaction Followed/Purpose	Stability	
Myrothecium verrucaria BOD	Enzyme adsorbed or electrochemically deposited on Au polycrystalline electrode	0.1 M acetate buffer pH 5, 25 °C	O_2 reduction	Adsorption mode: 5% initial activity after 100 potential cycles between 0 and 0.8 V vs. Ag/AgCl at 50 mV/s. Electrochemical deposition mode: 80% initial activity after 200 cycles	[228]
Myrothecium verrucaria BOD	Enzyme adsorbed on MgOC modified with 6 amino 2-naphtoïque acid	0.1 M acetate buffer pH 5, 24 °C	O_2 reduction	50% of the initial activity after 5 days.	[229]
Myrothecium verrucaria BOD	Enzyme covalently bound to reduced graphene oxide modified by 4-aminobenzoic acid to reduce its aggregation	0.1 M phosphate buffer pH 7	O_2 reduction	55 h half-life.	[230]
Myrothecium verrucaria BOD	Enzyme adsorbed on buckypaper CNTs	Phosphate buffer pH 6.5, 25 °C	O_2 reduction	10 h half-life	[231]
Magnaporthe orizae BOD	Molecular engineering for covalent immobilization in macroporous gold electrode	0.1 M phosphate buffer pH 7.2, 25 °C	O_2 reduction	75% of the initial activity after incubation for 5 days.	[232]
Myrothecium verrucaria BOD	*Mv* BOD entrapped with ABTS in polydopamine layer on CNTs	0.1 M phosphate buffer pH 7, 25 °C	O_2 reduction	*Mv* BOD stable for 22 h of operation at 0.1 V vs. Ag/AgCl.	[233]
Bacillus pumilus BOD	*Bp* BOD: entrapped in pyrenebetaine on CNTs	50 °C	O_2 reduction	Rapid drop in activity, almost 0 after 3 h of operation at 0.1 V vs. Ag/AgCl	[233]
Bacillus pumilus BOD	Enzyme entrapment in a carbon felt modified by pyrene-NH2 modified CNTs	0.2 M phosphate buffer pH 6,	O_2 reduction	7 days half-life at 25 °C upon continuous operation of a H_2/O_2 biofuel cell.	[198]
Pyrococcus furiosus HASE	Enzyme adsorbed onto a CNT modified carbon felt electrode	50 mM phosphate buffer pH 7, 50 °C	H_2 oxidation	Voltage retains 90% of the initial value 33 h operation in a hybrid fuel cell.	[234]
Desulfovibrio vulgaris HASE	Enzyme embedded in viologen-based redox hydrogel modified porous carbon cloth.	0.1 M phosphate buffer pH 7.4	H_2 oxidation	60% of initial activity after 20 h at 0.16 V vs. SHE.	[224]
Desulfomicrobium baculatum HASE	Oriented covalent immobilization of the enzyme on modified CNTs	50 mM phosphate buffer pH 7.6, 25 °C	H_2 oxidation	80% of initial activity after 1 h in the biofuel cell.	[235]

Table 1. Cont.

Enzyme	Bioelectrode Design	Electrochemical Conditions	Reaction Followed/Purpose	Stability	
Desulfovibrio desulfuricans HASE	Enzyme embedded in viologen-based redox hydrogel on carbon cloth-based electrode	0.1 M phosphate buffer pH 7.4	H_2 oxidation	50% of initial activity after 11 h at 0.16 V vs. SHE.	[236]
Desulfovibrio vulgaris HASE	Enzyme embedded in viologen-based redox hydrogel and protected by catalase-GOx layer.	0.1 M phosphate buffer pH 7.4	H_2 oxidation	60% initial activity after 8 h in the presence of 5% O_2 at 0.016 V vs. SHE.	[143]
Aquifex aeolicus HASE	Enzyme adsorption on CNT modified by π-π stacking with $PyrNH_2$	0.2 M phosphate buffer pH 7, 25 °C	H_2 oxidation	1 month half-life at 25 °C upon continuous operation of a H_2/O_2 biofuel cell.	[198]
[NiFeSe] Desulfovibrio vulgaris Hildenborough HASE	Enzyme entrapped in methylviologen-based redox polymer films deposited on carbon cloths	0.1 M phosphate buffer pH 7.4, room temperature	H_2 oxidation	25% and 50% of initial activity depending on the variant after 7 h in chronoamperometry experiments.	[237]

5.2.2. Engineering of Enzymes Seeking for Both Enhanced ET Efficiency and Stability

The general engineering procedures developed to enhance enzyme stability can be adapted for redox enzymes, including introduction of stronger bonds, removal of potential degradation sites, or oligomeric structure formation [238–240]. However, the most interesting strategies here are to obtain enhanced stability and ET efficiency at the same time. Then, the main targets for protein engineering will lie (i) in the vicinity of the active site or (ii) at specific points of the enzyme surface able to anchor it with the best orientation for the highest ET rate (Figure 11). For strategy (i), the vicinity of the heme cavity of peroxidase [241] or of the CuT1 site of LAC has been targeted [242,243]. Two recent reports illustrate this concept. Two alanine residues were mutated to one leucine and one valine. Not only did the introduction of hydrophobic residues increase the redox potential of the CuT1 site by 50 mV, but it also protected the CuT1 site from the solvent as well as enhanced temperature and pH stability of the variant. The mutant was stabilized at the electrode. It retained roughly 80% of its initial activity after 96 h incubation at pH 4.0 and presented a half-life of 60 min at 70 °C against 23 min for the wild type protein [203]. With the aim of electricity production directly from seawater recycling, recombinant forms of CotA LAC from *B. licheniformis* were produced [244]. A double mutation, one in a region close to the CuT1 and the other on the surface of the protein, was shown to induce a synergic effect in bioelectrocatalysis. Conformational changes in the double mutant were proved by CD and fluorescence that allow the variant to catalyze O_2 reduction in seawater, a very unusual property for classical LACs. Protein leaching under such high ionic strength conditions must however be solved.

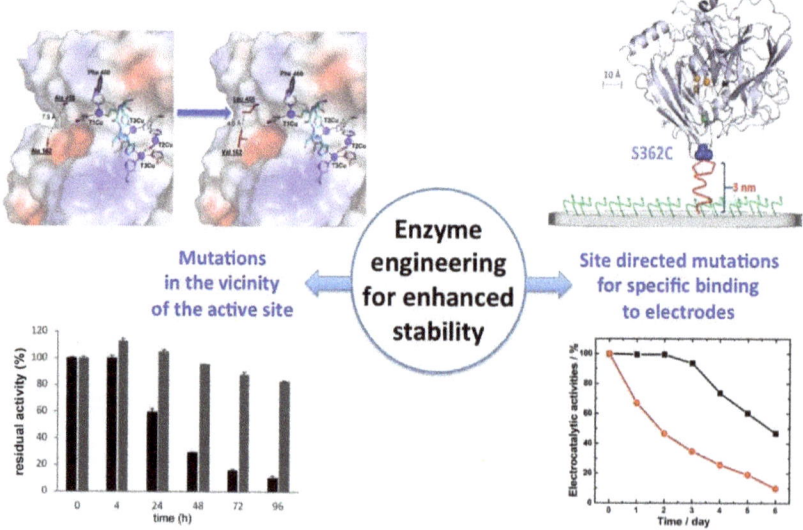

Figure 11. Enzyme engineering strategies for enhanced stability. (**Left**) Mutations in the vicinity of the active site; computer-guided mutagenesis and directed evolution of a fungal LAC in CuT1 vicinity induce enhanced stability at pH 4 compared to the WT. Reproduced with permission from [243]. (**Right**) Covalent immobilization of BOD through site-directed mutagenesis and reaction with maleimide groups on the electrode surface induce enhanced stability of the mutant (black curve) compared to the WT (red curve). Adapted with permission from [227].

For strategy (ii), covalent immobilization of enzymes at electrodes was realized by producing site-directed variants with cysteine residues located at different sites of the enzymes able to react with maleimide groups immobilized on the electrode. This elegant method combines controlled orientation of the protein at the electrode for ET, sufficient flexibility for activity, and expected enhanced stability. In the case of cellulose dehydrogenase, 40% of initial electroactivity was recovered after 2 months of storage while the wild type enzyme loses progressively its activity within 20 days [245,246]. In addition, controlled orientation allowed investigating mechanistic aspects of the ET. The immobilization method was extended to BOD from *Magnaporthe oryzae* showing an enhanced stability with almost no loss of electroactivity for 3 days [227] (Figure 11).

5.2.3. Conductive Porous Material as Redox Enzyme Host Matrices Favoring Fast ET

We discussed in the previous sections how porous materials can be beneficial for the stabilization of embedded enzymes. The further requirements for their use as host matrices of redox enzymes are that either the porous material is conductive to ensure DET, or redox mediators can be added either diffusing or co-immobilized within the matrix. We emphasized that MOFs are nowadays considered as very attractive materials. To ensure electron conduction within MOFs, multicomponent systems can be synthetized composed of MOFs, enzymes and conductive nanomaterials such as graphene nanosheets or CNTs [147]. A nice example is provided by Li et al. who encapsulated a LAC in ZIF-8 in the presence of CNTs (LAC@ZIF-8) [247]. Fluorescence measurements highlighted a negligible enzyme leaching. When compared to free enzyme or to enzyme embedded in MOF with no CNT, the LAC@ZIF-8 hybrid displayed improved thermal stability, enhanced resistance to solvents as well improved long-term storage stability. However, although DET was demonstrated, the catalytic efficiency was greatly enhanced by redox mediator addition, suggesting that a large enzyme population was not directly wired in the hybrid material. Mesoporous silica nanotubes are another example of a suitable matrix when coated with a graphene sheet layer to enhance its electrical conductivity. BOD entrapped in the material was able to reduce O_2 in the absence of redox mediators, and the bioelectrode stored in water at 4 °C for 15 days retained around 80% of initial current [248]. The stability of the bioelectrode was suggested to be linked to the size of the nano-channels (12 nm) being close to the size of the enzyme, limiting its mobility. Fructose dehydrogenase and HASE were immobilized in MgOC with pore size in the range 10–150 nm [249,250]. DET occurred between enzymes and the material regardless of the pore-size, but a pore diameter close to the enzyme size enhanced the thermal and long-term stabilities.

The relationship between pore size and stability is shared by non-redox and redox enzymes. However, for redox ones, the orientation of the enzyme toward the material wall must also be considered for a direct wiring. In pores presenting a diameter close to the enzyme size, the distance between electrical relays on the surface of enzymes and the material will be statistically minimized. Multi-point contacts are expected to occur at the same time. This means that both stability and ET rates should be enhanced simultaneously [238,251]. Porous gold obtained either by dealloying processes or anodization is increasingly used for immobilization of redox enzymes. Their pore sizes are compatible with efficient enzyme entrapment [144]. Using electrochemistry, a narrow distribution of orientation of BOD favorable for enhanced DET was measured [252]. A model based on close packing spheres explained why the curvature of pores was beneficial for DET. Unfortunately, no examination of the enzyme conformation in the porous gold material was provided. Effect of nanomaterial curvature was evocated in another work to explain LAC properties once immobilized in a CNT network in the presence of ethanol [253]. An improved direct catalytic current for O_2 reduction as well as reduced chloride inhibition was measured. CD and ATR-IR demonstrated a net difference in the secondary structure when the enzyme/CNT/ethanol system was immobilized at the electrode as compared with enzyme in solution. The structure stabilization suggested that ethanol favors contact between CNT and LAC in its native form with an orientation favorable for DET. Although not deeply investigated, CNT curvature was suggested to help in the stabilization of the immobilized dehydrated enzyme.

5.2.4. Redox Enzyme Encapsulation in Hydrogels for Bioelectrocatalysis

Entrapping redox enzymes into hydrogels is expected to provide a suitable hydrated environment able to maintain the enzyme in a functional active state. Some examples of use of a sole polymer for direct electrocatalysis involving redox enzymes were recently reported [254]. Conductive polyaniline (PANi) forms a nanostructured hydrogel once deposited on an electrode [255]. The PANi hydrogel exhibits a continuously connected hierarchical 3D network with pore diameter of 60 nm, allowing a short electron diffusion length. The PANi-formate dehydrogenase electrode was efficient for catalytic reduction of CO_2 into formate. Although no exhaustive study of stability was made, conversion of CO_2 was efficient for 12 h. However, hydrogel matrices are very often poorly conductive, and conductive nanostructures or redox mediators must be co-immobilized in the gel to ensure electron conduction. Chitosan is a natural polyamino-saccharide that forms thin hydrogel films at electrodes through controlled electrodeposition process. The amine groups are available for attachment to a functionalized electrode or for anchorage of the enzyme. CNTs mixed with chitosan render the material conductive. Hosting a LAC, the so-built bioelectrode retained 70% initial current after 60 days of continuous measurement [209,256]. The same authors reported an improved bioelectrode using genipin as a cross linker. The LAC-based electrode was implanted in rats and remained operational for 167 days in vivo [257].

Alternatively, redox mediators can act as electron relays between the electrode and the enzyme inside the hydrogel. In the search for enzyme stabilization, Matsumoto et al. prevented leaching of a biotinylated GDH by immobilizing it into a streptavidine-hydrogel prepared by click chemistry between sortase A and PEG [258]. In the presence of a ferrocene derivative as a redox mediator, the so-designed electrode showed improved stability when compared to the electrode with no hydrogel. One issue was the weak stability of the hydrogel itself on the electrode. This last result points out one more requirement for redox enzyme-based electrode stabilization: Not only the enzyme must be stable but the interaction between the electrode and the hybrid enzyme-host matrix must also be strong enough to avoid leaching of the whole biomaterial. To ensure stability of the bioelectrode, hydrogel can be combined within porous conductive material. As an illustration, a hydrogel formed by cross-linking bovine serum albumin, GOx and arginine was interlocked within the pores of a carbon cloth [259]. Leaching of enzyme was less than 0.5% of the total amount of embedded protein within 48 h. UV-Visible spectrometry and CD demonstrated that 90% of the enzyme secondary structure was preserved in the network. In the presence of ferricyanide as a redox mediator, the current density for glucose oxidation was maintained for 25 days.

Diffusing redox mediators are however undesirable in many cases, either because they limit the rate of reaction or because they can be toxic when leaching to the environment in which the enzyme operates. Redox hydrogels, introduced 30 years ago by Adam Heller, are now widely exploited for immobilization of a variety of enzymes [260]. These hydrogels combine the advantages of a hydrated matrix ensuring protection of the enzyme and rapid diffusion of the substrate with the presence of tethered redox moieties (they can be based on osmium, ruthenium, ferrocene, colbaltocene, quinone, or viologen derivatives) that can electrically wire the enzyme by electron hopping inside the gel [261–263]. The redox potential of the hydrogel matrix can be adapted to the desired reaction to be catalyzed by tuning the ligands of the redox moieties. Ferrocene entity was tethered to linear PEI and used to cross-link GOx [264]. Of initial current, 70% and 36% was retained after 21 days of storage and 6 h of continuous operation, respectively. This marked difference underlines the effect of applied potential on the stability of the bioelectrode, an issue that we will further discuss in Section 5.3.2. Os-tethered hydrogel embedding GDH was loaded in porous MgOC deposited by ink-drop-casting on an electrode [265]. The high catalytic currents for glucose oxidation were linked to the controlled pore diameter of the MgOC matrix that allows high loading of redox polymer and rapid substrate diffusion. Moreover, the combined advantage of hydrated redox polymer and porous structure allowed improved stability. Of initial catalytic current, 44% was recovered after 10 days of operation against 3% for the hydrogel-enzyme hybrid deposited on a glassy carbon electrode. Direct oxidation of glucose by *Corynascus thermophilus* cellobiose dehydrogenase was compared to the mediated one obtained when the CNT-based electrode was further modified by an Os-based redox hydrogel [200]. MET process generated higher catalytic currents than DET one proceeding at a lower potential through the FAD cofactor instead of the heme domain in the case of the DET process. Of the mediated electrocatalytic current, 30% remained after 9 days of CV cycling. In such configuration, it will be interesting in the future to get the comparative stability through the DET process. Last but not least, redox hydrogels have been shown in recent years to act as shields against O_2 for sensitive enzymes. This topic will be further developed below in Section 5.3.3.

5.2.5. When Aggregates or Partially Unfolded Enzymes Operate in Bioelectrocatalysis

We clearly stated in the introduction of this review that the secondary and tertiary structure of enzymes control their activity. Any change in the conformation was thus expected to modify the activity, yielding inactive states upon denaturation. We otherwise underlined that aggregated enzymes, for example in the form of CLEAs, could maintain the native conformation showing a stable, although often lower, enzymatic activity. Efforts to maintain stable enzyme electroactivity are mainly directed toward enzymes with unchanged conformations. However, and contrary to these strategies, examples of aggregates or partially unfolded forms of proteins have been reported to promote bioelectrocatalytic properties. Aggregated proteins immobilized on nanostructured carbon-based electrodes may enabled direct electrocatalysis showing enhanced stability compared to non-precipitated proteins [266,267]. Pyranose oxidase was immobilized on a CNT-based electrode via three different procedures: (i) covalent attachment (CA), (ii) enzyme coating (EC), (iii) enzyme precipitate coating (CLEA). After 34 days at room temperature, CLEA retained 65% of initial electroactivity toward glucose oxidation, while CA and EC maintained 9.2% and 26% of their initial activities, respectively. LAC aggregates were formed by computational design through production of a new dimeric interface that induces self-assembly of the protein into crystalline-like assemblies [268]. Combined with CNT incorporated in the aggregates, the bioelectrode showed high direct electroenzymatic rate for O_2 reduction with enhanced thermostability. LAC from *T. versicolor* was precipitated in the presence of Cu^{2+} yielding flower-like particles [207]. The Cu/LAC nanoflowers were integrated in carbonaceous nanomaterials and tested as catalysts for O_2 bioelectroreduction. A well-shaped direct electroenzymatic signal was observed, demonstrating the functionality of the electron pathway through CuT1 in the flower particles (Figure 12). The bioelectrode delivered 72% initial current after 6 days of daily 2 h discharge, and the H_2/O_2 fuel cell constructed based on the Cu/LAC flower-based biocathode maintained 85% initial power after 15 days of operation.

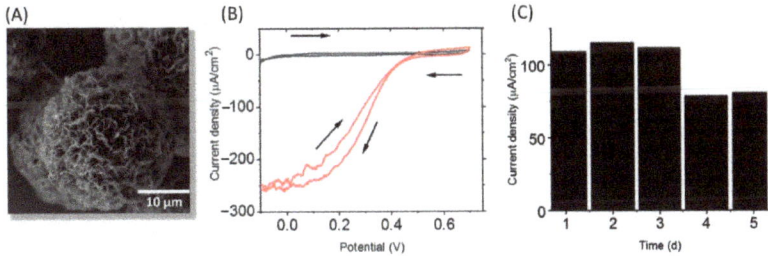

Figure 12. LAC assembled through Cu-flower like particles displays direct and stable electrocatalysis of O_2 reduction. (**A**) SEM image of a Cu/LAC nanoflower, (**B**) Catalytic reduction of O_2 (red curve) by LAC in Cu nanoflowers on CNTs, (**C**) Operational stability. Adapted with permission from [207].

It was also recently shown that the large sub-unit of a NiFe HASE was able to sustain H_2 oxidation by itself, showing that an incomplete structure still allows catalysis [269]. As another illustration, LAC from *Aspergillus* sp. was immobilized on Fe_2O_3 NPs [270]. Careful spectroscopic analysis, including FT-IR, fluorescence and EPR, demonstrated a partial unfolding of the protein which tends to a higher exposure of the CuT1. A large direct catalytic current in the range of 3 mA/cm^2 was observed, with an electrochemical CV shape denoting fast ET. However, it should be stressed that heterogeneity of enzyme population certainly exists at the electrochemical interface, and methods are now required to distinguish whether the electrochemical activity is actually linked to the unfolded proteins or to the small fraction of proteins remaining correctly folded.

5.3. Parameters Specific to Bioelectrocatalysis Stabilization

Bioelectrocatalysis implies not only the immobilization of the enzyme on a conductive support but also the application of a potential that the enzyme will sense. Hence, the stability of a bioelectrode will include the stability of the enzyme in the immobilized state and the stability of the electrocatalytic response under an electric field which is reflected in most cases by a measure of a current, a voltage or the impedance of the system. However, the variation of the electrochemical signal may have different origins. Progressive leaching of the enzymes from the electrochemical interface can explain a decrease in the catalytic current, while changes in the conformation or in the reorientation of the enzyme can explain modification in the catalytic response due to lower electron transfer rates. Methods probing the relationship between electroactivity and amount of proteins on the electrochemical interface, as well as the relationship between electroactivity and changes in enzyme conformation, are not readily available. Actually, many bioelectrocatalysis-related studies report a decrease of the catalytic current with time. This decrease has long been named as "film loss", suggesting that enzyme leaching from the electrode would be responsible for the current evolution. However, coupling SPR and quartz crystal microbalance with dissipation (QCM-D) to electrochemistry, it was shown that other phenomena than simple enzyme desorption must account for bioelectrode instability [271,272]. In the following, we would like to emphasize some critical parameters that will affect the overall electrocatalytic response stability and the methods developed to get more insights in the overall process.

5.3.1. Reorientation of the Immobilized Enzyme

We stated just before that one key issue in order to enhance the interfacial ET between a redox enzyme and an electrode is the orientation of the enzyme on the solid surface so that the electronic relay on the surface of the protein can be positioned at a distance compatible with fast ET. In many studies reporting enzyme orientation, hydrophobic or electrostatic interactions are used to control the enzyme positioning. These interactions are relevant because they mimic in vivo recognition with the substrate or between redox partners. LAC, as an example, is involved in transformation of many substrates such as phenolic compounds. In accordance, the Cu T1 is engaged in a hydrophobic pocket that can be recognized by the electrode modified by phenyl or anthracenyl groups, leading to an orientation favoring efficient electron transfer [273]. The recognition between HASE and its redox partner cytochrome c_3 is another example illustrating electrostatic parameters to be considered in ET. By NMR docking and by resolving the Poisson–Boltzmann equation, it was established that a functional complex is formed thanks to a small negatively charged region on the HASE surface that can interact with the positively charged cytochrome. Such local electrostatic interaction was able to overcome electrostatic repulsion of the overall positive charges of the interacting proteins [274,275]. The same situation arises in bioelectrocatalysis, considering the electrode as the redox partner. Local charges are reflected by charge heterogeneity on the surface of the enzyme and rationalization can be obtained through the calculation of the dipole moment direction. This latter will depend on the pH of the enzyme environment. Hence, it can be expected that an enzyme takes an orientation favorable to ET at one pH but rotates on the electrode when the pH of surrounding solution is modified. It is the most crucial when electrocatalysis involves proton consumption or production, as it is the case for many electroenzymatic reactions, thus inducing local changes in pH in the weakly buffered solution and potential modification of the catalytic signal.

Evidence of the rotation of BOD was demonstrated on gold electrodes modified by thiol-based SAMs [116]. Both the pH at which the enzyme was adsorbed and the pH of the electrolytic solution were varied. By doing so, not only the charge of the electrode but also the charge of the enzyme varied. Coupling electrochemistry to ellipsometry, SPR, and Phase Modulation Infrared Reflection Absorption Spectroscopy (PMIRRAS), it was shown that the global charge of the protein controls the amount of adsorbed molecules as a function of electrostatic interactions tuned by the pH of adsorption. It was also demonstrated that the local charge in the vicinity of the CuT1 controls the orientation of the enzyme, hence the catalytic current for O_2 reduction. Hence, when the amount of enzyme and its orientation were a priori fixed by adsorption at a fixed pH, the catalytic current recorded in buffers of various pHs nevertheless varied following rotation of the enzyme to adopt the most favorable orientation. It was clearly shown that orientation could be reversibly tuned by changing the pH of the electrolyte, shifting the catalysis from a slow ET rate to a fast one (Figure 13).

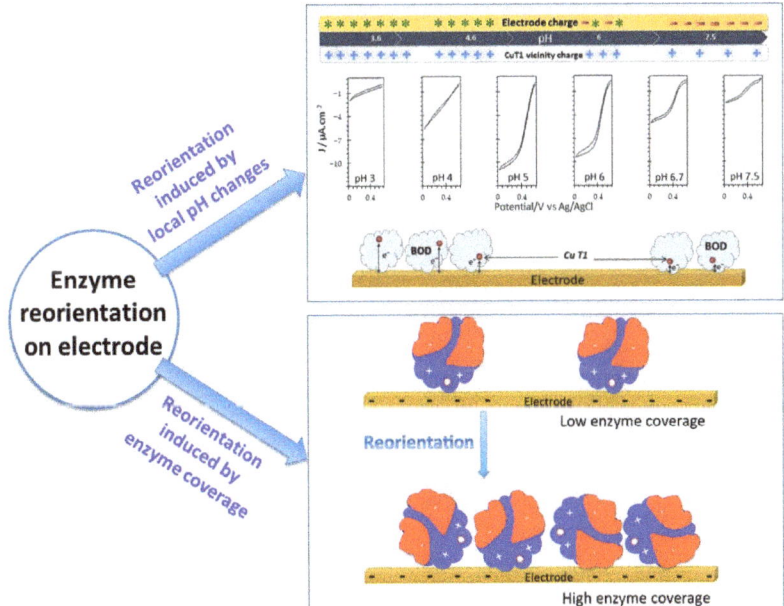

Figure 13. Reorientation of enzymes on electrode may alter enzyme stability. (**Top**) Tuning the pH of the electrolyte induces rotation of the adsorbed bilirubin oxidase on the electrode surface, hence induces modification in bioelectrocatalysis for O_2 reduction. Adapted with permission from [116]. (**Down**) Scheme of the reorientation of enzymes on an electrode induced by surface coverage. The white sphere represents the active site. Adapted with permission from [115].

Enzyme reorientation, leading to potential bioelectrocatalysis instability, may have another origin. In the course of protein adsorption, protein-protein interactions are increasingly dominating. Hence, while the adsorption of the first proteins in a favorable orientation for ET can be controlled by electrostatic interactions, the interactions between neighboring adsorbed proteins may induce reorientation to decrease the repulsive forces [115] (Figure 13). Coupling electrochemistry to QCM and interferometry for studying O_2 reduction by BOD on a gold electrode, Blanford et al. clearly demonstrated that an optimum surface coverage exists yielding the highest stable electrocatalytic current [276]. Although proof of enzyme reorientation was not provided, QCM-D allowed to conclude that this optimum was linked to a balance between rate of enzyme adsorption and deformation.

5.3.2. Effect of the Electric Field

Compared to other catalytic processes, the special feature of bioelectrocatalysis lies in the electric field imposed at the electrode interface. Although partially shielded by the ions present in the solution, the extent of the electric field can still reach the size of immobilized enzymes, especially in weakly supported electrolytes. It should be expected to have a major impact, but only few studies report the effect of the electric field on enzyme or bioelectrode stability. This lack of information most probably comes from the difficulty to determine which effect among protein leaching, denaturation or reconformation can be responsible for any variation in the electrocatalytic response. The recent use of coupled methods with electrochemistry will certainly lead to major advancements in that field. Adsorption of MCOs on planar gold electrodes was taken as a model for such studies. It was shown using electrochemistry coupled to QCM-D that cycling the potential for O_2 reduction by *M. verrucaria* BOD accelerated mass decrease compared to applying a fixed potential [272]. It was hypothesized that varying the electric field near the surface may distort the adsorbed protein, expelling water and contributing to enzyme denaturation. It was further shown with the same enzyme that continuous cycling induced more stability of the bioelectrode than cycling during a same period but at regular intervals, the electrode being at the open circuit potential (OCP) when not cycling [116]. This result suggests a much lower stability of the bioelectrode when the applied potential was close to the OCP. Considering the potential of zero charge (E_{pzc}) of the electrode, it could be concluded that the strong electric field generated when holding the potential far away from E_{pzc} induced bioelectrode destabilization. Kano's group rationalized these experimental data by modeling the electrostatic interactions between enzymes and electrodes in the electric double layer [275]. Taking the copper efflux oxidase (CueO) enzyme as a model enzyme adsorbed either at bare or butanethiol-modified gold electrodes, the authors demonstrated how the electrical double layer plays a key role in enzyme stability for bioelectrocatalysis. It can be argued that the diameter of the enzyme is much higher than the length of the electrical double layer (5 nm vs. 1 nm at a typical ionic strength for electrocatalysis). However, the domain of CueO facing the electrode is indeed located in the double layer and will face an electric field whose force depends on the applied potential and on the E_{pzc} of the electrode. Hence, at the bare gold electrode, the strong electric field induced by an applied potential much more positive than E_{pzc} of the electrode, in addition to the complementary charges chosen to favor DET orientation, leads to denaturation of the protein.

Heterogeneity of the electric field inside the pores of a porous matrix otherwise affects enzyme stabilization. A model developed in the group of Kano demonstrated the existence of an inner pore region where the electric field remains low, and where weak electrostatic forces help in protein stabilization [238,277,278]. The higher electric field at the entry of the pore should enhance the ET, while inducing less stability [279,280]. This effect known as the "edge effect" was recognized early in the history of bioelectrochemical studies using cytochrome *c* and graphite and may explain efficient bioelectrocatalysis at carbon nanofibers (CNFs) [281–284].

Finally, it should be mentioned the role of the electric field sensed by the enzyme within the course of mediated electrocatalysis against direct electrocatalysis. A relevant example is bioelectrocatalytic oxidation of hydrogen by HASEs. It is known that HASEs are reversibly inactivated at high potentials [285]. In a typical cyclic voltammetry experiment where the enzyme is adsorbed on an electrode, this inactivation results in a decrease in the catalytic current in the forward scan at a potential higher than ca. -0.4 V vs. Ag/AgCl at pH 7. Current recovers when the potential is reversed. It was shown that pushing the potential to even more positive values induces an irreversible inactivation of the enzyme [286]. Such inactivation process may have consequences on the operation of fuel cells using HASE as anode catalyst, requiring regular steps of reducing voltages to reactivate the enzyme. Alternatively, it was shown that embedding the hydrogenase in a viologen-based hydrogel protected the enzyme from high potential deactivation. In this MET process, the enzyme no longer senses the high potential of the electrode but the potential of the viologen redox couple which is sufficiently negative to prevent HASE inactivation [287–289]. Nevertheless, the use of redox mediators induces other detrimental effects towards enzyme stability, such as production of reactive oxygen species (ROS) that will be discussed in the next section.

5.3.3. Protection against Reactive Oxygen Species (ROS) Production

Two electron reduction of O_2 yields the formation of unstable superoxide $O_2^{\bullet -}$ which is the precursor for other reactive species including H_2O_2 and $^{\bullet}OH$. Incomplete oxygen reduction can be encountered in the course of certain enzymatic reactions, or when O_2 is directly reduced at an electrochemical interface. A relevant example of ROS produced by enzymatic reactions is the oxidation of sugars by the flavin-based GOx. Here, O_2 is a co-substrate of the enzyme and is reduced to H_2O_2 after the release of D-glucono-δ-lactone within glucose oxidation [290]. Identically, oxidation of lactate-by-lactate oxidase produces H_2O_2. To avoid any detrimental effect of H_2O_2, one solution is to electroenzymatically reduce this latter by peroxidases [291]. Alternatively, catalase has been used to decompose H_2O_2 [292]. Engineering of flavoproteins is also largely reported to decrease O_2 activity [293,294]. Such strategy is particularly useful when the flavoprotein is embedded in redox hydrogels, where the competition between reaction of the enzyme with the redox mediator or O_2 occurs within the polymer, decreasing the efficiency of the bioelectrocatalysis. An additional issue comes from the reduction of O_2 directly by the redox entities of the polymer.

Reactions other than the oxidation of sugar by GOx embedded in Os-based polymers can expose enzymes to ROS. Actually, low potential viologen-based redox polymers developed to protect HASEs against high potential inactivation also serve as shields against O_2 [224,237,288]. However, the viologen-based matrix itself may generate ROS in the course of the HASE protection mechanism, inducing its progressive decomposition [295]. To tackle this issue, a top polymer layer containing both GOx and catalase was used to ensure a complete HASE protection [143]. Alternatively, iodide was added in the whole system to induce dismutation of H_2O_2. An improvement of the half-life of the HASE in the presence of O_2 was demonstrated [295]. In the case of O_2-tolerant HASEs, their specific sensitivity/tolerance to O_2 has been shown to lead to an even greater sensitivity to ROS. Hence, it was reported that ROS formed by applying low potentials to a HASE-based carbon felt electrode in the presence of O_2 irreversibly deactivated the enzyme contrary to reversible O_2 inactivation [111,198] (Figure 14). In this case, a protective upper layer composed of a 3D porous carbon matrix increased the stability by 4–6 times against ROS compared to a 2D electrode. It was proposed that the 3D matrix could scavenge ROS before reaching the enzymes inside the pores.

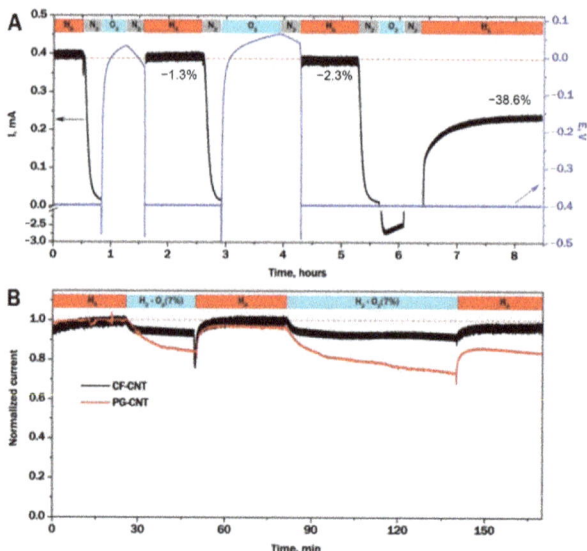

Figure 14. Chronoamperometric response for H_2 oxidation by *A. aeolicus* HASE embedded in a CNT-modified carbon felt electrode showing the inactivation of the enzyme by ROS electrochemically produced (**A**) and protection by a porous matrix (**B**). Reproduced with permission from [198].

6. Conclusions and Future Directions

Electrocatalysis involving redox enzymes takes advantage of the high selectivity and affinity of the biomolecules, as well as the large diversity of identified enzymes, hence a large variety of substrates. Possible applications range from implantable medical devices (insulin pump, neurological stimulator, artificial organs, etc.), wearable sensors, and environmental diagnostic to energy converting devices or CO_2 capture [16,17,296–299]. Many strategies are available to enhance enzyme stability in general. However, the specificity of redox enzymes that sense an electric field and require a controlled orientation to operate on electrochemical interfaces makes the stability of redox enzymes one of the most important issues to resolve before a large-scale commercial development of bioelectrocatalysis. The search for a good balance between stability and activity of enzymes is a prerequisite, even complexified by the search for efficient ET in the case of redox enzymes. Recent examples in the literature concerning MCOs revealed the difficulty of getting improved interfacial ET parameters or improved redox properties while maintaining both the stability and the activity of the enzyme [300,301]. No doubt that protein engineering, which is increasingly carried out with the aid of innovative solutions based on emerging genetic tools, will offer new opportunities of producing stable enzymes. There is also no doubt that the screening of the biodiversity will allow the discovery of redox enzymes with outstanding properties, which can be applied either directly in bioelectrochemical devices, or as model enzymes for protein engineering for high stability.

Alternative strategies can be envisioned in case enzyme stability is limiting the targeted device. The first one would be to refresh the bioelectrodes regularly, so as to renew inactivated enzymes by new active ones. Increasing salt concentration in the medium was reported to allow LAC desorption, rendering the electrode available for fresh enzyme adsorption [209]. Refreshment of the bioelectrode can also be done by adsorbing the enzyme on magnetic nanoparticles [302]. Applying a suitable magnetic field may allow assembling or release of the biocatalyst on the electrode. Cosnier et al. proposed to use enzymes and redox mediators encapsulated in glyconoparticles, both diffusing in solution [303]. This strategy was supposed to allow protein and mediators to freely rotate for an efficient ET and to offer in addition the possibility of easy refreshing of the biocatalyst by a simple exchange of the solutions. Although suitable to overcome enzyme instability, these procedures based on enzyme renewal, however, do not appear to be sustainable or cost effective. Nanozymes, or bioinspired catalysts, could be desirable alternatives to enzymes in the future [304–307]. Taking benefit of the fundamental understanding of bioelectrocatalysis, these synthetic catalysts may provide higher stability while operating the closest to the enzymatic catalysis.

Should redox enzymes or biosynthetic catalysts be used, novel methodologies must be developed in order to get new insights in the parameters that limit the stability of bioelectrocatalysis. In situ and *in operando* methods are certainly the most appealing within the objective of understanding and enhancing bioelectrode stability [308,309]. Recent reviews by Reisner et al. and by Crespilho et al. discussed various methods, and especially coupled methods, currently available to study bioelectrode interfaces [309,310]. Methods allowing mass or layer thickness quantification, i.e., SPR, QCM-D and ellipsometry, coupled to electrochemistry will give access to the relationship between enzyme loading or release and catalytic response [21,116,271,272,311]. Electrochemistry coupled to spectroscopies (IR (including SEIRA, PMIRRAS), XAS, fluorescence, Raman (including SERRS), EPR, CD, mass spectrometry) will help in the determination of conformational change once enzyme is immobilized or as a function of applied potential or current [312–317]. Microscopy coupled to electrochemistry will allow the mapping of the electrocatalysis [318–320]. Finally, getting more and more accurate experimental data on enzyme behavior on electrodes may be implemented in theoretical models that in return will help in bioelectrode rationalization [21,117,321–323].

Author Contributions: Writing, C.B.; review and editing, H.-M.M., A.d.P. and I.M.; writing, editing, and supervision, E.L. All authors have read and agreed to the published version of the manuscript.

Funding: This work was supported by ANR (ENZYMOR-ANR-16-CE05- 0024). H-M Man grant was funded by Région PACA, France, Centre National de la Recherche Scientifique, CNRS, France and Hyseas company, Cannes, France. C. Beaufils grant was funded by Aix-Marseille University and by "Agence Innovation Defense" of the French army ministry.

Conflicts of Interest: The authors declare no conflict of interest.

References

1. Woodley, J.M. New frontiers in biocatalysis for sustainable synthesis. *Curr. Opin. Green Sustain. Chem.* **2020**, *21*, 22–26. [CrossRef]
2. Reetz, M.T. What are the Limitations of Enzymes in Synthetic Organic Chemistry? *Chem. Rec.* **2016**, *16*, 2449–2459. [CrossRef] [PubMed]
3. Hauer, B. Embracing Nature's Catalysts: A Viewpoint on the Future of Biocatalysis. *ACS Catal.* **2020**, *10*, 8418–8427. [CrossRef]
4. Bommarius, A.S.; Paye, M.F. Stabilizing biocatalysts. *Chem. Soc. Rev.* **2013**, *42*, 6534–6565. [CrossRef] [PubMed]
5. Mandpe, P.; Prabhakar, B.; Gupta, H.; Shende, P. Glucose oxidase-based biosensor for glucose detection from biological fluids. *Sens. Rev.* **2020**, *40*, 497–511. [CrossRef]
6. Mano, N.; de Poulpiquet, A. O-2 Reduction in Enzymatic Biofuel Cells. *Chem. Rev.* **2018**, *118*, 2392–2468. [CrossRef] [PubMed]
7. Lojou, E. Hydrogenases as catalysts for fuel cells: Strategies for efficient immobilization at electrode interfaces. *Electrochim. Acta* **2011**, *56*, 10385–10397. [CrossRef]
8. Lubitz, W.; Ogata, H.; Rudiger, O.; Reijerse, E. Hydrogenases. *Chem. Rev.* **2014**, *114*, 4081–4148. [CrossRef]
9. Alfano, M.; Cavazza, C. Structure, function, and biosynthesis of nickel-dependent enzymes. *Protein Sci.* **2020**, *29*, 1071–1089. [CrossRef]
10. Najafpour, M.M.; Zaharieva, I.; Zand, Z.; Hosseini, S.M.; Kouzmanova, M.; Holynska, M.; Tranca, I.; Larkum, A.W.; Shen, J.R.; Allakhverdiev, S.I. Water-oxidizing complex in Photosystem II: Its structure and relation to Manganese-oxide based catalysts. *Coord. Chem. Rev.* **2020**, *409*. [CrossRef]
11. Morello, G.; Megarity, C.F.; Armstrong, F.A. The power of electrified nanoconfinement for energising, controlling and observing long enzyme cascades. *Nat. Commun.* **2021**, *12*, 340. [CrossRef] [PubMed]
12. Stolarczyk, K.; Rogalski, J.; Bilewicz, R. NAD(P)-dependent glucose dehydrogenase: Applications for biosensors, bioelectrodes, and biofuel cells. *Bioelectrochemistry* **2020**, *135*, 107574. [CrossRef] [PubMed]
13. Alvarez-Malmagro, J.; Garcia-Molina, G.; De Lacey, A.L. Electrochemical Biosensors Based on Membrane-Bound Enzymes in Biomimetic Configurations. *Sensors* **2020**, *20*, 3393. [CrossRef] [PubMed]

14. Bollella, P.; Katz, E. Bioelectrocatalysis at carbon nanotubes. In *Methods in Enzymology: Nanoarmoring of Enzymes with Carbon Nanotubes and Magnetic Nanoparticles*; Kumar, C.V., Pyle, A.M., Christianson, D.W., Eds.; Elsevier: Amsterdam, The Netherlands, 2020; Volume 630, pp. 215–247.
15. Wu, R.R.; Ma, C.L.; Zhu, Z.G. Enzymatic electrosynthesis as an emerging electrochemical synthesis platform. *Curr. Opin. Electrochem.* **2020**, *19*, 1–7. [CrossRef]
16. Yuan, M.W.; Kummer, M.J.; Minteer, S.D. Strategies for Bioelectrochemical CO_2 Reduction. *Chem. A Eur. J.* **2019**, *25*, 14258–14266. [CrossRef]
17. Mazurenko, I.; Wang, X.; de Poulpiquet, A.; Lojou, E. H-2/O-2 enzymatic fuel cells: From proof-of-concept to powerful devices. *Sustain. Energy Fuels* **2017**, *1*, 1475–1501. [CrossRef]
18. Xiao, X.X.; Xia, H.Q.; Wu, R.R.; Bai, L.; Yan, L.; Magner, E.; Cosnier, S.; Lojou, E.; Zhu, Z.G.; Liu, A.H. Tackling the Challenges of Enzymatic (Bio)Fuel Cells. *Chem. Rev.* **2019**, *119*, 9509–9558. [CrossRef] [PubMed]
19. Mazurenko, I.; de Poulpiquet, A.; Lojou, E. Recent developments in high surface area bioelectrodes for enzymatic fuel cells. *Curr. Opin. Electrochem.* **2017**, *5*, 74–84. [CrossRef]
20. Mazurenko, I.; Hitaishi, V.P.; Lojou, E. Recent advances in surface chemistry of electrodes to promote direct enzymatic bioelectrocatalysis. *Curr. Opin. Electrochem.* **2020**, *19*, 113–121. [CrossRef]
21. Hitaishi, V.P.; Clement, R.; Bourassin, N.; Baaden, M.; de Poulpiquet, A.; Sacquin-Mora, S.; Ciaccafava, A.; Lojou, E. Controlling Redox Enzyme Orientation at Planar Electrodes. *Catalysts* **2018**, *8*, 192. [CrossRef]
22. Iyer, P.V.; Ananthanarayan, L. Enzyme stability and stabilization—Aqueous and non-aqueous environment. *Process Biochemistry* **2008**, *43*, 1019–1032. [CrossRef]
23. Balcao, V.M.; Vila, M. Structural and functional stabilization of protein entities: State-of-the-art. *Adv. Drug Deliv. Rev.* **2015**, *93*, 25–41. [CrossRef] [PubMed]
24. Polizzi, K.M.; Bommarius, A.S.; Broering, J.M.; Chaparro-Riggers, J.F. Stability of biocatalysts. *Curr. Opin. Chem. Biol.* **2007**, *11*, 220–225. [CrossRef] [PubMed]
25. Gianfreda, L.; Scarfi, M.R. Enzyme Stabilization—State-Of-The-Art. *Mol. Cell. Biochem.* **1991**, *100*, 97–128. [CrossRef] [PubMed]
26. Stepankova, V.; Bidmanova, S.; Koudelakova, T.; Prokop, Z.; Chaloupkova, R.; Damborsky, J. Strategies for Stabilization of Enzymes in Organic Solvents. *ACS Catal.* **2013**, *3*, 2823–2836. [CrossRef]
27. Lalande, M.; Schwob, L.; Vizcaino, V.; Chirot, F.; Dugourd, P.; Schlatholter, T.; Poully, J.C. Direct Radiation Effects on the Structure and Stability of Collagen and Other Proteins. *Chembiochem* **2019**. [CrossRef]
28. Doster, W.; Settles, M. Protein-water displacement distributions. *Biochim. Biophys. Acta Proteins Proteom.* **2005**, *1749*, 173–186. [CrossRef]
29. Miyawaki, O. Hydration state change of proteins upon unfolding in sugar solutions. *Biochim. Biophys. Acta Proteins Proteom.* **2007**, *1774*, 928–935. [CrossRef] [PubMed]
30. Ganjalikhany, M.R.; Ranjbar, B.; Hosseinkhani, S.; Khalifeh, K.; Hassani, L. Roles of trehalose and magnesium sulfate on structural and functional stability of firefly luciferase. *J. Mol. Catal. B Enzym.* **2010**, *62*, 127–132. [CrossRef]
31. Haque, I.; Singh, R.; Moosavi-Movahedi, A.A.; Ahmad, F. Effect of polyol osmolytes on Delta G(D), the Gibbs energy of stabilisation of proteins at different pH values. *Biophys. Chem.* **2005**, *117*, 1–12. [CrossRef] [PubMed]
32. Tiwari, A.; Bhat, R. Stabilization of yeast hexokinase A by polyol osmolytes: Correlation with the physicochemical properties of aqueous solutions. *Biophys. Chem.* **2006**, *124*, 90–99. [CrossRef] [PubMed]
33. Wlodarczyk, S.R.; Costa-Silva, T.A.; Pessoa, A.; Madeira, P.; Monteiro, G. Effect of osmolytes on the activity of anti-cancer enzyme L-Asparaginase II from Erwinia chrysanthemi. *Process Biochem.* **2019**, *81*, 123–131. [CrossRef]
34. Chen, G.; Zhang, Q.P.; Lu, Q.Y.; Feng, B. Protection effect of polyols on Rhizopus chinensis lipase counteracting the deactivation from high pressure and high temperature treatment. *Int. J. Biol. Macromol.* **2019**, *127*, 555–562. [CrossRef]
35. Pazhang, M.; Mardi, N.; Mehrnejad, F.; Chaparzadeh, N. The combinatorial effects of osmolytes and alcohols on the stability of pyrazinamidase: Methanol affects the enzyme stability through hydrophobic interactions and hydrogen bonds. *Int. J. Biol. Macromol.* **2018**, *108*, 1339–1347. [CrossRef]
36. Khan, M.V.; Ishtikhar, M.; Rabbani, G.; Zaman, M.; Abdelhameed, A.S.; Khan, R.H. Polyols (Glycerol and Ethylene glycol) mediated amorphous aggregate inhibition and secondary structure restoration of metalloproteinase-conalbumin (ovotransferrin). *Int. J. Biol. Macromol.* **2017**, *94*, 290–300. [CrossRef]
37. Kaushik, J.K.; Bhat, R. Why is trehalose an exceptional protein stabilizer? An analysis of the thermal stability of proteins in the presence of the compatible osmolyte trehalose. *J. Biol. Chem.* **2003**, *278*, 26458–26465. [CrossRef]
38. Silva, C.; Martins, M.; Jing, S.; Fu, J.J.; Cavaco-Paulo, A. Practical insights on enzyme stabilization. *Crit. Rev. Biotechnol.* **2018**, *38*, 335–350. [CrossRef]
39. Scharnagl, C.; Reif, M.; Friedrich, J. Local compressibilities of proteins: Comparison of optical experiments and simulations for horse heart cytochrome-c. *Biophys. J.* **2005**, *89*, 64–75. [CrossRef]
40. Mangiagalli, M.; Carvalho, H.; Natalello, A.; Ferrario, V.; Pennati, M.L.; Barbiroli, A.; Lotti, M.; Pleiss, J.; Brocca, S. Diverse effects of aqueous polar co-solvents on Candida antarctica lipase B. *Int. J. Biol. Macromol.* **2020**, *150*, 930–940. [CrossRef]
41. Mukherjee, A.; Sarkar, S.; Gupta, S.; Banerjee, S.; Senapati, S.; Chakrabarty, R.; Gachhui, R. DMSO strengthens chitin deacetylase-chitin interaction: Physicochemical, kinetic, structural and catalytic insights. *Carbohydr. Polym.* **2019**, *223*. [CrossRef] [PubMed]

42. Woll, A.K.; Hubbuch, J. Investigation of the reversibility of freeze/thaw stress-induced protein instability using heat cycling as a function of different cryoprotectants. *Bioprocess Biosyst. Eng.* **2020**, *43*, 1309–1327. [CrossRef] [PubMed]
43. Dal Magro, L.; Kornecki, J.F.; Klein, M.P.; Rodrigues, R.C.; Fernandez-Lafuente, R. Stability/activity features of the main enzyme components of rohapect 10L. *Biotechnol. Prog.* **2019**, *35*. [CrossRef] [PubMed]
44. Nolan, V.; Collin, A.; Rodriguez, C.; Perillo, M.A. Effect of Polyethylene Glycol-Induced Molecular Crowding on the Enzymatic Activity and Thermal Stability of beta-Galactosidase from Kluyveromyces lactis. *J. Agric. Food Chem.* **2020**, *68*, 8875–8882. [CrossRef] [PubMed]
45. Jain, E.; Flanagan, M.; Sheth, S.; Patel, S.; Gan, Q.; Patel, B.; Montano, A.M.; Zustiak, S.P. Biodegradable polyethylene glycol hydrogels for sustained release and enhanced stability of rhGALNS enzyme. *Drug Deliv. Transl. Res.* **2020**, *10*, 1341–1352. [CrossRef]
46. Sasahara, K.; McPhie, P.; Minton, A.P. Effect of dextran on protein stability and conformation attributed to macromolecular crowding. *J. Mol. Biol.* **2003**, *326*, 1227–1237. [CrossRef]
47. Kuchler, A.; Yoshimoto, M.; Luginbuhl, S.; Mavelli, F.; Walde, P. Enzymatic reactions in confined environments. *Nat. Nanotechnol.* **2016**, *11*, 409–420. [CrossRef]
48. Li, J.H.; Zheng, H.Y.; Feng, C.J. Effect of Macromolecular Crowding on the FMN-Herne Intraprotein Electron Transfer in Inducible NO Synthase. *Biochemistry* **2019**, *58*, 3087–3096. [CrossRef]
49. Shahid, S.; Hassan, M.I.; Islam, A.; Ahmad, F. Size-dependent studies of macromolecular crowding on the thermodynamic stability, structure and functional activity of proteins: In vitro and in silico approaches. *Biochim. Et Biophys. Acta-Gen. Subj.* **2017**, *1861*, 178–197. [CrossRef]
50. Christiansen, A.; Wang, Q.; Samiotakis, A.; Cheung, M.S.; Wittung-Stafshede, P. Factors Defining Effects of Macromolecular Crowding on Protein Stability: An in Vitro/in Silico Case Study Using Cytochrome c. *Biochemistry* **2010**, *49*, 6519–6530. [CrossRef]
51. Timasheff, S.N. Protein-solvent preferential interactions, protein hydration, and the modulation of biochemical reactions by solvent components. *Proc. Natl. Acad. Sci. USA* **2002**, *99*, 9721–9726. [CrossRef]
52. Hasan, S.; Isar, M.; Naeem, A. Macromolecular crowding stabilises native structure of alpha-chymotrypsinogen-A against hexafluoropropanol-induced aggregates. *Int. J. Biol. Macromol.* **2020**. [CrossRef] [PubMed]
53. Ghosh, S.; Shahid, S.; Raina, N.; Ahmad, F.; Hassan, M.I.; Islam, A. Molecular and macromolecular crowding-induced stabilization of proteins: Effect of dextran and its building block alone and their mixtures on stability and structure of lysozyme. *Int. J. Biol. Macromol.* **2020**, *150*, 1238–1248. [CrossRef] [PubMed]
54. Baykal, E.; Vardar, G.; Attar, A.; Yapaoz, M.A. Complexes of glucose oxidase with chitosan and dextran possessing enhanced stability. *Prep. Biochem. Biotechnol.* **2020**, *50*, 572–577. [CrossRef] [PubMed]
55. Das, N.; Sen, P. Size-dependent macromolecular crowding effect on the thermodynamics of protein unfolding revealed at the single molecular level. *Int. J. Biol. Macromol.* **2019**, *141*, 843–854. [CrossRef] [PubMed]
56. Mittal, S.; Chowhan, R.K.; Singh, L.R. Macromolecular crowding: Macromolecules friend or foe. *Biochim. Et Biophys. Acta-Gen. Subj.* **2015**, *1850*, 1822–1831. [CrossRef] [PubMed]
57. Lavelle, L.; Fresco, J.R. Stabilization of nucleic acid triplexes by high concentrations of sodium and ammonium salts follows the Hofmeister series. *Biophys. Chem.* **2003**, *105*, 681–699. [CrossRef]
58. Okur, H.I.; Hladilkova, J.; Rembert, K.B.; Cho, Y.; Heyda, J.; Dzubiella, J.; Cremer, P.S.; Jungwirth, P. Beyond the Hofmeister Series: Ion-Specific Effects on Proteins and Their Biological Functions. *J. Phys. Chem. B* **2017**, *121*, 1997–2014. [CrossRef] [PubMed]
59. Hofmeister, F. Zur Lehre von der Wirkung der Salze. *Arch. Exp. Pathol. Pharmakol.* **1888**, *24*, 247–260. [CrossRef]
60. Salis, A.; Ninham, B.W. Models and mechanisms of Hofmeister effects in electrolyte solutions, and colloid and protein systems revisited. *Chem. Soc. Rev.* **2014**, *43*, 7358–7377. [CrossRef]
61. Ghobadi, R.; Divsalar, A. Enzymatic behavior of bovine liver catalase in aqueous medium of sugar based deep eutectic solvents. *J. Mol. Liq.* **2020**, *310*. [CrossRef]
62. Usoltsev, D.; Sitnikova, V.; Kajava, A.; Uspenskaya, M. FTIR Spectroscopy Study of the Secondary Structure Changes in Human Serum Albumin and Trypsin under Neutral Salts. *Biomolecules* **2020**, *10*, 606. [CrossRef] [PubMed]
63. Sedlak, E.; Sedlakova, D.; Marek, J.; Hancar, J.; Garajova, K.; Zoldak, G. Ion-Specific Protein/Water Interface Determines the Hofmeister Effect on the Kinetic Stability of Glucose Oxidase. *J. Phys. Chem. B* **2019**, *123*, 7965–7973. [CrossRef] [PubMed]
64. Garajova, K.; Balogova, A.; Dusekova, E.; Sedlakova, D.; Sedlak, E.; Varhac, R. Correlation of lysozyme activity and stability in the presence of Hofmeister series anions. *Biochim. Biophys. Acta Proteins Proteom.* **2017**, *1865*, 281–288. [CrossRef] [PubMed]
65. Nemoto, M.; Sugihara, K.; Adachi, T.; Murata, K.; Shiraki, K.; Tsujimura, S. Effect of Electrolyte Ions on the Stability of Flavin Adenine Dinucleotide-Dependent Glucose Dehydrogenase. *Chemelectrochem* **2019**, *6*, 1028–1031. [CrossRef]
66. Hani, F.M.; Cole, A.E.; Altman, E. The ability of salts to stabilize proteins in vivo or intracellularly correlates with the Hofmeister series of ions. *Int. J. Biochem. Mol. Biol.* **2019**, *10*, 23–31.
67. Banerjee, S.; Arora, A.; Vijayaraghavan, R.; Patti, A.F. Extraction and crosslinking of bromelain aggregates for improved stability and reusability from pineapple processing waste. *Int. J. Biol. Macromol.* **2020**, *158*, 318–326. [CrossRef]
68. Kulkarni, N.H.; Muley, A.B.; Bedade, D.K.; Singhal, R.S. Cross-linked enzyme aggregates of arylamidase from Cupriavidus oxalaticus ICTDB921: Process optimization, characterization, and application for mitigation of acrylamide in industrial wastewater. *Bioprocess Biosyst. Eng.* **2020**, *43*, 457–471. [CrossRef]

69. Mayolo-Deloisa, K.; Gonzalez-Gonzalez, M.; Simental-Martinez, J.; Rito-Palomares, M. Aldehyde PEGylation of laccase from Trametes versicolor in route to increase its stability: Effect on enzymatic activity. *J. Mol. Recognit.* **2015**, *28*, 173–179. [CrossRef]
70. El-Sayed, A.S.A.; Shindia, A.A.; Abou Zeid, A.A.; Yassin, A.M.; Sitohy, M.Z.; Sitohy, B. Aspergillus nidulans thermostable arginine deiminase-Dextran conjugates with enhanced molecular stability, proteolytic resistance, pharmacokinetic properties and anticancer activity. *Enzym. Microb. Technol.* **2019**, *131*. [CrossRef]
71. Veronese, F.M. Peptide and protein PEGylation: A review of problems and solutions. *Biomaterials* **2001**, *22*, 405–417. [CrossRef]
72. Huang, H.F.; Fang, L.; Xue, L.; Zhang, T.; Kim, K.B.; Hou, S.R.; Zheng, F.; Zhan, C.G. PEGylation but Not Fc-Fusion Improves in Vivo Residence Time of a Thermostable Mutant of Bacterial Cocaine Esterase. *Bioconjugate Chem.* **2019**, *30*, 3021–3027. [CrossRef]
73. Cummings, C.; Murata, H.; Koepsel, R.; Russell, A.J. Tailoring enzyme activity and stability using polymer-based protein engineering. *Biomaterials* **2013**, *34*, 7437–7443. [CrossRef] [PubMed]
74. Chung, S.F.; Kim, C.F.; Kwok, S.Y.; Tam, S.Y.; Chen, Y.W.; Chong, H.C.; Leung, S.L.; So, P.K.; Wong, K.Y.; Leung, Y.C.; et al. Mono-PEGylation of a Thermostable Arginine-Depleting Enzyme for the Treatment of Lung Cancer. *Int. J. Mol. Sci.* **2020**, *21*, 4234. [CrossRef] [PubMed]
75. Kovaliov, M.; Zhang, B.; Konkolewicz, D.; Szczesniak, K.; Jurga, S.; Averick, S. Polymer grafting from a metallo-centered enzyme improves activity in non-native environments. *Poly. Int.* **2020**. [CrossRef]
76. Ritter, D.W.; Newton, J.M.; McShane, M.J. Modification of PEGylated enzyme with glutaraldehyde can enhance stability while avoiding intermolecular crosslinking. *RSC Adv.* **2014**, *4*, 28036–28040. [CrossRef]
77. Sharma, A.; Gupta, G.; Ahmad, T.; Mansoor, S.; Kaur, B. Enzyme Engineering: Current Trends and Future Perspectives. *Food Rev. Int.* **2021**. [CrossRef]
78. Korendovych, I.V. Rational and Semirational Protein Design. *Methods Mol. Biol.* **2018**, *1685*, 15–23. [CrossRef]
79. de Morais, M.A.B.; Polo, C.C.; Domingues, M.N.; Persinoti, G.F.; Pirolla, R.A.S.; de Souza, F.H.M.; Correa, J.B.D.; dos Santos, C.R.; Murakami, M.T. Exploring the Molecular Basis for Substrate Affinity and Structural Stability in Bacterial GH39 beta-Xylosidases. *Front. Bioeng. Biotechnol.* **2020**, *8*. [CrossRef]
80. Ito, Y.; Ikeuchi, A.; Imamura, T. Advanced evolutionary molecular engineering to produce thermostable cellulase by using a small but efficient library. *Protein Eng. Des. Sel.* **2013**, *26*, 73–79. [CrossRef] [PubMed]
81. Minges, H.; Schnepel, C.; Bottcher, D.; Weiss, M.S.; Spross, J.; Bornscheuer, U.T.; Sewald, N. Targeted Enzyme Engineering Unveiled Unexpected Patterns of Halogenase Stabilization. *Chemcatchem* **2020**, *12*, 818–831. [CrossRef]
82. Bulut, H.; Yuksel, B.; Gul, M.; Eren, M.; Karatas, E.; Kara, N.; Yilmazer, B.; Kocyigit, A.; Labrou, N.E.; Binay, B. Conserved Amino Acid Residues that Affect Structural Stability ofCandida boidiniiFormate Dehydrogenase. *Appl. Biochem. Biotechnol.* **2020**. [CrossRef]
83. Alvarez, C.E.; Bovdilova, A.; Hoppner, A.; Wolff, C.C.; Saigo, M.; Trajtenberg, F.; Zhang, T.; Buschiazzo, A.; Nagel-Steger, L.; Drincovich, M.F.; et al. Molecular adaptations of NADP-malic enzyme for its function in C-4 photosynthesis in grasses. *Nat. Plants* **2019**, *5*, 755–765. [CrossRef] [PubMed]
84. Zhang, Y.; Yang, J.; Yu, X.; Hu, X.; Zhang, H. Engineering Leuconostoc mesenteroides dextransucrase by inserting disulfide bridges for enhanced thermotolerance. *Enzym. Microb. Technol.* **2020**, *139*. [CrossRef]
85. Gomez-Fernandez, B.J.; Risso, V.A.; Sanchez-Ruiz, J.M.; Alcalde, M. Consensus Design of an Evolved High-Redox Potential Laccase. *Front. Bioeng. Biotechnol.* **2020**, *8*. [CrossRef] [PubMed]
86. Zhao, Z.X.; Lan, D.M.; Tan, X.Y.; Hollmann, F.; Bornscheuer, U.T.; Yang, B.; Wang, Y.H. How To Break the Janus Effect of H2O2 in Biocatalysis? Understanding Inactivation Mechanisms To Generate more Robust Enzymes. *ACS Catal.* **2019**, *9*, 2916–2921. [CrossRef]
87. Liu, Q.; Xun, G.H.; Feng, Y. The state-of-the-art strategies of protein engineering for enzyme stabilization. *Biotechnol. Adv.* **2019**, *37*, 530–537. [CrossRef]
88. Roger, M.; Castelle, C.; Guiral, M.; Infossi, P.; Lojou, E.; Giudici-Orticoni, M.T.; Ilbert, M. Mineral respiration under extreme acidic conditions: From a supramolecular organization to a molecular adaptation in Acidithiobacillus ferrooxidans. *Biochem. Soc. Trans.* **2012**, *40*, 1324–1329. [CrossRef]
89. Hassan, N.; Rafiq, M.; Rehman, M.; Sajjad, W.; Hasan, F.; Abdullah, S. Fungi in acidic fire: A potential source of industrially important enzymes. *Fungal Biol. Rev.* **2019**, *33*, 58–71. [CrossRef]
90. Jin, M.; Gai, Y.B.; Guo, X.; Hou, Y.P.; Zeng, R.Y. Properties and Applications of Extremozymes from Deep-Sea Extremophilic Microorganisms: A Mini Review. *Mar. Drugs* **2019**, *17*, 656. [CrossRef]
91. Guiral, M.; Prunetti, L.; Aussignargues, C.; Ciaccafava, A.; Infossi, P.; Ilbert, M.; Lojou, E.; Giudici-Orticoni, M.T. The hyperthermophilic bacterium Aquifex aeolicus: From respiratory pathways to extremely resistant enzymes and biotechnological applications. *Adv. Microb. Physiol.* **2012**, *61*, 125–194. [CrossRef] [PubMed]
92. Furukawa, R.; Toma, W.; Yamazaki, K.; Akanuma, S. Ancestral sequence reconstruction produces thermally stable enzymes with mesophilic enzyme-like catalytic properties. *Sci. Rep.* **2020**, *10*, 15493. [CrossRef] [PubMed]
93. Abdollahi, P.; Ghane, M.; Babaeekhou, L. Isolation and Characterization of Thermophilic Bacteria from Gavmesh Goli Hot Spring in Sabalan Geothermal Field, Iran: *Thermomonas hydrothermalis* and *Bacillus altitudinis* Isolates as a Potential Source of Thermostable Protease. *Geomicrobiol. J.* **2020**. [CrossRef]
94. Hilden, K.; Hakala, T.K.; Lundell, T. Thermotolerant and thermostable laccases. *Biotechnol. Lett.* **2009**, *31*, 1117–1128. [CrossRef] [PubMed]

95. Sterner, R.; Liebl, W. Thermophilic adaptation of proteins. *Crit. Rev. Biochem. Mol. Biol.* **2001**, *36*, 39–106. [CrossRef]
96. Maiello, F.; Gallo, G.; Coelho, C.; Sucharski, F.; Hardy, L.; Wurtele, M. Crystal structure of Thermus thermophilus methylenetetrahydrofolate dehydrogenase and determinants of thermostability. *PLoS ONE* **2020**, *15*. [CrossRef]
97. McManus, T.J.; Wells, S.A.; Walker, A.B. Salt bridge impact on global rigidity and thermostability in thermophilic citrate synthase. *Phys. Biol.* **2020**, *17*. [CrossRef]
98. Xie, Z.H.; Zhai, L.X.; Meng, D.; Tian, Q.P.; Guan, Z.B.; Cai, Y.J.; Liao, X.R. Improving the catalytic thermostability ofBacillus altitudinisW3 omega-transaminase by proline substitutions. *3 Biotech* **2020**, *10*. [CrossRef]
99. Kargar, F.; Mortezavi, M.; Torkzadeh-Mahani, M.; Lotfi, S.; Shakeri, S. Evaluation of Luciferase Thermal Stability by Arginine Saturation in the Flexible Loops. *Curr. Proteom.* **2020**, *17*, 30–39. [CrossRef]
100. Enguita, F.J.; Martins, L.O.; Henriques, A.O.; Carrondo, M.A. Crystal structure of a bacterial endospore coat component—A laccase with enhanced thermostability properties. *J. Biol. Chem.* **2003**, *278*, 19416–19425. [CrossRef]
101. Brininger, C.; Spradlin, S.; Cobani, L.; Evilia, C. The more adaptive to change, the more likely you are to survive: Protein adaptation in extremophiles. *Semin. Cell Dev. Biol.* **2018**, *84*, 158–169. [CrossRef] [PubMed]
102. Cea, P.A.; Araya, G.; Vallejos, G.; Recabarren, R.; Alzate-Morales, J.; Babul, J.; Guixe, V.; Castro-Fernandez, V. Characterization of hydroxymethylpyrimidine phosphate kinase from mesophilic and thermophilic bacteria and structural insights into their differential thermal stability. *Arch. Biochem. Biophys.* **2020**, *688*. [CrossRef] [PubMed]
103. Maffucci, I.; Laage, D.; Stirnemann, G.; Sterpone, F. Differences in thermal structural changes and melting between mesophilic and thermophilic dihydrofolate reductase enzymes. *Phys. Chem. Chem. Phys.* **2020**, *22*, 18361–18373. [CrossRef] [PubMed]
104. Hou, J.; Yang, X.Y.; Xu, Q.; Cui, H.L. Characterization of a novel Cu-containing dissimilatory nitrite reductase from the haloarchaeon Halorussus sp. YCN54. *Extremophiles* **2020**, *24*, 403–411. [CrossRef] [PubMed]
105. Graziano, G.; Merlino, A. Molecular bases of protein halotolerance. *Biochim. Biophys. Acta Proteins Proteom.* **2014**, *1844*, 850–858. [CrossRef] [PubMed]
106. Qin, H.M.; Gao, D.K.; Zhu, M.L.; Li, C.; Zhu, Z.L.; Wang, H.B.; Liu, W.D.; Tanokura, M.; Lu, F.P. Biochemical characterization and structural analysis of ulvan lyase from marine Alteromonas sp. reveals the basis for its salt tolerance. *Int. J. Biol. Macromol.* **2020**, *147*, 1309–1317. [CrossRef] [PubMed]
107. Zheng, C.; Li, Z.; Yang, H.; Zhang, T.; Niu, H.; Liu, D.; Wang, J.; Ying, H. Computation-aided rational design of a halophilic choline kinase for cytidine diphosphate choline production in high-salt condition. *J. Biotechnol.* **2019**, *290*, 59–66. [CrossRef] [PubMed]
108. Liu, Z.; Yuan, M.; Zhang, X.; Liang, Q.; Yang, M.; Mou, H.; Zhu, C. A thermostable glucose oxidase from Aspergillus heteromophus CBS 117.55 with broad pH stability and digestive enzyme resistance. *Protein Expr. Purif.* **2020**, *176*. [CrossRef]
109. Yi, Z.; Cai, Z.; Zeng, B.; Zeng, R.; Zhang, G. Identification and Characterization of a Novel Thermostable and Salt-Tolerant beta-1,3 Xylanase from Flammeovirga pacifica Strain WPAGA1. *Biomolecules* **2020**, *10*, 1287. [CrossRef]
110. Zhu, T.; Li, R.; Sun, J.; Cui, Y.; Wu, B. Characterization and efficient production of a thermostable, halostable and organic solvent-stable cellulase from an oil reservoir. *Int. J. Biol. Macromol.* **2020**, *159*, 622–629. [CrossRef]
111. Monsalve, K.; Mazurenko, I.; Gutierrez-Sanchez, C.; Ilbert, M.; Infossi, P.; Frielingsdorf, S.; Giudici-Orticoni, M.T.; Lenz, O.; Lojou, E. Impact of Carbon Nanotube Surface Chemistry on Hydrogen Oxidation by Membrane-Bound Oxygen-Tolerant Hydrogenases. *Chemelectrochem* **2016**, *3*, 2179–2188. [CrossRef]
112. Oteri, F.; Baaden, M.; Lojou, E.; Sacquin-Mora, S. Multiscale Simulations Give Insight into the Hydrogen In and Out Pathways of NiFe -Hydrogenases from Aquifex aeolicus and Desulfovibrio fructosovorans. *J. Phys. Chem. B* **2014**, *118*, 13800–13811. [CrossRef]
113. Coglitore, D.; Janot, J.M.; Balme, S. Protein at liquid solid interfaces: Toward a new paradigm to change the approach to design hybrid protein/solid-state materials. *Adv. Colloid Interface Sci.* **2019**, *270*, 278–292. [CrossRef]
114. Arai, T.; Norde, W. The behavior of some model proteins at solid liquid interfaces 1. adsorption from single protein solutions. *Adv. Colloid Interface Sci.* **1990**, *51*, 1–15. [CrossRef]
115. Rabe, M.; Verdes, D.; Seeger, S. Understanding protein adsorption phenomena at solid surfaces. *Adv. Colloid Interface Sci.* **2011**, *162*, 87–106. [CrossRef] [PubMed]
116. Hitaishi, V.P.; Mazurenko, I.; Harb, M.; Clement, R.; Taris, M.; Castano, S.; Duche, D.; Lecomte, S.; Ilbert, M.; de Poulpiquet, A.; et al. Electrostatic-Driven Activity, Loading, Dynamics, and Stability of a Redox Enzyme on Functionalized-Gold Electrodes for Bioelectrocatalysis. *ACS Catal.* **2018**, *8*, 12004–12014. [CrossRef]
117. Bourassin, N.; Baaden, M.; Lojou, E.; Sacquin-Mora, S. Implicit Modeling of the Impact of Adsorption on Solid Surfaces for Protein Mechanics and Activity with a Coarse-Grained Representation. *J. Phys. Chem. B* **2020**, *124*, 8516–8523. [CrossRef] [PubMed]
118. Pankratov, D.; Sotres, J.; Barrantes, A.; Arnebrant, T.; Shleev, S. Interfacial Behavior and Activity of Laccase and Bilirubin Oxidase on Bare Gold Surfaces. *Langmuir* **2014**, *30*, 2943–2951. [CrossRef] [PubMed]
119. Kienle, D.F.; Falatach, R.M.; Kaar, J.L.; Schwartz, D.K. Correlating Structural and Functional Heterogeneity of Immobilized Enzymes. *ACS Nano* **2018**, *12*, 8091–8103. [CrossRef] [PubMed]
120. Zhang, Y.F.; Ge, J.; Liu, Z. Enhanced Activity of Immobilized or Chemically Modified Enzymes. *ACS Catal.* **2015**, *5*, 4503–4513. [CrossRef]
121. Garcia-Garcia, P.; Guisan, J.M.; Fernandez-Lorente, G. A mild intensity of the enzyme-support multi-point attachment promotes the optimal stabilization of mesophilic multimeric enzymes: Amine oxidase from Pisum sativum. *J. Biotechnol.* **2020**, *318*, 39–44. [CrossRef]

122. Weltz, J.S.; Kienle, D.F.; Schwartz, D.K.; Kaar, J.L. Reduced Enzyme Dynamics upon Multipoint Covalent Immobilization Leads to Stability-Activity Trade-off. *J. Am. Chem. Soc.* **2020**, *142*, 3463–3471. [CrossRef]
123. Mateo, C.; Palomo, J.M.; Fernandez-Lorente, G.; Guisan, J.M.; Fernandez-Lafuente, R. Improvement of enzyme activity, stability and selectivity via immobilization techniques. *Enzym. Microb. Technol.* **2007**, *40*, 1451–1463. [CrossRef]
124. Hitaishi, V.P.; Mazurenko, I.; Vengasseril Murali, A.; de Poulpiquet, A.; Coustillier, G.; Delaporte, P.; Lojou, E. Nanosecond Laser-Fabricated Monolayer of Gold Nanoparticles on ITO for Bioelectrocatalysis. *Front. Chem.* **2020**, *8*. [CrossRef] [PubMed]
125. Sigurdardottir, S.B.; Lehmann, J.; Grivel, J.C.; Zhang, W.J.; Kaiser, A.; Pinelo, M. Alcohol dehydrogenase on inorganic powders: Zeta potential and particle agglomeration as main factors determining activity during immobilization. *Colloids Surf. B-Biointerfaces* **2019**, *175*, 136–142. [CrossRef] [PubMed]
126. Johnson, B.J.; Algar, W.R.; Malanoski, A.P.; Ancona, M.G.; Medintz, I.L. Understanding enzymatic acceleration at nanoparticle interfaces: Approaches and challenges. *Nano Today* **2014**, *9*, 102–131. [CrossRef]
127. Sharifi, M.; Sohrabi, M.J.; Hosseinali, S.H.; Hasan, A.; Kani, P.H.; Talaei, A.J.; Karim, A.Y.; Nanakali, N.M.Q.; Salihi, A.; Aziz, F.M.; et al. Enzyme immobilization onto the nanomaterials: Application in enzyme stability and prodrug-activated cancer therapy. *Int. J. Biol. Macromol.* **2020**, *143*, 665–676. [CrossRef]
128. Zhang, Y.; Tang, Z.; Wang, J.; Wu, H.; Lin, C.-T.; Lin, Y. Apoferritin nanoparticle: A novel and biocompatible carrier for enzyme immobilization with enhanced activity and stability. *J. Mater. Chem.* **2011**, *21*, 17468–17475. [CrossRef]
129. Yang, Y.L.; Zhu, G.X.; Wang, G.C.; Li, Y.L.; Tang, R.K. Robust glucose oxidase with a Fe3O4@C-silica nanohybrid structure. *J. Mater. Chem. B* **2016**, *4*, 4726–4731. [CrossRef] [PubMed]
130. Suo, H.B.; Gao, Z.; Xu, L.L.; Xu, C.; Yu, D.H.; Xiang, X.R.; Huang, H.; Hu, Y. Synthesis of functional ionic liquid modified magnetic chitosan nanoparticles for porcine pancreatic lipase immobilization. *Mater. Sci. Eng. C Mater. Biol. Appl.* **2019**, *96*, 356–364. [CrossRef] [PubMed]
131. Ansari, S.A.; Husain, Q.; Qayyum, S.; Azam, A. Designing and surface modification of zinc oxide nanoparticles for biomedical applications. *Food Chem. Toxicol.* **2011**, *49*, 2107–2115. [CrossRef]
132. Grewal, J.; Ahmad, R.; Khare, S.K. Development of cellulase-nanoconjugates with enhanced ionic liquid and thermal stability for in situ lignocellulose saccharification. *Bioresour. Technol.* **2017**, *242*, 236–243. [CrossRef]
133. Shang, W.; Nuffer, J.H.; Dordick, J.S.; Siegel, R.W. Unfolding of ribonuclease A on silica nanoparticle surfaces. *Nano Lett.* **2007**, *7*, 1991–1995. [CrossRef]
134. Song, Y.; Zhong, D.; Luo, D.; Huang, M.; Huang, Z.; Tan, H.; Sun, L.; Wang, L. Effect of particle size on conformation and enzymatic activity of EcoRI adsorbed on CdS nanoparticles. *Colloids Surf. B-Biointerfaces* **2014**, *114*, 269–276. [CrossRef] [PubMed]
135. Asuri, P.; Karajanagi, S.S.; Yang, H.C.; Yim, T.J.; Kane, R.S.; Dordick, J.S. Increasing protein stability through control of the nanoscale environment. *Langmuir* **2006**, *22*, 5833–5836. [CrossRef] [PubMed]
136. Vertegel, A.A.; Siegel, R.W.; Dordick, J.S. Silica nanoparticle size influences the structure and enzymatic activity of adsorbed lysozyme. *Langmuir* **2004**, *20*, 6800–6807. [CrossRef] [PubMed]
137. Gagner, J.E.; Lopez, M.D.; Dordick, J.S.; Siegel, R.W. Effect of gold nanoparticle morphology on adsorbed protein structure and function. *Biomaterials* **2011**, *32*, 7241–7252. [CrossRef]
138. Bilal, M.; Asgher, M.; Cheng, H.R.; Yan, Y.J.; Iqbal, H.M.N. Multi-point enzyme immobilization, surface chemistry, and novel platforms: A paradigm shift in biocatalyst design. *Crit. Rev. Biotechnol.* **2019**, *39*, 202–219. [CrossRef]
139. Vila, N.; Andre, E.; Ciganda, R.; Ruiz, J.; Astruc, D.; Walcarius, A. Molecular Sieving with Vertically Aligned Mesoporous Silica Films and Electronic Wiring through Isolating Nanochannels. *Chem. Mater.* **2016**, *28*, 2511–2514. [CrossRef]
140. Abdelbar, M.F.; Shams, R.S.; Morsy, O.M.; Hady, M.A.; Shoueir, K.; Abdelmonem, R. Highly ordered functionalized mesoporous silicate nanoparticles reinforced poly (lactic acid) gatekeeper surface for infection treatment. *Int. J. Biol. Macromol.* **2020**, *156*, 858–868. [CrossRef]
141. Rodriguez-Abetxuko, A.; Sanchez-deAlcazar, D.; Munumer, P.; Beloqui, A. Tunable Polymeric Scaffolds for Enzyme Immobilization. *Front. Bioeng. Biotechnol.* **2020**, *8*. [CrossRef]
142. Inagaki, M.; Toyoda, M.; Soneda, Y.; Tsujimura, S.; Morishita, T. Templated mesoporous carbons: Synthesis and applications. *Carbon* **2016**, *107*, 448–473. [CrossRef]
143. Ruff, A.; Szczesny, J.; Markovic, N.; Conzuelo, F.; Zacarias, S.; Pereira, I.A.C.; Lubitz, W.; Schuhmann, W. A fully protected hydrogenase/polymer-based bioanode for high-performance hydrogen/glucose biofuel cells. *Nat. Commun.* **2018**, *9*. [CrossRef]
144. Stine, K.J. Enzyme Immobilization on Nanoporous Gold: A Review. *Biochem. Insights* **2017**, *10*. [CrossRef] [PubMed]
145. Baruch-Shpigler, Y.; Avnir, D. Enzymes in a golden cage. *Chem. Sci.* **2020**, *11*, 3965–3977. [CrossRef]
146. Ge, J.; Lei, J.D.; Zare, R.N. Protein-inorganic hybrid nanoflowers. *Nat. Nanotechnol.* **2012**, *7*, 428–432. [CrossRef] [PubMed]
147. Liang, W.; Wied, P.; Carraro, F.; Sumby, C.J.; Nidetzky, B.; Tsung, C.-K.; Falcaro, P.; Doonan, C.J. Metal-Organic Framework-Based Enzyme Biocomposites. *Chem. Rev.* **2021**. [CrossRef]
148. Lian, X.Z.; Fang, Y.; Joseph, E.; Wang, Q.; Li, J.L.; Banerjee, S.; Lollar, C.; Wang, X.; Zhou, H.C. Enzyme-MOF (metal-organic framework) composites. *Chem. Soc. Rev.* **2017**, *46*, 3386–3401. [CrossRef] [PubMed]
149. Li, P.; Moon, S.Y.; Guelta, M.A.; Harvey, S.P.; Hupp, J.T.; Farha, O.K. Encapsulation of a Nerve Agent Detoxifying Enzyme by a Mesoporous Zirconium Metal-Organic Framework Engenders Thermal and Long-Term Stability. *J. Am. Chem. Soc.* **2016**, *138*, 8052–8055. [CrossRef] [PubMed]

150. Feng, Y.X.; Zhong, L.; Hou, Y.; Jia, S.R.; Cui, J.D. Acid-resistant enzyme@MOF nanocomposites with mesoporous silica shells for enzymatic applications in acidic environments. *J. Biotechnol.* **2019**, *306*, 54–61. [CrossRef]
151. He, H.M.; Han, H.B.; Shi, H.; Tian, Y.Y.; Sun, F.X.; Song, Y.; Li, Q.S.; Zhu, G.S. Construction of Thermophilic Lipase-Embedded Metal Organic Frameworks via Biomimetic Mineralization: A Biocatalyst for Ester Hydrolysis and Kinetic Resolution. *ACS Appl. Mater. Interfaces* **2016**, *8*, 24517–24524. [CrossRef]
152. Wu, X.L.; Yang, C.; Ge, J.; Liu, Z. Polydopamine tethered enzyme/metal-organic framework composites with high stability and reusability. *Nanoscale* **2015**, *7*, 18883–18886. [CrossRef] [PubMed]
153. Wu, X.L.; Yang, C.; Ge, J. Green synthesis of enzyme/metal-organic framework composites with high stability in protein denaturing solvents. *Bioresour. Bioprocess.* **2017**, *4*. [CrossRef] [PubMed]
154. Knedel, T.O.; Ricklefs, E.; Schlusener, C.; Urlacher, V.B.; Janiak, A.C. Laccase Encapsulation in ZIF-8 Metal-Organic Framework Shows Stability Enhancement and Substrate Selectivity. *Chemistryopen* **2019**, *8*, 1337–1344. [CrossRef] [PubMed]
155. Song, J.Y.; He, W.T.; Shen, H.; Zhou, Z.X.; Li, M.Q.; Su, P.; Yang, Y. Construction of multiple enzyme metal-organic frameworks biocatalyst via DNA scaffold: A promising strategy for enzyme encapsulation. *Chem. Eng. J.* **2019**, *363*, 174–182. [CrossRef]
156. Gao, Y.; Doherty, C.M.; Mulet, X. A Systematic Study of the Stability of Enzyme/Zeolitic Imidazolate Framework-8 Composites in Various Biologically Relevant Solutions. *Chemistryselect* **2020**, *5*, 13766–13774. [CrossRef]
157. Nadar, S.S.; Rathod, V.K. Magnetic-metal organic framework (magnetic-MOF): A novel platform for enzyme immobilization and nanozyme applications. *Int. J. Biol. Macromol.* **2018**, *120*, 2293–2302. [CrossRef] [PubMed]
158. Gan, J.; Bagheri, A.R.; Aramesh, N.; Gul, I.; Franco, M.; Almulaiky, Y.Q.; Bilal, M. Covalent organic frameworks as emerging host platforms for enzyme immobilization and robust biocatalysis—A review. *Int. J. Biol. Macromol.* **2021**, *167*, 502–515. [CrossRef]
159. Chapman, R.; Stenzel, M.H. All Wrapped up: Stabilization of Enzymes within Single Enzyme Nanoparticles. *J. Am. Chem. Soc.* **2019**, *141*, 2754–2769. [CrossRef]
160. Chen, Y.Z.; Luo, Z.G.; Lu, X.X. Construction of Novel Enzyme-Graphene Oxide Catalytic Interface with Improved Enzymatic Performance and Its Assembly Mechanism. *ACS Appl. Mater. Interfaces* **2019**, *11*, 11349–11359. [CrossRef]
161. Wu, L.; Lu, X.; Niu, K.; Ma, D.; Chen, J. Tyrosinase nanocapsule based nano-biosensor for ultrasensitive and rapid detection of bisphenol A with excellent stability in different application scenarios. *Biosens. Bioelectron.* **2020**, *165*. [CrossRef]
162. Shen, X.T.; Yang, M.; Cui, C.X.; Cao, H. In situ immobilization of glucose oxidase and catalase in a hybrid interpenetrating polymer network by 3D bioprinting and its application. *Colloids Surf. A Physicochem. Eng. Asp.* **2019**, *568*, 411–418. [CrossRef]
163. Zhang, S.T.; Wu, Z.F.; Chen, G.; Wang, Z. An Improved Method to Encapsulate Laccase from Trametes versicolor with Enhanced Stability and Catalytic Activity. *Catalysts* **2018**, *8*, 286. [CrossRef]
164. Wan, L.; Chen, Q.S.; Liu, J.B.; Yang, X.H.; Huang, J.; Li, L.; Guo, X.; Zhang, J.; Wang, K.M. Programmable Self-Assembly of DNA-Protein Hybrid Hydrogel for Enzyme Encapsulation with Enhanced Biological Stability. *Biomacromolecules* **2016**, *17*, 1543–1550. [CrossRef]
165. Liang, H.; Jiang, S.H.; Yuan, Q.P.; Li, G.F.; Wang, F.; Zhang, Z.J.; Liu, J.W. Co-immobilization of multiple enzymes by metal coordinated nucleotide hydrogel nanofibers: Improved stability and an enzyme cascade for glucose detection. *Nanoscale* **2016**, *8*, 6071–6078. [CrossRef] [PubMed]
166. Yamaguchi, H.; Kiyota, Y.; Miyazaki, M. Techniques for Preparation of Cross-Linked Enzyme Aggregates and Their Applications in Bioconversions. *Catalysts* **2018**, *8*, 174. [CrossRef]
167. Sheldon, R.A.; van Pelt, S. Enzyme immobilisation in biocatalysis: Why, what and how. *Chem. Soc. Rev.* **2013**, *42*, 6223–6235. [CrossRef] [PubMed]
168. Sheldon, R.A.; van Pelt, S.; Kanbak-Aksu, S.; Rasmussen, J.A.; Janssen, M.H.A. Cross-Linked Enzyme Aggregates (CLEAs) in Organic Synthesis. *Aldrichimica Acta* **2013**, *46*, 81–93.
169. Qian, J.; Zhao, C.; Ding, J.; Chen, Y.; Guo, H. Preparation of nano-enzyme aggregates by crosslinking lipase with sodium tripolyphosphate. *Process Biochem.* **2020**, *97*, 19–26. [CrossRef]
170. Nadar, S.S.; Muley, A.B.; Ladole, M.R.; Joshi, P.U. Macromolecular cross-linked enzyme aggregates (M-CLEAs) of alpha-amylase. *Int. J. Biol. Macromol.* **2016**, *84*, 69–78. [CrossRef] [PubMed]
171. Wang, S.; Zheng, D.; Yin, L.; Wang, F. Preparation, activity and structure of cross-linked enzyme aggregates (CLEAs) with nanoparticle. *Enzym. Microb. Technol.* **2017**, *107*, 22–31. [CrossRef] [PubMed]
172. Ren, S.Z.; Li, C.H.; Jiao, X.B.; Jia, S.R.; Jiang, Y.J.; Bilal, M.; Cui, J.D. Recent progress in multienzymes co-immobilization and multienzyme system applications. *Chem. Eng. J.* **2019**, *373*, 1254–1278. [CrossRef]
173. Monteiro, R.R.C.; dos Santos, J.C.S.; Alcantara, A.R.; Fernandez-Lafuente, R. Enzyme-Coated Micro-Crystals: An Almost Forgotten but Very Simple and Elegant Immobilization Strategy. *Catalysts* **2020**, *10*, 891. [CrossRef]
174. Zerva, A.; Pentari, C.; Topakas, E. Crosslinked Enzyme Aggregates (CLEAs) of Laccases from Pleurotus citrinopileatus Induced in Olive Oil Mill Wastewater (OOMW). *Molecules* **2020**, *25*, 2221. [CrossRef]
175. Kopp, W.; da Costa, T.P.; Pereira, S.C.; Jafelicci, M., Jr.; Giordano, R.C.; Marques, R.F.C.; Araujo-Moreira, F.M.; Giordano, R.L.C. Easily handling penicillin G acylase magnetic cross-linked enzymes aggregates: Catalytic and morphological studies. *Process Biochem.* **2014**, *49*, 38–46. [CrossRef]
176. Cui, J.D.; Zhang, S.; Sun, L.M. Cross-Linked Enzyme Aggregates of Phenylalanine Ammonia Lyase: Novel Biocatalysts for Synthesis of L-Phenylalanine. *Appl. Biochem. Biotechnol.* **2012**, *167*, 835–844. [CrossRef] [PubMed]

177. Cui, J.D.; Jia, S.R. Optimization protocols and improved strategies of cross-linked enzyme aggregates technology: Current development and future challenges. *Crit. Rev. Biotechnol.* **2015**, *35*, 15–28. [CrossRef]
178. Tandjaoui, N.; Tassist, A.; Abouseoud, M.; Couvert, A.; Amrane, A. Preparation and characterization of cross-linked enzyme aggregates (CLEAs) of Brassica rapa peroxidase. *Biocatal. Agric. Biotechnol.* **2015**, *4*, 208–213. [CrossRef]
179. Dinh, T.H.; Jang, N.Y.; McDonald, K.A.; Won, K. Cross-linked aggregation of glutamate decarboxylase to extend its activity range toward alkaline pH. *J. Chem. Technol. Biotechnol.* **2015**, *90*, 2100–2105. [CrossRef]
180. Gupta, K.; Jana, A.K.; Kumar, S.; Jana, M.M. Solid state fermentation with recovery of Amyloglucosidase from extract by direct immobilization in cross linked enzyme aggregate for starch hydrolysis. *Agric. Biotechnol.* **2015**, *4*, 486–492. [CrossRef]
181. Lucena, G.N.; dos Santos, C.C.; Pinto, G.C.; Piazza, R.D.; Guedes, W.N.; Jafelicci Junior, M.; de Paula, A.V.; Marques, R.F.C. Synthesis and characterization of magnetic cross-linked enzyme aggregate and its evaluation of the alternating magnetic field (AMF) effects in the catalytic activity. *J. Magn. Magn. Mater.* **2020**, *516*. [CrossRef]
182. Primozic, M.; Kravanja, G.; Knez, Z.; Crnjac, A.; Leitgeb, M. Immobilized laccase in the form of (magnetic) cross-linked enzyme aggregates for sustainable diclofenac (bio) degradation. *J. Clean. Prod.* **2020**, *275*. [CrossRef]
183. Xu, M.Q.; Wang, S.S.; Li, L.N.; Gao, J.; Zhang, Y.W. Combined Cross-Linked Enzyme Aggregates as Biocatalysts. *Catalysts* **2018**, *8*, 460. [CrossRef]
184. Xu, M.Q.; Li, F.L.; Yu, W.Q.; Li, R.F.; Zhang, Y.W. Combined cross-linked enzyme aggregates of glycerol dehydrogenase and NADH oxidase for high efficiency in situ NAD(+) regeneration. *Int. J. Biol. Macromol.* **2020**, *144*, 1013–1021. [CrossRef] [PubMed]
185. Sellami, K.; Couvert, A.; Nasrallah, N.; Maachi, R.; Tandjaoui, N.; Abouseoud, M.; Amrane, A. Bio-based and cost effective method for phenolic compounds removal using cross-linked enzyme aggregates. *J. Hazard. Mater.* **2021**, *403*. [CrossRef] [PubMed]
186. Mai-Lan, P.; Polakovic, M. Microbial cell surface display of oxidoreductases: Concepts and applications. *Int. J. Biol. Macromol.* **2020**, *165*, 835–841. [CrossRef]
187. Ugwuodo, C.J.; Nwagu, T.N. Stabilizing enzymes by immobilization on bacterial spores: A review of literature. *Int. J. Biol. Macromol.* **2021**, *166*, 238–250. [CrossRef] [PubMed]
188. Hsieh, H.-Y.; Lin, C.-H.; Hsu, S.-Y.; Stewart, G.C. A Bacillus Spore-Based Display System for Bioremediation of Atrazine. *Appl. Environ. Microbiol.* **2020**, *86*. [CrossRef]
189. Tang, X.J.; Liang, B.; Yi, T.Y.; Manco, G.; Palchetti, I.; Liu, A.H. Cell surface display of organophosphorus hydrolase for sensitive spectrophotometric detection of p-nitrophenol substituted organophosphates. *Enzym. Microb. Technol.* **2014**, *55*, 107–112. [CrossRef]
190. Padkina, M.V.; Sambuk, E.V. Prospects for the Application of Yeast Display in Biotechnology and Cell Biology (Review). *Appl. Biochem. Microbiol.* **2018**, *54*, 337–351. [CrossRef]
191. Chen, H.; Chen, Z.; Ni, Z.; Tian, R.; Zhang, T.; Jia, J.; Chen, K.; Yang, S. Display of Thermotoga maritima MSB8 nitrilase on the spore surface of Bacillus subtilis using out coat protein CotG as the fusion partner. *J. Mol. Catal. B Enzym.* **2016**, *123*, 73–80. [CrossRef]
192. Lozancic, M.; Hossain, A.S.; Mrsa, V.; Teparic, R. Surface Display-An Alternative to Classic Enzyme Immobilization. *Catalysts* **2019**, *9*, 728. [CrossRef]
193. Marcus, R.A. On the theory of oxidation-reduction reactions involving electron transfer. 1. *J. Chem. Phys.* **1956**, *24*, 966–978. [CrossRef]
194. Richardson, D.J.; Butt, J.N.; Fredrickson, J.K.; Zachara, J.M.; Shi, L.; Edwards, M.J.; White, G.; Baiden, N.; Gates, A.J.; Marritt, S.J.; et al. The porin-cytochrome' model for microbe-to-mineral electron transfer. *Mol. Microbiol.* **2012**, *85*, 201–212. [CrossRef] [PubMed]
195. Hitaishi, V.P.; Clement, R.; Quattrocchi, L.; Parent, P.; Duche, D.; Zuily, L.; Ilbert, M.; Lojou, E.; Mazurenko, I. Interplay between Orientation at Electrodes and Copper Activation of Thermus thermophilus Laccase for O-2 Reduction. *J. Am. Chem. Soc.* **2020**, *142*, 1394–1405. [CrossRef] [PubMed]
196. Mazurenko, I.; Monsalve, K.; Rouhana, J.; Parent, P.; Laffon, C.; Le Goff, A.; Szunerits, S.; Boukherroub, R.; Giudici-Orticoni, M.T.; Mano, N.; et al. How the Intricate Interactions between Carbon Nanotubes and Two Bilirubin Oxidases Control Direct and Mediated O-2 Reduction. *ACS Appl. Mater. Interfaces* **2016**, *8*, 23074–23085. [CrossRef] [PubMed]
197. Oteri, F.; Ciaccafava, A.; de Poulpiquet, A.; Baaden, M.; Lojou, E.; Sacquin-Mora, S. The weak, fluctuating, dipole moment of membrane-bound hydrogenase from Aquifex aeolicus accounts for its adaptability to charged electrodes. *Phys. Chem. Chem. Phys.* **2014**, *16*, 11318–11322. [CrossRef]
198. Mazurenko, I.; Monsalve, K.; Infossi, P.; Giudici-Orticoni, M.T.; Topin, F.; Mano, N.; Lojou, E. Impact of substrate diffusion and enzyme distribution in 3D-porous electrodes: A combined electrochemical and modelling study of a thermostable H-2/O-2 enzymatic fuel cell. *Energy Environ. Sci.* **2017**, *10*, 1966–1982. [CrossRef]
199. Atalah, J.; Zhou, Y.; Espina, G.; Blamey, J.M.; Ramasamy, R.P. Improved stability of multicopper oxidase-carbon nanotube conjugates using a thermophilic laccase. *Catal. Sci. Technol.* **2018**, *8*, 1272–1276. [CrossRef]
200. Zafar, M.N.; Aslam, I.; Ludwig, R.; Xu, G.B.; Gorton, L. An efficient and versatile membraneless bioanode for biofuel cells based on Corynascus thermophilus cellobiose dehydrogenase. *Electrochim. Acta* **2019**, *295*, 316–324. [CrossRef]
201. Malinowski, S.; Wardak, C.; Jaroszynska-Wolinska, J.; Herbert, P.A.F.; Panek, R. Cold Plasma as an Innovative Construction Method of Voltammetric Biosensor Based on Laccase. *Sensors* **2018**, *18*, 4086. [CrossRef] [PubMed]

202. Bathinapatla, A.; Kanchi, S.; Sabela, M.I.; Ling, Y.C.; Bisetty, K.; Inamuddin. Experimental and Computational Studies of a Laccase Immobilized ZnONPs/GO-Based Electrochemical Enzymatic Biosensor for the Detection of Sucralose in Food Samples. *Food Anal. Methods* **2020**, *13*, 2014–2027. [CrossRef]
203. Aleksejeva, O.; Mateljak, I.; Ludwig, R.; Alcalde, M.; Shleev, S. Electrochemistry of a high redox potential laccase obtained by computer-guided mutagenesis combined with directed evolution. *Electrochem. Commun.* **2019**, *106*. [CrossRef]
204. Mohtar, L.G.; Aranda, P.; Messina, G.A.; Nazaren, M.A.; Pereira, S.V.; Raba, J.; Bertolino, F.A. Amperometric biosensor based on laccase immobilized onto a nanostructured screen-printed electrode for determination of polyphenols in propolis. *Microchem. J.* **2019**, *144*, 13–18. [CrossRef]
205. Chen, T.; Xu, Y.H.; Peng, Z.; Li, A.H.; Liu, J.Q. Simultaneous Enhancement of Bioactivity and Stability of Laccase by Cu^{2+}/PAA/PPEGA Matrix for Efficient Biosensing and Recyclable Decontamination of Pyrocatechol. *Anal. Chem.* **2017**, *89*, 2065–2072. [CrossRef] [PubMed]
206. Palanisamy, S.; Ramaraj, S.K.; Chen, S.M.; Yang, T.C.K.; Yi-Fan, P.; Chen, T.W.; Velusamy, V.; Selvam, S. A novel Laccase Biosensor based on Laccase immobilized Graphene-Cellulose Microfiber Composite modified Screen-Printed Carbon Electrode for Sensitive Determination of Catechol. *Sci. Rep.* **2017**, *7*. [CrossRef]
207. Lee, J.Y.Y.; Elouarzaki, K.; Sabharwal, H.S.; Fisher, A.C.; Lee, J.-M. A hydrogen/oxygen hybrid biofuel cell comprising an electrocatalytically active nanoflower/laccase-based biocathode. *Catal. Sci. Technol.* **2020**, *10*, 6235–6243. [CrossRef]
208. Zhang, Y.N.; Li, X.; Li, D.W.; Wei, Q.F. A laccase based biosensor on AuNPs-MoS2 modified glassy carbon electrode for catechol detection. *Colloids Surf. B Biointerfaces* **2020**, *186*. [CrossRef]
209. El Ichi-Ribault, S.; Zebda, A.; Tingry, S.; Petit, M.; Suherman, A.L.; Boualam, A.; Cinquin, P.; Martin, D.K. Performance and stability of chitosan-MWCNTs-laccase biocathode: Effect of MWCNTs surface charges and ionic strength. *J. Electroanal. Chem.* **2017**, *799*, 26–33. [CrossRef]
210. Yang, Y.; Zeng, H.; Huo, W.S.; Zhang, Y.H. Direct Electrochemistry and Catalytic Function on Oxygen Reduction Reaction of Electrodes Based on Two Kinds of Magnetic Nano-particles with Immobilized Laccase Molecules. *J. Inorg. Organomet. Polym. Mater.* **2017**, *27*, 201–214. [CrossRef]
211. Zappi, D.; Masci, G.; Sadun, C.; Tortolini, C.; Antonelli, M.L.; Bollella, P. Evaluation of new cholinium-amino acids based room temperature ionic liquids (RTILs) as immobilization matrix for electrochemical biosensor development: Proof-of-concept with Trametes Versicolor laccase. *Microchem. J.* **2018**, *141*, 346–352. [CrossRef]
212. Zhang, Y.; Lv, Z.Y.; Zhou, J.; Fang, Y.; Wu, H.; Xin, F.X.; Zhang, W.M.; Ma, J.F.; Xu, N.; He, A.Y.; et al. Amperometric Biosensors Based on Recombinant Bacterial Laccase CotA for Hydroquinone Determination. *Electroanalysis* **2020**, *32*, 142–148. [CrossRef]
213. Xu, R.; Cui, J.Y.; Tang, R.Z.; Li, F.T.; Zhang, B.R. Removal of 2,4,6-trichlorophenol by laccase immobilized on nano-copper incorporated electrospun fibrous membrane-high efficiency, stability and reusability. *Chem. Eng. J.* **2017**, *326*, 647–655. [CrossRef]
214. Wang, A.Q.; Ding, Y.P.; Li, L.; Duan, D.D.; Mei, Q.W.; Zhuang, Q.; Cui, S.Q.; He, X.Y. A novel electrochemical enzyme biosensor for detection of 17 beta-estradiol by mediated electron-transfer system. *Talanta* **2019**, *192*, 478–485. [CrossRef]
215. Yashas, S.R.; Sandeep, S.; Shivakumar, B.P.; Swamy, N.K. A matrix of perovskite micro-seeds and polypyrrole nanotubes tethered laccase/graphite biosensor for sensitive quantification of 2,4-dichlorophenol in wastewater. *Anal. Methods* **2019**, *11*, 4511–4519. [CrossRef]
216. Castrovilli, M.C.; Bolognesi, P.; Chiarinelli, J.; Avaldi, L.; Cartoni, A.; Calandra, P.; Tempesta, E.; Giardi, M.T.; Antonacci, A.; Arduini, F.; et al. Electrospray deposition as a smart technique for laccase immobilisation on carbon black-nanomodified screen-printed electrodes. *Biosens. Bioelectron.* **2020**, *163*. [CrossRef]
217. Lee, Y.G.; Liao, B.X.; Weng, Y.C. Ascorbic acid sensor using a PVA/laccase-Au-NPs/Pt electrode. *RSC Adv.* **2018**, *8*, 37872–37879. [CrossRef]
218. Lou, C.Q.; Jing, T.; Zhou, J.Y.; Tian, J.Z.; Zheng, Y.J.; Wang, C.; Zhao, Z.Y.; Lin, J.; Liu, H.; Zhao, C.Q.; et al. Laccase immobilized polyaniline/magnetic graphene composite electrode for detecting hydroquinone. *Int. J. Biol. Macromol.* **2020**, *149*, 1130–1138. [CrossRef] [PubMed]
219. Chen, T.; Xu, Y.H.; Wei, S.; Li, A.H.; Huang, L.; Liu, J.Q. A signal amplification system constructed by bi-enzymes and bi-nanospheres for sensitive detection of norepinephrine and miRNA. *Biosens. Bioelectron.* **2019**, *124*, 224–232. [CrossRef] [PubMed]
220. Mazlan, S.Z.; Lee, Y.H.; Abu Hanifah, S. A New Laccase Based Biosensor for Tartrazine. *Sensors* **2017**, *17*, 2859. [CrossRef] [PubMed]
221. Moraes, J.T.; Salamanca-Neto, C.A.R.; Svorc, L.; Schirmann, J.G.; Barbosa-Dekker, A.M.; Dekker, R.F.H.; Sartori, E.R. Laccase from Botryosphaeria rhodina MAMB-05 as a biological component in electrochemical biosensing devices. *Anal. Methods* **2019**, *11*, 717–720. [CrossRef]
222. Sedenho, G.C.; Hassan, A.; Macedo, L.J.A.; Crespilho, F.N. Stabilization of bilirubin oxidase in a biogel matrix for high-performance gas diffusion electrodes. *J. Power Sources* **2021**, *482*. [CrossRef]
223. Gentil, S.; Carriere, M.; Cosnier, S.; Gounel, S.; Mano, N.; Le Goff, A. Direct Electrochemistry of Bilirubin Oxidase from Magnaporthe orizae on Covalently-Functionalized MWCNT for the Design of High-Performance Oxygen-Reducing Biocathodes. *Chem. A Eur. J.* **2018**, *24*, 8404–8408. [CrossRef]
224. Szczesny, J.; Markovic, N.; Conzuelo, F.; Zacarias, S.; Pereira, I.A.C.; Lubitz, W.; Plumere, N.; Schuhmann, W.; Ruff, A. A gas breathing hydrogen/air biofuel cell comprising a redox polymer/hydrogenase-based bioanode. *Nat. Commun.* **2018**, *9*. [CrossRef] [PubMed]

225. Markovic, N.; Conzuelo, F.; Szczesny, J.; Garcia, M.B.G.; Santos, D.H.; Ruff, A.; Schuhmann, W. An Air-breathing Carbon Cloth-based Screen-printed Electrode for Applications in Enzymatic Biofuel Cells. *Electroanalysis* **2019**, *31*, 217–221. [CrossRef]
226. Trifonov, A.; Stemmer, A.; Tel-Vered, R. Carbon-coated magnetic nanoparticles as a removable protection layer extending the operation lifetime of bilirubin oxidase-based bioelectrode. *Bioelectrochemistry* **2021**, *137*, 107640. [CrossRef] [PubMed]
227. Al-Lolage, F.A.; Bartlett, P.N.; Gounel, S.; Staigre, P.; Mano, N. Site-Directed Immobilization of Bilirubin Oxidase for Electrocatalytic Oxygen Reduction. *ACS Catal.* **2019**, *9*, 2068–2078. [CrossRef]
228. Takimoto, D.; Tsujimura, S. Oxygen Reduction Reaction Activity and Stability of Electrochemically Deposited Bilirubin Oxidase. *Chem. Lett.* **2018**, *47*, 1269–1271. [CrossRef]
229. Tsujimura, S.; Oyama, M.; Funabashi, H.; Ishii, S. Effects of pore size and surface properties of MgO-templated carbon on the performance of bilirubin oxidase-modified oxygen reduction reaction cathode. *Electrochim. Acta* **2019**, *322*. [CrossRef]
230. Tang, J.; Yan, X.M.; Huang, W.; Engelbrekt, C.; Duus, J.O.; Ulstrup, J.; Xiao, X.X.; Zhang, J.D. Bilirubin oxidase oriented on novel type three-dimensional biocathodes with reduced graphene aggregation for biocathode. *Biosens. Bioelectron.* **2020**, *167*. [CrossRef]
231. Walgama, C.; Pathiranage, A.; Akinwale, M.; Montealegre, R.; Niroula, J.; Echeverria, E.; McIlroy, D.N.; Harriman, T.A.; Lucca, D.A.; Krishnan, S. Buckypaper–Bilirubin Oxidase Biointerface for Electrocatalytic Applications: Buckypaper Thickness. *ACS Appl. Bio Mater.* **2019**, *2*, 2229–2236. [CrossRef]
232. Zhang, L.; Carucci, C.; Reculusa, S.; Goudeau, B.; Lefrancois, P.; Gounel, S.; Mano, N.; Kuhn, A. Rational Design of Enzyme-Modified Electrodes for Optimized Bioelectrocatalytic Activity. *Chemelectrochem* **2019**, *6*, 4980–4984. [CrossRef]
233. Mukha, D.; Cohen, Y.; Yehezkeli, O. Bismuth Vanadate/Bilirubin Oxidase Photo(bio)electrochemical Cells for Unbiased, Light-Triggered Electrical Power Generation. *Chemsuschem* **2020**, *13*, 2684–2692. [CrossRef]
234. Wang, Y.; Song, Y.; Ma, C.; Kang, Z.; Zhu, Z. A heterologously-expressed thermostable Pyrococcus furiosus cytoplasmic [NiFe]-hydrogenase I used as the catalyst of H2/air biofuel cells. *Int. J. Hydrogen Energy* **2021**, *46*, 3035–3044. [CrossRef]
235. Gentil, S.; Mansor, S.M.C.; Jamet, H.; Cosnier, S.; Cavazza, C.; Le Goff, A. Oriented Immobilization of NiFeSe Hydrogenases on Covalently and Noncovalently Functionalized Carbon Nanotubes for H-2/Air Enzymatic Fuel Cells. *ACS Catal.* **2018**, *8*, 3957–3964. [CrossRef]
236. Szczesny, J.; Birrell, J.A.; Conzuelo, F.; Lubitz, W.; Ruff, A.; Schuhmann, W. Redox-Polymer-Based High-Current-Density Gas-Diffusion H-2-Oxidation Bioanode Using FeFe Hydrogenase fromDesulfovibrio desulfuricansin a Membrane-free Biofuel Cell. *Angew. Chem. Int. Ed.* **2020**, *59*, 16506–16510. [CrossRef]
237. Ruff, A.; Szczesny, J.; Vega, M.; Zacarias, S.; Matias, P.M.; Gounel, S.; Mano, N.; Pereira, I.A.C.; Schuhmann, W. Redox-Polymer-Wired NiFeSe Hydrogenase Variants with Enhanced O-2 Stability for Triple-Protected High-Current-Density H-2-Oxidation Bioanodes. *Chemsuschem* **2020**, *13*, 3627–3635. [CrossRef] [PubMed]
238. Adachi, T.; Kitazumi, Y.; Shirai, O.; Kano, K. Direct Electron Transfer-Type Bioelectrocatalysis of Redox Enzymes at Nanostructured Electrodes. *Catalysts* **2020**, *10*, 236. [CrossRef]
239. Pereira, A.R.; Luz, R.A.S.; Lima, F.; Crespilho, F.N. Protein Oligomerization Based on Bronsted Acid Reaction. *ACS Catal.* **2017**, *7*, 3082–3088. [CrossRef]
240. Bulutoglu, B.; Macazo, F.C.; Bale, J.; King, N.; Baker, D.; Minteer, S.D.; Banta, S. Multimerization of an Alcohol Dehydrogenase by Fusion to a Designed Self-Assembling Protein Results in Enhanced Bioelectrocatalytic Operational Stability. *ACS Appl. Mater. Interfaces* **2019**, *11*, 20022–20028. [CrossRef] [PubMed]
241. Barbosa, C.; Silveira, C.M.; Silva, D.; Brissos, V.; Hildebrandt, P.; Martins, L.O.; Todorovic, S. Immobilized dye-decolorizing peroxidase (DyP) and directed evolution variants for hydrogen peroxide biosensing. *Biosens. Bioelectron.* **2020**, *153*. [CrossRef] [PubMed]
242. Mate, D.M.; Alcalde, M. Laccase engineering: From rational design to directed evolution. *Biotechnol. Adv.* **2015**, *33*, 25–40. [CrossRef]
243. Mateljak, I.; Monza, E.; Lucas, M.F.; Guallar, V.; Aleksejeva, O.; Ludwig, R.; Leech, D.; Shleev, S.; Alcalde, M. Increasing Redox Potential, Redox Mediator Activity, and Stability in a Fungal Laccase by Computer-Guided Mutagenesis and Directed Evolution. *ACS Catal.* **2019**, *9*, 4561–4572. [CrossRef]
244. Lopes, P.; Koschorreck, K.; Pedersen, J.N.; Ferapontov, A.; Lorcher, S.; Pedersen, J.S.; Urlacher, V.B.; Ferapontova, E.E. Bacillus Licheniformis CotA Laccase Mutant: ElectrocatalyticReduction of O-2 from 0.6 V (SHE) at pH 8 and in Seawater. *Chemelectrochem* **2019**, *6*, 2043–2049. [CrossRef]
245. Al-Lolage, F.A.; Meneghello, M.; Ma, S.; Ludwig, R.; Bartlett, P.N. A Flexible Method for the Stable, Covalent Immobilization of Enzymes at Electrode Surfaces. *Chemelectrochem* **2017**, *4*, 1528–1534. [CrossRef]
246. Meneghello, M.; Al-Lolage, F.A.; Ma, S.; Ludwig, R.; Bartlett, P.N. Studying Direct Electron Transfer by Site-Directed Immobilization of Cellobiose Dehydrogenase. *Chemelectrochem* **2019**, *6*, 700–713. [CrossRef] [PubMed]
247. Li, X.; Li, D.W.; Zhang, Y.N.; Lv, P.F.; Feng, Q.; Wei, Q.F. Encapsulation of enzyme by metal-organic framework for single-enzymatic biofuel cell-based self-powered biosensor. *Nano Energy* **2020**, *68*. [CrossRef]
248. Itoh, T.; Shibuya, Y.; Yamaguchi, A.; Hoshikawa, Y.; Tanaike, O.; Tsunoda, T.; Hanaoka, T.A.; Hamakawa, S.; Mizukami, F.; Hayashi, A.; et al. High-performance bioelectrocatalysts created by immobilization of an enzyme into carbon-coated composite membranes with nano-tailored structures. *J. Mater. Chem. A* **2017**, *5*, 20244–20251. [CrossRef]
249. Funabashi, H.; Murata, K.; Tsujimura, S. Effect of Pore Size of MgO-templated Carbon on the Direct Electrochemistry of D-fructose Dehydrogenase. *Electrochemistry* **2015**, *83*, 372–375. [CrossRef]

250. Mazurenko, I.; Clement, R.; Byrne-Kodjabachian, D.; de Poulpiquet, A.; Tsujimura, S.; Lojou, E. Pore size effect of MgO-templated carbon on enzymatic H-2 oxidation by the hyperthermophilic hydrogenase from Aquifex aeolicus. *J. Electroanal. Chem.* **2018**, *812*, 221–226. [CrossRef]
251. Wanibuchi, M.; Kitazumi, Y.; Shirai, O.; Kano, K. Enhancement of the Direct Electron Transfer-type Bioelectrocatalysis of Bilirubin Oxidase at the Interface between Carbon Particles. *Electrochemistry* **2021**, *89*, 43–48. [CrossRef]
252. Takahashi, Y.; Wanibuchi, M.; Kitazumi, Y.; Shirai, O.; Kano, K. Improved direct electron transfer-type bioelectrocatalysis of bilirubin oxidase using porous gold electrodes. *J. Electroanal. Chem.* **2019**, *843*, 47–53. [CrossRef]
253. Wu, F.; Su, L.; Yu, P.; Mao, L.Q. Role of Organic Solvents in Immobilizing Fungus Laccase on Single Walled Carbon Nanotubes for Improved Current Response in Direct Bioelectrocatalysis. *J. Am. Chem. Soc.* **2017**, *139*, 1565–1574. [CrossRef] [PubMed]
254. Hickey, D.P.; Lim, K.; Cai, R.; Patterson, A.R.; Yuan, M.W.; Sahin, S.; Abdellaoui, S.; Minteer, S.D. Pyrene hydrogel for promoting direct bioelectrochemistry: ATP-independent electroenzymatic reduction of N-2. *Chem. Sci.* **2018**, *9*, 5172–5177. [CrossRef] [PubMed]
255. Kuk, S.K.; Gopinath, K.; Singh, R.K.; Kim, T.D.; Lee, Y.; Choi, W.S.; Lee, J.K.; Park, C.B. NADH-Free Electroenzymatic Reduction of CO_2 by Conductive Hydrogel-Conjugated Formate Dehydrogenase. *ACS Catal.* **2019**, *9*, 5584–5589. [CrossRef]
256. El Ichi, S.; Zebda, A.; Laaroussi, A.; Reverdy-Bruas, N.; Chaussy, D.; Belgacem, M.N.; Cinquin, P.; Martin, D.K. Chitosan improves stability of carbon nanotube biocathodes for glucose biofuel cells. *Chem. Commun.* **2014**, *50*, 14535–14538. [CrossRef]
257. El Ichi, S.; Zebda, A.; Alcaraz, J.P.; Laaroussi, A.; Boucher, F.; Boutonnat, J.; Reverdy-Bruas, N.; Chaussy, D.; Belgacem, M.N.; Cinquin, P.; et al. Bioelectrodes modified with chitosan for long-term energy supply from the body. *Energy Environ. Sci.* **2015**, *8*, 1017–1026. [CrossRef]
258. Matsumoto, T.; Isogawa, Y.; Tanaka, T.; Kondo, A. Streptavidin-hydrogel prepared by sortase A-assisted click chemistry for enzyme immobilization on an electrode. *Biosens. Bioelectron.* **2018**, *99*, 56–61. [CrossRef]
259. Ghimire, A.; Pattamrnattel, A.; Maher, C.E.; Kasi, R.M.; Kumar, C.V. Three-Dimensional, Enzyme Biohydrogel Electrode for Improved Bioelectrocatalysis. *ACS Appl. Mater. Interfaces* **2017**, *9*, 42556–42565. [CrossRef]
260. Heller, A. Electrical connection of enzyme redox centers to electrodes. *J. Phys. Chem.* **1992**, *96*, 3579–3587. [CrossRef]
261. Cadoux, C.; Milton, R.D. Recent Enzymatic Electrochemistry for Reductive Reactions. *Chemelectrochem* **2020**, *7*, 1974–1986. [CrossRef]
262. Ruth, J.C.; Milton, R.D.; Gu, W.Y.; Spormann, A.M. Enhanced Electrosynthetic Hydrogen Evolution by Hydrogenases Embedded in a Redox-Active Hydrogel. *Chem. A Eur. J.* **2020**, *26*, 7323–7329. [CrossRef]
263. Xiao, X.X.; Conghaile, P.O.; Leech, D.; Magner, E. Use of Polymer Coatings to Enhance the Response of Redox-Polymer-Mediated Electrodes. *Chemelectrochem* **2019**, *6*, 1344–1349. [CrossRef]
264. Diaz-Gonzalez, J.C.M.; Escalona-Villalpando, R.A.; Arriaga, L.G.; Minteer, S.D.; Casanova-Moreno, J.R. Effects of the cross-linker on the performance and stability of enzymatic electrocatalytic films of glucose oxidase and dimethylferrocene-modified linear poly(ethyleneimine). *Electrochim. Acta* **2020**, *337*. [CrossRef]
265. Tsujimura, S.; Takeuchi, S. Toward an ideal platform structure based on MgO-templated carbon for flavin adenine dinucleotide-dependent glucose dehydrogenase-Os polymer-hydrogel electrodes. *Electrochim. Acta* **2020**, *343*. [CrossRef]
266. Huang, X.C.; Zhang, L.L.; Zhang, Z.; Guo, S.; Shang, H.; Li, Y.B.; Liu, J. Wearable biofuel cells based on the classification of enzyme for high power outputs and lifetimes. *Biosens. Bioelectron.* **2019**, *124*, 40–52. [CrossRef]
267. Kim, J.H.; Hong, S.G.; Wee, Y.; Hu, S.; Kwon, Y.; Ha, S.; Kim, J. Enzyme precipitate coating of pyranose oxidase on carbon nanotubes and their electrochemical applications. *Biosens. Bioelectron.* **2017**, *87*, 365–372. [CrossRef]
268. Garcia, K.E.; Babanova, S.; Scheffler, W.; Hans, M.; Baker, D.; Atanassov, P.; Banta, S. Designed Protein Aggregates Entrapping Carbon Nanotubes for Bioelectrochemical Oxygen Reduction. *Biotechnol. Bioeng.* **2016**, *113*, 2321–2327. [CrossRef]
269. Caserta, G.; Lorent, C.; Ciaccafava, A.; Keck, M.; Breglia, R.; Greco, C.; Limberg, C.; Hildebrandt, P.; Cramer, S.P.; Zebger, I.; et al. The large subunit of the regulatory NiFe -hydrogenase from Ralstonia eutropha—A minimal hydrogenase? *Chem. Sci.* **2020**, *11*, 5453–5465. [CrossRef]
270. Rodriguez-Padron, D.; Puente-Santiago, A.R.; Caballero, A.; Balu, A.M.; Romero, A.A.; Luque, R. Highly efficient direct oxygen electro-reduction by partially unfolded laccases immobilized on waste-derived magnetically separable nanoparticles. *Nanoscale* **2018**, *10*, 3961–3968. [CrossRef] [PubMed]
271. Gutierrez-Sanchez, C.; Ciaccafava, A.; Blanchard, P.Y.; Monsalve, K.; Giudici-Orticoni, M.T.; Lecomte, S.; Lojou, E. Efficiency of Enzymatic O-2 Reduction by Myrothecium verrucaria Bilirubin Oxidase Probed by Surface Plasmon Resonance, PMIRRAS, and Electrochemistry. *ACS Catal.* **2016**, *6*, 5482–5492. [CrossRef]
272. Singh, K.; McArdle, T.; Sullivan, P.R.; Blanford, C.F. Sources of activity loss in the fuel cell enzyme bilirubin oxidase. *Energy Environ. Sci.* **2013**, *6*, 2460–2464. [CrossRef]
273. Zelechowska, K.; Stolarczyk, K.; Lyp, D.; Rogalski, J.; Roberts, K.P.; Bilewicz, R.; Biernat, J.F. Aryl and N-arylamide carbon nanotubes for electrical coupling of laccase to electrodes in biofuel cells and biobatteries. *Biocybern. Biomed. Eng.* **2013**, *33*, 235–245. [CrossRef]
274. Yahata, N.; Saitoh, T.; Takayama, Y.; Ozawa, K.; Ogata, H.; Higuchi, Y.; Akutsu, H. Redox interaction of cytochrome c(3) with NiFe hydrogenase from Desulfovibrio vulgaris Miyazaki F. *Biochemistry* **2006**, *45*, 1653–1662. [CrossRef]

275. Sugimoto, Y.; Kitazumi, Y.; Shirai, O.; Nishikawa, K.; Higuchi, Y.; Yamamoto, M.; Kano, K. Electrostatic roles in electron transfer from NiFe hydrogenase to cytochrome c(3) from Desulfovibrio vulgaris Miyazaki F. *Biochim. Et Biophys. Acta Proteins Proteom.* **2017**, *1865*, 481–487. [CrossRef]
276. McArdle, T.; McNamara, T.P.; Fei, F.; Singh, K.; Blanford, C.F. Optimizing the Mass-Specific Activity of Bilirubin Oxidase Adlayers through Combined Electrochemical Quartz Crystal Microbalance and Dual Polarization Interferometry Analyses. *ACS Appl. Mater. Interfaces* **2015**, *7*, 25270–25280. [CrossRef] [PubMed]
277. Kitazumi, Y.; Shirai, O.; Yamamoto, M.; Kano, K. Numerical simulation of diffuse double layer around microporous electrodes based on the Poisson-Boltzmann equation. *Electrochim. Acta* **2013**, *112*, 171–175. [CrossRef]
278. Olloqui-Sariego, J.L.; Calvente, J.J.; Andreu, R. Immobilizing redox enzymes at mesoporous and nanostructured electrodes. *Curr. Opin. Electrochem.* **2021**, *26*, 100658. [CrossRef]
279. Sakai, K.; Xia, H.-Q.; Kitazumi, Y.; Shirai, O.; Kano, K. Assembly of direct-electron-transfer-type bioelectrodes with high performance. *Electrochim. Acta* **2018**, *271*, 305–311. [CrossRef]
280. Sakai, K.; Kitazumi, Y.; Shirai, O.; Kano, K. Nanostructured Porous Electrodes by the Anodization of Gold for an Application as Scaffolds in Direct-electron-transfer-type Bioelectrocatalysis. *Electrochim. Acta* **2018**, *34*, 1317–1322. [CrossRef]
281. Armstrong, F.A.; Bond, A.M.; Hill, H.A.O.; Oliver, B.N.; Psalti, I.S.M. Electrochemistry of cytochrome-c, plastocyanin, and ferredoxin at edge-plane and basal-plane graphite-electrodes interpreted via a model based on electron-transfer at electroactive sites of microscopic dimensions in size. *J. Am. Chem. Soc.* **1989**, *111*, 9185–9189. [CrossRef]
282. Komori, K.; Huang, J.; Mizushima, N.; Ko, S.; Tatsuma, T.; Sakai, Y. Controlled direct electron transfer kinetics of fructose dehydrogenase at cup-stacked carbon nanofibers. *Phys. Chem. Chem. Phys.* **2017**, *19*, 27795–27800. [CrossRef] [PubMed]
283. Kavetskyy, T.; Smutok, O.; Demkiv, O.; Mat'ko, I.; Svajdlenkova, H.; Sausa, O.; Novak, I.; Berek, D.; Cechova, K.; Pecz, M.; et al. Microporous carbon fibers as electroconductive immobilization matrixes: Effect of their structure on operational parameters of laccase-based amperometric biosensor. *Mater. Sci. Eng. C Mater. Biol. Appl.* **2020**, *109*. [CrossRef] [PubMed]
284. de Poulpiquet, A.; Marques-Knopf, H.; Wernert, V.; Giudici-Orticoni, M.T.; Gadiou, R.; Lojou, E. Carbon nanofiber mesoporous films: Efficient platforms for bio-hydrogen oxidation in biofuel cells. *Phys. Chem. Chem. Phys.* **2014**, *16*, 1366–1378. [CrossRef] [PubMed]
285. Pandelia, M.E.; Fourmond, V.; Tron-Infossi, P.; Lojou, E.; Bertrand, P.; Leger, C.; Giudici-Orticoni, M.T.; Lubitz, W. Membrane-Bound Hydrogenase I from the Hyperthermophilic Bacterium Aquifex aeolicus: Enzyme Activation, Redox Intermediates and Oxygen Tolerance. *J. Am. Chem. Soc.* **2010**, *132*, 6991–7004. [CrossRef]
286. Ciaccafava, A.; De Poulpiquet, A.; Techer, V.; Giudici-Orticoni, M.T.; Tingry, S.; Innocent, C.; Lojou, E. An innovative powerful and mediatorless H-2/O-2 biofuel cell based on an outstanding bioanode. *Electrochem. Commun.* **2012**, *23*, 25–28. [CrossRef]
287. Oughli, A.A.; Velez, M.; Birrell, J.A.; Schuhmann, W.; Lubitz, W.; Plumere, N.; Rudiger, O. Viologen-modified electrodes for protection of hydrogenases from high potential inactivation while performing H-2 oxidation at low overpotential. *Dalton Trans.* **2018**, *47*, 10685–10691. [CrossRef]
288. Plumere, N.; Rudiger, O.; Oughli, A.A.; Williams, R.; Vivekananthan, J.; Poller, S.; Schuhmann, W.; Lubitz, W. A redox hydrogel protects hydrogenase from high-potential deactivation and oxygen damage. *Nat. Chem.* **2014**, *6*, 822–827. [CrossRef]
289. Ruff, A.; Szczesny, J.; Zacarias, S.; Pereira, I.A.C.; Plumere, N.; Schuhmann, W. Protection and Reactivation of the NiFeSe Hydrogenase from Desulfovibrio vulgaris Hildenborough under Oxidative Conditions. *ACS Energy Lett.* **2017**, *2*, 964–968. [CrossRef]
290. Mano, N. Engineering glucose oxidase for bioelectrochemical applications. *Bioelectrochemistry* **2019**, *128*, 218–240. [CrossRef]
291. Elouarzaki, K.; Bourourou, M.; Holzinger, M.; Le Goff, A.; Marks, R.S.; Cosnier, S. Freestanding HRP-GOx redox buckypaper as an oxygen-reducing biocathode for biofuel cell applications. *Energy Environ. Sci.* **2015**, *8*, 2069–2074. [CrossRef]
292. Lopez, F.; Zerria, S.; Ruff, A.; Schuhmann, W. An O-2 Tolerant Polymer/Glucose Oxidase Based Bioanode as Basis for a Self-powered Glucose Sensor. *Electroanalysis* **2018**, *30*, 1311–1318. [CrossRef]
293. Sahin, S.; Wongnate, T.; Chuaboon, L.; Chaiyen, P.; Yu, E.H. Enzymatic fuel cells with an oxygen resistant variant of pyranose-2-oxidase as anode biocatalyst. *Biosens. Bioelectron.* **2018**, *107*, 17–25. [CrossRef]
294. Tremey, E.; Stines-Chaumeil, C.; Gounel, S.; Mano, N. Designing an O-2-Insensitive Glucose Oxidase for Improved Electrochemical Applications. *Chemelectrochem* **2017**, *4*, 2520–2526. [CrossRef]
295. Li, H.; Münchberg, U.; Oughli, A.A.; Buesen, D.; Lubitz, W.; Freier, E.; Plumeré, N. Suppressing hydrogen peroxide generation to achieve oxygen-insensitivity of a [NiFe] hydrogenase in redox active films. *Nat. Commun.* **2020**, *11*, 920. [CrossRef]
296. Zebda, A.; Alcaraz, J.P.; Vadgama, P.; Shleev, S.; Minteer, S.D.; Boucher, F.; Cinquin, P.; Martin, D.K. Challenges for successful implantation of biofuel cells. *Bioelectrochemistry* **2018**, *124*, 57–72. [CrossRef] [PubMed]
297. Xu, Z.H.; Liu, Y.C.; Williams, I.; Li, Y.; Qian, F.Y.; Wang, L.; Lei, Y.; Li, B.K. Flat enzyme-based lactate biofuel cell integrated with power management system: Towards long term in situ power supply for wearable sensors. *Appl. Energy* **2017**, *194*, 71–80. [CrossRef]
298. Lv, J.; Jeerapan, I.; Tehrani, F.; Yin, L.; Silva-Lopez, C.A.; Jang, J.H.; Joshuia, D.; Shah, R.; Liang, Y.Y.; Xie, L.Y.; et al. Sweat-based wearable energy harvesting-storage hybrid textile devices. *Energy Environ. Sci.* **2018**, *11*, 3431–3442. [CrossRef]
299. Ruff, A.; Conzuelo, F.; Schuhmann, W. Bioelectrocatalysis as the basis for the design of enzyme-based biofuel cells and semi-artificial biophotoelectrodes. *Nat. Catal.* **2020**, *3*, 214–224. [CrossRef]

300. Clement, R.; Wang, X.; Biaso, F.; Ilbert, M.; Mazurenko, I.; Lojou, E. Mutations in the coordination spheres of T1 Cu affect Cu2+-activation of the laccase from Thermus thermophilus. *Biochimie* **2021**. [CrossRef]
301. Zhang, L.L.; Cui, H.Y.; Zou, Z.; Garakani, T.M.; Novoa-Henriquez, C.; Jooyeh, B.; Schwaneberg, U. Directed Evolution of a Bacterial Laccase (CueO) for Enzymatic Biofuel Cells. *Angew. Chem. Int. Ed.* **2019**, *58*, 4562–4565. [CrossRef]
302. Herkendell, K.; Stemmer, A.; Tel-Vered, R. Extending the operational lifetimes of all-direct electron transfer enzymatic biofuel cells by magnetically assembling and exchanging the active biocatalyst layers on stationary electrodes. *Nano Res.* **2019**, *12*, 767–775. [CrossRef]
303. Hammond, J.L.; Gross, A.J.; Giroud, F.; Travelet, C.; Borsali, R.; Cosnier, S. Solubilized Enzymatic Fuel Cell (SEFC) for Quasi-Continuous Operation Exploiting Carbohydrate Block Copolymer Glyconanoparticle Mediators. *ACS Energy Lett.* **2019**, *4*, 142–148. [CrossRef]
304. Wang, D.D.; Jana, D.L.; Zhao, Y.L. Metal-Organic Framework Derived Nanozymes in Biomedicine. *Acc. Chem. Res.* **2020**, *53*, 1389–1400. [CrossRef]
305. Navyatha, B.; Singh, S.; Nara, S. AuPeroxidase nanozymes: Promises and applications in biosensing. *Biosens. Bioelectron.* **2021**, *175*, 112882. [CrossRef] [PubMed]
306. Ahmed, M.E.; Dey, A. Recent developments in bioinspired modelling of NiFe- and FeFe-hydrogenases. *Curr. Opin. Electrochem.* **2019**, *15*, 155–164. [CrossRef]
307. Laureanti, J.A.; O'Hagan, M.; Shaw, W.J. Chicken fat for catalysis: A scaffold is as important for molecular complexes for energy transformations as it is for enzymes in catalytic function. *Sustain. Energy Fuels* **2019**, *3*, 3260–3278. [CrossRef]
308. Yang, Y.; Xiong, Y.; Zeng, R.; Lu, X.; Krumov, M.; Huang, X.; Xu, W.; Wang, H.; DiSalvo, F.J.; Brock, J.D.; et al. Operando Methods in Electrocatalysis. *ACS Catal.* **2021**, *11*, 1136–1178. [CrossRef]
309. de Souza, J.C.P.; Macedo, L.J.A.; Hassan, A.; Sedenho, G.C.; Modenez, I.A.; Crespilho, F.N. In Situ and Operando Techniques for Investigating Electron Transfer in Biological Systems. *Chemelectrochem* **2020**. [CrossRef]
310. Kornienko, N.; Ly, K.H.; Robinson, W.E.; Heidary, N.; Zhang, J.Z.; Reisner, E. Advancing Techniques for Investigating the Enzyme-Electrode Interface. *Acc. Chem. Res.* **2019**, *52*, 1439–1448. [CrossRef]
311. Singh, K.; Blanford, C.F. Electrochemical Quartz Crystal Microbalance with Dissipation Monitoring: A Technique to Optimize Enzyme Use in Bioelectrocatalysis. *Chemcatchem* **2014**, *6*, 921–929. [CrossRef]
312. Sezer, M.; Kielb, P.; Kuhlmann, U.; Mohrmann, H.; Schulz, C.; Heinrich, D.; Schlesinger, R.; Heberle, J.; Weidinger, I.M. Surface Enhanced Resonance Raman Spectroscopy Reveals Potential Induced Redox and Conformational Changes of Cytochrome c Oxidase on Electrodes. *J. Phys. Chem. B* **2015**, *119*, 9586–9591. [CrossRef] [PubMed]
313. Dagys, M.; Laurynenas, A.; Ratautas, D.; Kulys, J.; Vidziunaite, R.; Talaikis, M.; Niaura, G.; Marcinkeviciene, L.; Meskys, R.; Shleev, S. Oxygen electroreduction catalysed by laccase wired to gold nanoparticles via the trinuclear copper cluster. *Energy Environ. Sci.* **2017**, *10*, 498–502. [CrossRef]
314. Olejnik, P.; Pawlowska, A.; Palys, B. Application of Polarization Modulated Infrared Reflection Absorption Spectroscopy for electrocatalytic activity studies of laccase adsorbed on modified gold electrodes. *Electrochim. Acta* **2013**, *110*, 105–111. [CrossRef]
315. Macedo, L.J.A.; Hassan, A.; Sedenho, G.C.; Crespilho, F.N. Assessing electron transfer reactions and catalysis in multicopper oxidases with operando X-ray absorption spectroscopy. *Nat. Commun.* **2020**, *11*. [CrossRef] [PubMed]
316. Abdiaziz, K.; Salvadori, E.; Sokol, K.P.; Reisner, E.; Roessler, M.M. Protein film electrochemical EPR spectroscopy as a technique to investigate redox reactions in biomolecules. *Chem. Commun* **2019**, *55*, 8840–8843. [CrossRef]
317. Harris, T.; Heidary, N.; Kozuch, J.; Frielingsdorf, S.; Lenz, O.; Mroginski, M.A.; Hildebrandt, P.; Zebger, I.; Fischer, A. In Situ Spectroelectrochemical Studies into the Formation and Stability of Robust Diazonium-Derived Interfaces on Gold Electrodes for the Immobilization of an Oxygen-Tolerant Hydrogenase. *ACS Appl. Mater. Interfaces* **2018**, *10*, 23380–23391. [CrossRef]
318. Abad, J.M.; Tesio, A.Y.; Martinez-Perinan, E.; Pariente, F.; Lorenzo, E. Imaging resolution of biocatalytic activity using nanoscale scanning electrochemical microscopy. *Nano Res.* **2018**, *11*, 4232–4244. [CrossRef]
319. Tassy, B.; Dauphin, A.L.; Man, H.M.; Le Guenno, H.; Lojou, E.; Bouffier, L.; de Poulpiquet, A. In Situ Fluorescence Tomography Enables a 3D Mapping of Enzymatic O-2 Reduction at the Electrochemical Interface. *Anal. Chem.* **2020**, *92*, 7249–7256. [CrossRef] [PubMed]
320. Zigah, D.; Lojou, E.; de Poulpiquet, A. Micro- and Nanoscopic Imaging of Enzymatic Electrodes: A Review. *Chemelectrochem* **2019**, *6*, 5524–5546. [CrossRef]
321. Barrozo, A.; Orio, M. Molecular Electrocatalysts for the Hydrogen Evolution Reaction: Input from Quantum Chemistry. *Chemsuschem* **2019**, *12*, 4905–4915. [CrossRef] [PubMed]
322. Qiu, S.Y.; Li, Q.Y.; Xu, Y.J.; Shen, S.H.; Sun, C.H. Learning from nature: Understanding hydrogenase enzyme using computational approach. *Wiley Interdiscip. Rev. Comput. Mol. Sci.* **2020**, *10*. [CrossRef]
323. Yang, S.J.; Liu, J.; Quan, X.B.; Zhou, J. Bilirubin Oxidase Adsorption onto Charged Self-Assembled Monolayers: Insights from Multiscale Simulations. *Langmuir* **2018**, *34*, 9818–9828. [CrossRef] [PubMed]

Review

Rational Surface Modification of Carbon Nanomaterials for Improved Direct Electron Transfer-Type Bioelectrocatalysis of Redox Enzymes

Hongqi Xia [1,2] and Jiwu Zeng [1,2,*]

1. Key Laboratory of South Subtropical Fruit Biology and Genetic Resource Utilization (MOA), Institute of Fruit Tree Research, Guangdong Academy of Agricultural Sciences, Guangzhou 510640, China; hqixia@sina.cn
2. Guangdong Province Key Laboratory of Tropical and Subtropical Fruit Tree Research, Guangzhou 510640, China
* Correspondence: zengjiwu@gdaas.cn

Received: 27 October 2020; Accepted: 4 December 2020; Published: 10 December 2020

Abstract: Interfacial electron transfer between redox enzymes and electrodes is a key step for enzymatic bioelectrocatalysis in various bioelectrochemical devices. Although the use of carbon nanomaterials enables an increasing number of redox enzymes to carry out bioelectrocatalysis involving direct electron transfer (DET), the role of carbon nanomaterials in interfacial electron transfer remains unclear. Based on the recent progress reported in the literature, in this mini review, the significance of carbon nanomaterials on DET-type bioelectrocatalysis is discussed. Strategies for the oriented immobilization of redox enzymes in rationally modified carbon nanomaterials are also summarized and discussed. Furthermore, techniques to probe redox enzymes in carbon nanomaterials are introduced.

Keywords: direct electron transfer; bioelectrocatalysis; orientation; carbon nanomaterials; surface modification

1. Introduction

Bioelectrocatalysis, which couples an enzymatic catalysis with an electrode reaction and thereby transforms the chemical energy of the reactant into electrical energy (or vice versa), plays an important role in various applications, including biological fuel cells [1], biosensors [2], and bio-electrosynthesis [3]. Redox enzymes in solutions usually show significantly high catalytic efficiency toward their natural substrates. However, enzymes immobilized on the surface of a solid electrode usually do not enable sufficient electron transfer kinetics between the redox-active sites and the support owing to electrical insulation of the redox-active site by the surrounding polypeptides.

In general, mechanisms of electron transfer between enzymes and electrodes are classified into mediated electron transfer (MET) and direct electron transfer (DET) [4] (Figure 1). In an MET-type system, an extrinsic redox-active species, referred to as a mediator, is utilized to shuttle electrons between the enzyme redox site and an electrode [5]. In this case, the redox enzyme catalyzes the oxidation or reduction of the mediator as a co-substrate. The reverse transformation (regeneration) of the mediator occurs reversibly or quasi-reversibly on the electrode surface. The use of small, low-molecular-weight electron mediators that require low overpotentials can be beneficial because they can enable rapid electron transfer between an enzyme and an electrode with low power loss. However, the cost, stability, selectivity, and ability to exchange electrons in the immobilized state of such mediators must also be considered. In contrast, in a DET-type system, fast electron transfer to or from a solid electrode occurs through the intrinsic electron relay system in the protein [6] (such as a series of ion-sulfur clusters [7], hemes [8], copper atoms [9], or some amino acid residues [10,11]). Accordingly, the electrode surface acts as a co-substrate of the redox enzyme, and the enzymatic and electrode reactions proceed simultaneously. From this viewpoint, a DET-type bioelectrocatalysis constructed with only an enzyme and an electrode is mostly an ideal system that provides high selectivity for electrochemical biosensors and a high cell voltage for enzymatic biofuel cells without a proton exchange membrane.

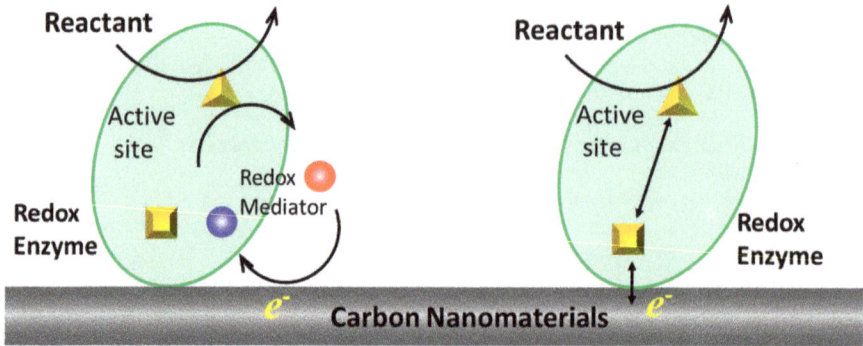

Figure 1. Schematic of electron transfer routes of (**left**) MET-type and (**right**) DET-type bioelectrocatalysis.

Over the past few decades, significant efforts have been devoted to improving the performance of DET-type bioelectrocatalysis, including protein engineering, electrode materials, and bio-interfacial engineering [12,13]. The interfacial electron transfer rate constant (k), according to Macrus's theory [14,15], is governed by the potential difference, re-organization energy, and most importantly, the distance between the active center of the enzymes and the electrode surface. The electron transfer rate constant between the redox site and the electrode decreases exponentially with increasing distance. Because redox enzymes are sizable proteins with anisotropic properties and their active sites are not always located in the central region, the distance between the active site and the electrodes should be varied according to the orientation of the enzymes when immobilizing at the electrode surfaces. Therefore, controlling the orientation of the enzyme on the electrode surface is a key factor in realizing a fast interfacial electron transfer [16,17].

Carbon nanomaterials, such as carbon nanotubes (CNTs), graphene, carbon nanoparticles, and their combinations thereof, have been widely utilized in DET-type bioelectrocatalysis owing to their high conductivity, chemical stability, and low cost. CNTs are nanowires constituted from one or more layers of seamlessly rolled graphene (single-walled and multi-walled CNTs, respectively) with large specific surface areas (in a precise sense, with large values for surface-to-weight ratio) and behave electrically as metals or as semiconductors [18]. Moreover, purified CNTs are usually electrochemically inert and do not exhibit voltametric response in the potential window commonly used [19]. In contrary to CNTs, graphene is a two-dimensional sheet of sp^2 bonded carbon atoms possess unique properties, like ballistic conductivity, high specific surface area, and rapid heterogeneous electron transfer [20]. For example, it has been reported that the specific surface area of graphene sheets is frequently larger than that of single-walled CNTs. Unlike structurally more defined CNTs and graphene, structurally less-defined carbon nanoparticles are considered more economical than other carbon nanomaterials because they are produced commercially in bulk quantities and also often formed as waste byproducts during the formation of other nanomaterials, such as CNTs [21]. In addition to these intrinsic properties mentioned above, carbon nanomaterials can be easily modified for different purposes through various approaches, including diazonium grafting, amine electrochemical oxidation, and π-π interactions. An increasing number of redox enzymes, including glucose dehydrogenase [22], fructose dehydrogenase (FDH) [23], cellobiose dehydrogenase [24], formate dehydrogenase (FoDH) [25], hydrogenase (H_2ase), bilirubin oxidase (BOD), laccase (Lac) [26], cooper efflux oxidase (CueO) [27], and peroxidase [28], have been reported to be capable of DET at a suitable carbon nanomaterial-based electrode. Notably, although many researchers in the previous decades have claimed that glucose oxidase, a well-known enzyme utilized in glucose biosensors and biofuel cells, achieves a DET capability in carbon nanomaterials, recent studies have concluded that no evidence supports the idea that native glucose oxidase undergoes DET in carbon nanotubes or graphene [29–31]. Understanding the mechanisms and roles of carbon nanomaterials in direct bioelectrocatalysis is essential to identifying whether a DET reaction occurs and for preparing a suitable platform for DET-type bioelectrocatalysis with a high performance.

In this mini review, we start with the significance of carbon nanomaterials in promoting direct bioelectrocatalysis of redox enzymes. The effects of the pore distribution and surface chemical properties of carbon nanomaterials on DET-type bioelectrocatalysis are emphasized. Following a summarization of the surface modification of carbon nanomaterials, the oriented immobilization of redox enzymes in rationally functionalized carbon nanomaterials via different strategies is reviewed. Techniques to probe the redox enzymes on the carbon nanomaterial surfaces are also described.

2. Significance of Carbon Nanomaterials on Direct Bioelectrocatalysis

2.1. Effect of Pore Distribution

The primary motivation of using carbon nanomaterials for direct electrocatalysis, at the early stage, is ascribed to the high enzyme loading because of its large specific surface area. Although we agree that enzyme loading is essential to the performance of direct bioelectrocatalysis, several recent studies have proposed that the microstructure and the surface physiochemistry of carbon nanomaterials also play important roles in DET-type bioelectrocatalysis. For example, carbon cryogels with a controlled average pore radii of 5–40 nm have been utilized for FDH immobilization [32]. It has been found that, although the estimated Brunauer–Emmett–Teller (BET) specific surface area of a carbon cryogel with an average pore radius of 40 nm is the smallest, the DET-type bioelectrocatalytic current of fructose oxidation in a carbon cryogel with an average pore radius of 40 nm is the highest (~5 mA cm^{-2}), compared to those of carbon cryogels with radii of 5, 11, 16, and 26 nm, which have higher BET specific surface areas. Similar results have been obtained for MgO-templated carbon nanomaterial electrodes using different redox enzymes, including FDH [33], BOD [34], and H$_2$ase [35]. Rather than a high specific surface area, carbon nanomaterials with suitable pore size distributions are apparently more essential for the DET-type bioelectrocatalysis of redox enzymes.

A theoretical model of a randomly oriented spherical enzyme adsorbed in a planar or three-dimensional electrode was proposed [36,37] (Figure 2). In that model, spherical pores with a radius close to that of the enzyme improves the interfacial electron transfer kinetics thanks to increased probability of orientations with a short distance during the interfacial electron transfer (Figure 2). This effect is referred to as the curvature effect of porous structures [12,13]. These porous structures can be found at the surface of mesoporous carbon materials or can be formed through primary carbon nanoparticle aggregation. For example, in a comparative study, three types of carbon materials—Ketjen Black EC300J (KB), Vulcan XC-72R (Vulcan), and high-purity exfoliated graphite J-SP (JSP)—were utilized as scaffolds for the DET-type bioelectrocatalysis of BOD [38]. The results show that the micropores at the JSP surface are highly effective for the DET-reaction of BOD, whereas gaps between several primary particles in the KB and Vulcan aggregates play important roles as scaffolds for *Mv*BOD.

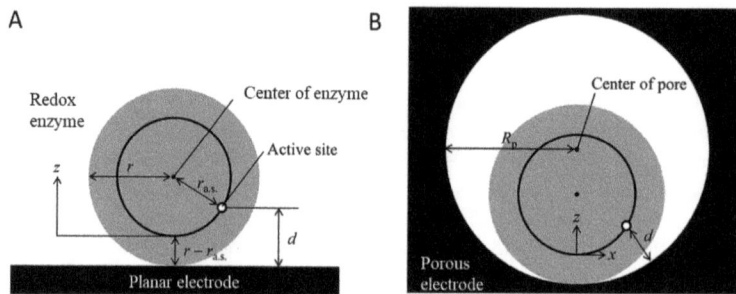

Figure 2. Schematic model of an adsorbed enzyme on (**A**) planar and (**B**) porous electrodes. Reprinted from [37]. Copyright (2016), with permission from American Chemical Society.

Notably, the optimization of carbon nanomaterials for DET-type bioelectrocatalysis requires balancing the ratio of macro-, meso-, and micropores for the construction of a hierarchical three-dimensional structure effective for the mass transport of an electrolyte solution and substrates as well as the penetration of enzymes into the nanostructures, and enhancing the interfacial electron transfer kinetics of macromolecular redox enzymes. The dispersion parameter of H_2ase in MgO-templated carbon with a pore size of 35 nm is notably smaller than that in MgO-templated carbons with a pore size of 150 nm [35], which can be explained by the curvature effect [36,37], as mentioned above. However, the H_2-oxidation bioelectrocatalytic current produced by H_2ase in MgO-templated carbons with a pore size of 150 nm (~450 µA cm^{-2}) is higher than that in MgO-templated carbons with a pore size of 35 nm (~50 µA cm^{-2}) owing to greater enzyme penetration in carbon with larger pores [35]. Furthermore, a hierarchical dual MgO-templated carbon, with a mixture of 33% macropores (150 nm) and 67% mesopores (40 nm), was prepared for the DET-type bioelectrocatalysis of BOD [39]. The results show that the oxygen reduction current in dual MgO-templated carbon reached to around 10 mA cm^{-2}, which is much higher than that in a MgO-templated carbon electrode with only meso- or macro-porous structures (~5.6 mA cm^{-2} or ~5.1 mA cm^{-2}). These results clearly show the importance of the pore distribution in carbon nanomaterials in DET-type bioelectrocatalysis.

2.2. Effects of Surface Chemistry

Carbon nanomaterials are frequently formed with hydrophobic and chemically inactivate surfaces, whereas functional groups, such as hydroxyl, carboxyl, and carbonyl, can be formed at the surfaces of carbon nanomaterials, depending on the preparation and treatment approaches. For example, carboxyl groups are frequently found in oxidized CNTs, which can be produced by treatment with acid or peroxides. By contrast, redox enzyme surfaces usually consist of polar amino acids, some of which can even be charged in a solution. Intricate interactions between carbon nanomaterials and redox enzymes have been proposed to play important roles in the orientation and interfacial electron transfer rate of redox enzymes [40]. For example, an improved electron transfer has been reported in BOD on UV-ozone treated carbon black [41] and carbon nanofiber [42] surfaces. UV-ozone treatment was proposed to produce hydrophobically and negatively oxygenated functional groups in carbon nanomaterials. In addition, the O_2-reduction current produced from the DET-type bioelectrocatalyst of *Tr*Lac at the CNT decreases with increasing N dopant ratio [43]. Strong electrostatic interactions between N-doped CNTs, which have a positive surface potential, and laccase has been proposed to cause the denaturation of redox enzymes and/or decrease the DET reaction rate. Furthermore, the zeta potential (i.e., surface charge density) of graphene oxide has been reported to affect the enzyme adsorption, direct electron transfer, and catalytic current [44]. In a recent study, three types of carbon nanotubes with different lengths were utilized as platforms for the DET-type bioelectrocatalysis of three redox enzymes, viz. BOD from *Myrothecium verrucaria* (*Mv*BOD), CueO from *Escherichia coli*, and H_2ase from *Desulfovibrio vulgaris* Miyazaki F (*Dv*H_2ase) [45]. As a result, diffusion-controlled O_2

reduction reaction with catalytic current density of ~8 mA cm^{-2} in an O$_2$-saturated neutral buffer was realized by BOD in CNTs of a length of 1 µm, but the catalytic current densities decreased as the length of CNTs increased [45] (Figure 3). However, in the cases of CueO and H$_2$ase, the catalytic current (O$_2$ reduction for CueO and H$_2$ oxidation for DvH$_2$ase) increases as the length of CNTs increased. This tendency can be weakened by the addition of Ca(NO$_3$)$_2$ to the electrolyte. The CNT surface is negatively charged in neutral solutions owing to the dissociation of the carboxyl group on the CNT surface, and the number of carboxy groups increases with a decreasing CNT length. It was concluded through this research that the electrostatic interaction between the region close to the active site of the enzymes and CNTs is one of the most important factors controlling the enzyme adsorption for DET-type bioelectrocatalysis.

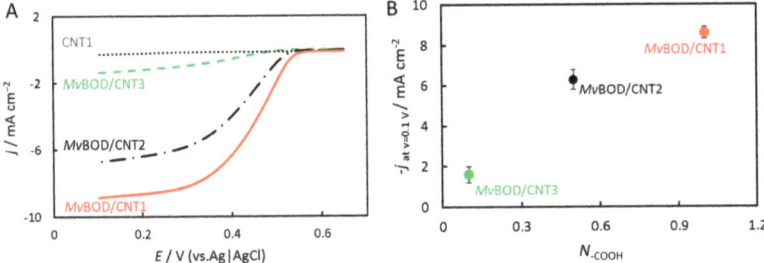

Figure 3. (**A**) Rotating disk linear scan voltammograms (LSVs) of O$_2$ reduction for MvBOD/CNT1 (**solid line**), MvBOD/CNT2 (**dashed–dotted line**), and MvBOD/CNT3 (**broken line**). The dotted line represents an LSV on a CNT1 without enzyme. All measurements were carried out in O$_2$-saturated phosphate buffer (0.1 M, pH 7.0, and 25 °C) at ω = 4000 rpm and υ = 5 mV s^{-1}. (**B**) Relationships between the steady state current density and the relative amount of carboxy groups at different CNT surfaces (N_{-COOH}). Reprint from [45]. Copyright (2017), with permission from Elsevier.

3. Oriented Immobilization of Redox Enzymes in Modified Carbon Nanomaterials

Because the surface physiochemistry of carbon nanomaterials plays a key role in the orientation of redox enzymes, strategies for the rational surface modification of carbon nanomaterials to provide suitable platforms for DET-type bioelectrocatalysis of redox enzymes are highly expected. In general, functional groups at carbon nanomaterial surfaces can be introduced by different approaches, including diazonium electroreduction [16], amine electrooxidation [47], π-stacking [48], and *in-situ* chemical modification [49] (Figure 4). We do not discuss these approaches in detail herein, focusing instead on strategies to control the orientation of redox enzymes in modified carbon nanomaterials for enhanced DET-type bioelectrocatalysis.

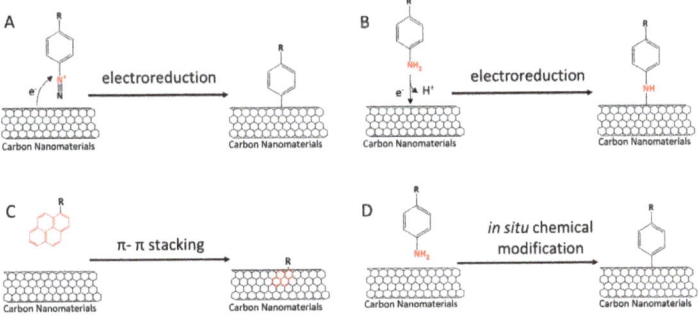

Figure 4. Typical strategies for surface modification of carbon nanomaterials. CNTs were utilized here as representatives of carbon nanomaterials.

3.1. Modification of Carbon Nanomaterials with Charged Compounds

Electrostatic effects play major roles in enzyme-substrate interactions and in DET-type bioelectrocatalysis. For example, it has been found that the electrode potential may affect the adsorption and orientation, and thus, the performance of DET-type bioelectrocatalysis of redox enzymes [50]. In addition, improved O_2-reduction reactions have been achieved through the adsorption of MvBOD on negatively charged surfaces, which were produced by the electrooxidation of 4-aminobenzoic acid at the KB [12,47,51], reduced graphene oxide [52], or MgO-templated mesoporous carbon [34] modified electrodes. Specially, the O_2 reduction-current densities produced by MvBOD at 4-aminobenzoic acid functionalized KB modified electrodes is more than 2 times higher than that at pristine KB modified electrodes [12,47] (Figure 5A). The net charge of MvBOD is negative at pH 7, whereas the surface close to the type I copper, the site that is believed to accept electrons from the electrode in the DET reaction case, is positive at pH 7. The electrostatic interaction between the 4-aminobenzoic acid-functionalized carbon surface and the positively charged T1 site is proposed to provide an effective orientation for the DET-type bioelectrocatalysis of MvBOD [47]. Such enhancement can also be found through the adsorption of MvBOD at naphthoate-functionalized CNT-modified electrodes [53], which were prepared using the electrochemical reduction of 6-carboxynaphthalenediazonium or by the in-situ chemical modification of 6-amino-2-naphthoic acid onto CNTs. Similar phenomena have also been reported in *Magnaporthe orizae* BOD (MoBOD), which illustrated improved oxygen reduction reactions at 6-carboxynaphthalenediazonium functionalized CNTs [54]. In contrast to MvBOD and MoBOD, BOD from *Bacillus pumilus* (BpBOD) presents a negatively charged T1 site. As expected, the O_2-reduction current produced by BpBOD at a positively charged 1-pyrenemethylamine hydrochloride-functionalized CNT electrode is ~0.4 mA cm^{-2}, which is much higher than that at a pristine CNT electrode (~0.05 mA cm^{-2}) [55]. Similarly, CueO [56] and DvH$_2$ase [51], which have negatively charged active sites for interfacial electron transfer, show enhanced DET-type bioelectrocatalysis at a p-phenylenediamine-functionalized KB electrode. Consequently, improved DET-type bioelectrocatalysis based on an electrostatic interaction has been applied to the construction of a H_2/Air(O_2) enzymatic biological fuel cell, which consists of a MvBOD-adsorbed 4-aminobenzoic acid-modified KB biocathode and DvH$_2$ase adsorbed p-phenylenediamine-functionalized KB bioanode, with a high-power density of 6.1 mW cm^{-2} [51], which is the highest output of DET-type biological fuel cells reported thus far.

3.2. Modification of Carbon Nanomaterials with Polycyclic Compounds

Contrary to the charged surface of redox enzymes, the active site of redox enzymes is usually surrounded by polypeptides to form a substrate-binding "pocket" that is often non-charged and hydrophobic. Based on this knowledge, by modifying an electrode surface with polycyclic aromatic compounds, the hydrophobic interactions (π-π stacking) between the active center "pocket" of the enzyme and the modifier may induce a favorable orientation. Anthracene and its derivatives are the most common groups that are modified into carbon nanomaterials for the oriented immobilization of redox enzymes. The concept was first proposed by Armstrong and his coworkers, who reported an improved O_2-reduction reaction by the adsorption of laccase at the anthracene-functionalized pyrolytic graphite "edge" electrode [57]. This approach has been further developed for the orientated-immobilization of laccase at CNTs functionalized with anthracene-2-carbonyl chloride [58] or 1-(2-anthraquinonylaminomethyl) pyrene [59]. For example, the optimized O_2-reduction current density produced by TvLac at an anthracene-2-carbonyl chloride functionalized CNT electrode is 140 µA cm^{-2}, while no obvious catalytic current can be observed by TvLac at a pristine CNT electrode [58]. In these studies, polycyclic anthracene groups tailored to carbon nanomaterials were expected to insert the hydrophobic substrate-binding "pocket" to control the orientation of redox enzymes for an efficient interfacial electron transfer. Interestingly, CNTs modified by 1-[bis(2-anthraquinonyl)aminomethyl] pyrene (PyrAQ$_2$) with two anthraquinone groups showed greater efficiency for improvement of DET-type bioelectrocatalysis of laccase than that modified by 1-(2-anthraquinonylaminomethyl) pyrene (PyrAQ) with only one anthraquinone group [60].

The O_2-reduction current density produced by *Tv*Lac at PyrAQ$_2$-functionalized CNT electrodes reached to 0.9 mA cm^{-2}, which is much higher than that at PyrAQ-functionalized CNT electrodes (0.35 mA cm^{-2}) [60]. The increased anthraquinone groups seem to increase the probability of laccases with a controlled orientation. In addition to anthracene, surfaces functionalized by pyrene [61] and adamantane [62] have been found to be efficient in controlling the orientation of Lac for the improvement of DET-type bioelectrocatalysis. (Figure 5B) Furthermore, such improved DET-type bioelectrocatalysis using a similar approach has also been reported when using FDH in anthracene-functionalized CNTs [63] and *Desulfomicrobium baculatum* H$_2$ase (*Db*H$_2$ase) in anthraquinone- or adamantane-functionalized CNTs [64].

3.3. Modification of Carbon Nanomaterials with Substrate Mimics

Compared to nonspecific electrostatic and hydrophobic interactions, interactions between redox enzymes and their natural substrates are usually highly specific. A more effective approach is the direct functionalization of carbon nanomaterials with a natural substrate (or its analogues) of redox enzymes to control the orientation of the redox enzymes. A significant improvement in the O_2-reduction current has been realized through the adsorption of *Mv*BOD in bilirubin, the natural substrate of *Mv*BOD, which is a functionalized KB electrode [65]. (Figure 5C) A computational study that combines the density functional theory with docking simulations showed that the modification of the electrode surface with bilirubin provides an optimal orientation of BOD toward the support and that bilirubin facilitates the interfacial electron transfer by decreasing the distance between the electrode surface and the T1 Cu atom [66]. Similar improvements in DET-type bioelectrocatalysis can be realized through the adsorption of *Mv*BOD at 2,5-dimethyl-1-phenyl-1H-pyrrole-3-carbaldehyde [67], syringaldazine [68], or protoporphyrin [69], analogues of bilirubin, and functionalized CNT electrodes. Syringaldazine-functionalized CNTs were also shown to be positively effective in improving the interfacial electron transfer of laccase [68]. Moreover, an improved DET-type bioelectrocatalysis of FDH at a methoxy-functionalized electrode surface [70] based on the specific interactions between FDH and methoxy substituents [8] has been reported. The fructose oxidation current produced by FDH at 2,4-dimethoxyaniline-functionalized KB electrodes reached to 23 ± 2 mA cm^{-2}, which is the largest measured to date. A recent study proposed that 4-mercaptopyridine functionalized gold nanoparticle-embedded KB can act as a favorable platform for DET-type bioelectrocatalysis of FoDH [71], probably owing to the attractive interactions between the pyridine moiety and the FeS site of FoDH. These results suggest that site-specific interactions are also effective in inducing a suitable orientation of redox enzymes for DET-type bioelectrocatalysis.

3.4. Oriented Immobilization of Engineered Enzymes in Functionalized Carbon Nanomaterials

The conventional method for oriented immobilization of redox enzymes is dependent on the natural properties of redox enzymes. Engineered redox enzymes with specific residues for oriented immobilization in modified carbon nanomaterials have recently been reported. Bartlett et al. constructed a site-specific variant of *Mo*BOD S362C, in which Ser362 at the *Mo*BOD surface close to the Tl Cu was replaced by a Cys residue [72]. The distance between the T1 site and Ser362 is ~1.33 nm. A thiol-maleimide click reaction between the S362C*Mo*BOD variant and maleimide-functionalized MWCNT was employed to construct a stable bioelectrode. A clear DET-type bioelectrocatalytic O_2 reduction wave with a higher current and more sigmoidal shape was obtained in a S362C*Mo*BOD-modified electrode than that in a native *Mo*BOD-modified electrode. Such improvements in DET-type bioelectrocatalytic performance can be explained by the oriented immobilization of the enzyme with a short distance between the active site and the electrodes. A site-specific surface modification of a fungal Lac with a single covalently bound pyrene group close to the T1 site showed improved DET-type bioelectrocatalysis in the CNTs [73]. (Figure 5D) The site-specific modified pyrene was attached to the CNT through a π-π stacking interaction to control the orientation of Lac with the T1 site face to the CNT. As a result, a maximum O_2-reduction current density produced by site-specific pyrene-modified Lac at a CNT electrode is 1.15 mA cm^{-2}, which is four times higher than that produced by native Lac at a CNT electrode (0.27 mA cm^{-2}). A further improved maximum O_2-reduction current density of 2.75 mA cm^{-2} was realized by immobilizing site-specific pyrene-modified Lac at β-cyclodextrin-modified gold nanoparticles(β-CD-AuNPs)-functionalized CNT [73]. The β-CD-AuNPs offer host−guest interactions between pyrene groups and β-CD moieties on the AuNPs surface to increase the number of effectively wired enzymes. A recent study on the oriented immobilization of site-specific alkylated Lac in azido-modified CNTs through copper-catalyzed click reactions was reported by the same group [74]. Interestingly, it has been found that the variant with an alkyne group in the position close to the T2/T3 site showed a higher O_2-reaction current (~3 mA cm^{-2}) than that with an alkyne group close to the T1 site (~1.8 mA cm^{-2}). Although several studies have reported a direct electron transfer from the electrode to the T2/T3 cluster of Lac in gold electrodes [75], more experimental evidence is needed to support this finding.

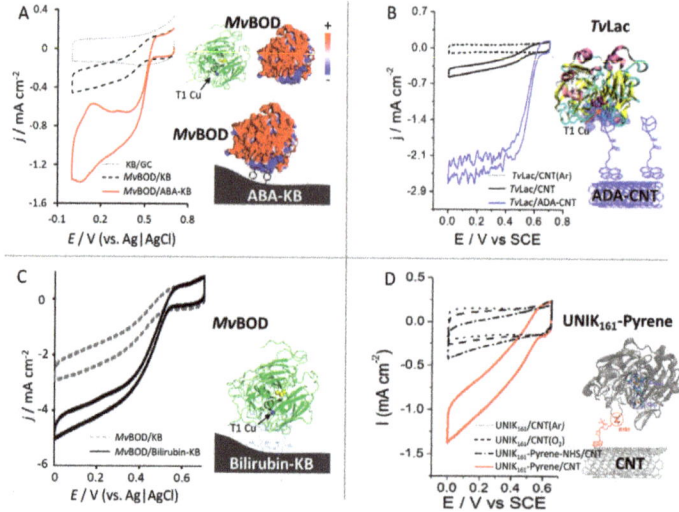

Figure 5. Examples of oriented immobilization of redox enzymes in modified carbon nanomaterials. (**A**) Cyclic voltammograms (CVs) of *Mv*BOD adsorbed in pristine KB (dashed line) and 4-aminobenzoic acid-functionalized KB (solid line) electrodes in an air-saturated phosphate buffer (0.1 M, pH 7.0) with a rotating rate of 2000 rpm. The dotted line shows the cyclic voltammogram of a bare KB electrode without *Mv*BOD. Reprinted from [47] Copyright (2016) with permission from Elsevier. (**B**) CVs of *Tv*Lac adsorbed in pristine CNTs (black solid line) and 1-pyrenebutyric acid adamantyl amide-functionalized CNTs (ADA-CNT, blue solid line) in a stirred oxygen-saturated McIlvaine buffer (pH 5). The dotted line shows the cyclic voltammogram of the *Tv*Lac adsorbed CNT electrode in an argon-saturated buffer. Reprinted from [62] Copyright (2016) with permission from American Chemical Society. (**C**) CVs of *Mv*BOD adsorbed in pristine KB (dashed line) and bilirubin functionalized KB (solid lines) electrodes in an O_2-saturated phosphate buffer (0.1 M, pH 7) with a rotating rate of 4000 rpm. The inset shows the structure of bilirubin. Reprinted from [65] Copyright (2014) with permission from Royal Society of Chemistry. (**D**) CVs of the bare $UNIK_{161}$ (dotted line), $UNIK_{161}$–NHS–pyrene (dashed-dotted line), and $UNIK_{161}$–pyrene (solid line) adsorbed in CNT electrodes in O_2-saturated 0.1 M phosphate buffer (pH 5.0). The dotted line shows bare UNIK161 adsorbed CNT electrodes in an argon-saturated buffer. $UNIK_{161}$ is a variant of the LAC3 enzyme (GENEBANK AAR00925.1) containing lysine, methionine, and histidine residues replacing arginine R161 (R > K161) and lysines K40 (K > M40) and K71 (K > H71) in the native sequence, respectively; in addition, UNIK161–NHS–pyrene indicates that the enzyme was modified with pyrene–NHS, which was randomly attached to $UNIK_{161}$ through an activated ester coupling. $UNIK_{161}$–pyrene is produced through the highly selective modifications of K161 with pyrene Rreprinted from [73] Copyright (2016) with permission from American Chemical Society.

4. Techniques to Probe the Redox Enzymes in Carbon Nanomaterials

Although electrochemical investigations (e.g., rotating disk electrodes) coupled with computational simulations [66,76] and a theoretical analysis of current–potential curves [77,78] for an investigation into DET-type bioelectrocatalysis have been well studied during the past 10 years, techniques to directly probe the behavior of redox enzymes in carbon nanomaterials are still highly desired [79]. Several techniques, particularly a quartz crystal microbalance (QCM) and infrared (IR) spectroscopy, used to analyze the enzyme adsorption, conformation, and orientation of redox enzymes in carbon nanomaterials have been proposed and developed during the past decades.

4.1. Quartz Crystal Microbalance

A quartz crystal microbalance (QCM) is a powerful technique commonly used to measure the adsorption of redox enzymes onto an electrode surface. In general, a QCM instrument operates by measuring the resonance frequency of a piezoelectric quartz chip. According to the Sauerbrey equation [80], the frequency of a piezoelectric quartz chip is typically proportional to the mass on the electrode (assuming a rigid thin film). By coating the piezoelectric quartz chip with carbon nanomaterials, the adsorption behaviors of redox enzymes at the corresponding surfaces can be obtained. Furthermore, considering the large size and structural flexibility of redox enzymes, QCM with dissipation (QCM-D) monitoring to further understand the rigidity of enzyme binding, desorption, reorientation, and conformational changes has been proposed [81]. For example, much greater decreases in frequency and increases in the dissipation factor have been found for the adsorption of TvLac onto adamantane-functionalized CNT electrodes than onto pristine CNT electrodes [62]. The results indicate higher enzyme loading and stable immobilization of laccase in adamantane-functionalized CNTs than in pristine CNT electrodes. A similar study reported by the same group also showed increased frequency decreases and dissipation factor increases for the adsorption of DbH$_2$ase onto adamantane- and anthraquinone-functionalized CNT electrodes [63]. In combination with electrochemical measurements, these studies have concluded that the functionalization of CNTs with adamantane or anthraquinone groups increase the effective adsorption of TvLac and DbH$_2$ase for DET-type bioelectrocatalysis owing to the hydrophobic interaction between the active site pocket of redox enzymes and a functionalized polycyclic modifier. A recent study revealed that upon adsorption of TrLac onto naphthalene-functionalized CNTs, dissipation increased and resonance frequency decreased rapidly, followed by slow dissipation decrease [82].(Figure 6A) The first rapid dissipation increase and resonance frequency decrease are typical behaviors of protein adsorption, whereas the subsequent slow decrease in dissipation was proposed to be caused by the rearrangement of adsorbed TrLac, which is important for an electric connection of TrLac to the naphthalene-functionalized CNTs by cross-comparing the results of an open circuit potential monitoring and dissipation response.

4.2. Infrared Spectroscopy

Infrared (IR) spectroscopy is a well-established experimental technique for analyzing the secondary structures of proteins [83]. IR vibrational spectra contain a wealth of information about the structure and environment of amino acid side chains, as well as protein conformation and the polypeptide backbone. For in-situ investigation of the redox enzymes at the electrode surfaces, reflection-based methods, particularly attenuated total-reflectance infrared (ATR-IR) spectroscopy, are the most commonly used. The strongest features in the ATR-IR spectra of enzymes are typically amide bands (e.g., amide I at ~1650 cm^{-1} and amide II at ~1550 cm^{-1}) that arise from the amide bonds in the polypeptide backbone and sensitively encode the secondary structure of the enzyme. Mao et al. developed an ATR-IR spectroscopy-based method for investigating the effects of organic solvents on the immobilization of TvLac on CNTs [84]. Six typical secondary structures in TvLac with and without ethanol treatment adsorbed on CNTs were resolved and compared. It has been found that the content of high-frequency (HF) β-sheets, which corresponds to the flexible β-extend, dropped to just above zero after immobilization in the CNTs, while it increased by ~200% in ethanol-treated *Tv*Lac. In addition, a stronger amide II peak was obtained with ethanol-treated *Tv*Lac-CNT electrodes, showing an increased amount of detectable N–H in-plane bending events. The ratio of the amide I/II peak intensity was reduced at the ethanol-treated *Tv*Lac-CNT electrode, compared to that at the non-treated *Tv*Lac-CNT electrodes, indicating a significant orientation change of the adsorbed *Tv*Lac. With improved O_2-reduction currents, ethanol-treated *Tv*Lac facilitating a favorable "end-on" orientation at the CNT surfaces for fast direct electron transfer was proposed. (Figure 6B) Based on a similar ATR-IR technique, the orientations of the small Lac in carbon nanomaterials were investigated [85]. Consequently, the effects of the surface curvature of carbon nanomaterials on the orientation of adsorbed small Lac in the view of DET-type bioelectrocatalysis was proposed.

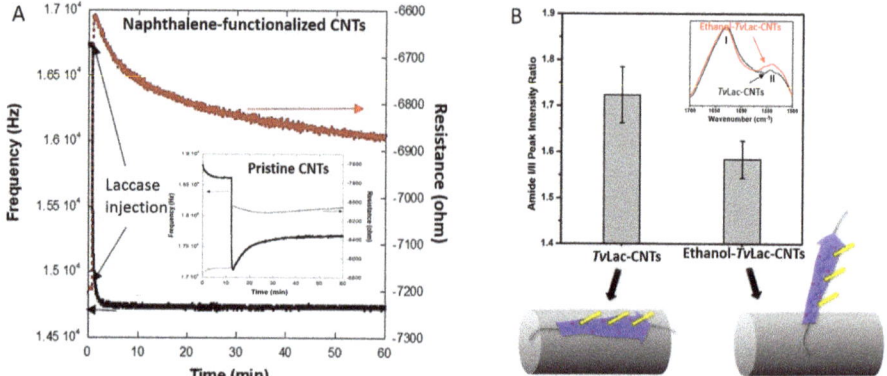

Figure 6. (**A**) QCM-D response of naphthalene-functionalized CNTs modified gold surface before and after *Tr*Lac injection. The inset shows the QCM-D response of pristine CNTs modified gold surface before and after *Tr*Lac injection. Reprinted from [82] Copyright (2020) with permission from Elsevier. (**B**) Two orientations of β-strand at the surface of CNTs predicted from the Amide I/II peak intensity ratios. Inset: Amide I and II band regions on difference IR spectra (corrected by a linear baseline between 1700 and 1500 cm^{-1}) of the untreated *Tv*Lac-CNT film (black line) and Ethanol-treated *Tv*Lac-CNT film (red line). Yellow arrows represent incident light. Amide bonds (ball and sticks) are in the plane of β-strand. Reprinted from [85] Copyright (2017) with permission from American Chemical Society.

5. Conclusions

DET-type bioelectrocatalysis is an ideal system that employs redox enzymes as biocatalysts for energy conversion under extremely mild conditions. The properties of redox enzymes lead to reactions with high specificity and high catalytic efficiency, which are suitable for various electrochemical devices, including biosensors, biofuel cells, and bioreactors. The major challenges for DET bioelectrocatalysis are the interfacial electron transfer between redox enzymes and electrodes. Carbon nanomaterials with high conductivity, large specific surface areas, and three-dimensional nanostructures have been widely utilized for the construction of bioelectrodes with the capability of DET.

In this review, we focused on the rational surface modification of carbon nanomaterials for improving DET-type bioelectrocatalysis based on an understanding of the effects of carbon nanomaterials on the interfacial electron transfer between redox enzymes and electrodes. Although a large specific surface area is important for increasing the enzyme loading, a suitable pore size distribution and specific surface chemistry of carbon nanomaterials have recently been proposed to play important roles in the promotion of interfacial electron transfer between redox enzymes and electrodes. Carbon nanomaterials, with hierarchical porosity that balance the enzyme adsorption, electron transfer, and mass transfer, are expected to be suitable for high-performance DET-type bioelectrocatalysis. By contrast, intricate interactions between redox enzymes and carbon nanomaterials play an important role in the orientation of redox enzymes. From this perspective, rational surface modification based on an understanding of the interaction between enzymes and the specific modifier to control the orientation of the redox enzymes for improved DET-type bioelectrocatalysis has been developed during the past decades. Typically, charged compounds, polycyclic aromatics, and substrate mimics have usually been utilized for the construction of functionalized carbon nanomaterials for controlling the orientation of redox enzymes. Unlike the random surface properties of native enzymes, recently developed engineered enzymes with specific sites that can be immobilized onto modified carbon nanomaterials with controlled orientations have also attracted increased attention. To further understand the situation of redox enzymes in carbon nanomaterials, various techniques other than the use of electrochemical methods alone are highly desired. In particular, QCM-D and ATR-IR spectroscopy have been utilized to investigate the adsorption, conformational change, and orientation distribution of redox enzymes in carbon nanomaterials.

Although significant progress has been made in the past decades, several important issues regarding the DET-type bioelectrocatalysis are still waiting to be tackled. Firstly, detail mechanism of DET reaction is still instinct. Advanced technology, for example coupling-technique as well as single molecule analysis, is desired to deeply understand the process of electron transfer between a redox enzyme and electrode surfaces. Further development in theoretical discussion on this issue is also expected. Secondly, redox enzyme that capable of DET-type bioelectrocatalysis is limited in number. Native redox enzymes have usually large and sophisticated three-dimensional molecular structures. Besides to electrode surface modification discussed in this review, protein engineering to reform redox enzymes, for example shortening the distance between active site and enzyme surfaces by elimination of the domains that are not related to the electron transfer, is proposed to improve the performance of DET-type bioelectrocatalysis effectively. Thirdly, stability is continued to be an important issue in DET-type bioelectrocatalysis for practical applications. Strategies including optimization of enzyme immobilization approaches, tuning enzyme properties and using suitable electrode materials have been proposed recently to improve the stability and increase the lifetime of redox enzymes at electrode surfaces for rapid interfacial electron transfer.

Author Contributions: H.X.: Original manuscript preparation and revision; J.Z.: Revision and discussion. All authors have read and agreed to the published version of the manuscript.

Funding: This research is partially supported by Guangdong Basic and Applied Basic Research Foundation (No. 2019A1515111183) and President Foundation of Institute of Fruit Tree Research, Guangdong Academy of Agricultural Science (No. 202005).

Conflicts of Interest: The authors declare no conflict of interest.

References

1. Xiao, X.; Xia, H.Q.; Wu, R.; Bai, L.; Yan, L.; Magner, E.; Cosnier, S.; Lojou, E.; Zhu, Z.; Liu, A. Tackling the Challenges of Enzymatic (Bio)Fuel Cells. *Chem. Rev.* **2019**, *119*, 9509. [CrossRef]
2. Adachi, T.; Kitazumi, Y.; Shirai, O.; Kano, K. Development Perspective of Bioelectrocatalysis-Based Biosensors. *Sensors* **2020**, *20*, 4826. [CrossRef]
3. Wu, R.; Ma, C.; Zhu, Z. Enzymatic Electrosynthesis as an Emerging Electrochemical Synthesis Platform. *Curr. Opin. Electrochem.* **2020**, *19*, 1–7. [CrossRef]
4. Karimian, N.; Hashemi, P.; Khanmohammadi, A.; Afkhami, A.; Bagheri, H. The Principles and Recent Applications of Bioelectrocatalysis. *Anal. Bioanal. Chem. Res.* **2020**, *7*, 281.
5. Kano, K.; Ikeda, T. Fundamentals and Practices of Mediated Bioelectrocatalysis. *Anal. Sci.* **2000**, *16*, 1013. [CrossRef]
6. Karyakin, A.A. Principles of Direct (Mediator Free) Bioelectrocatalysis. *Bioelectrochemistry* **2012**, *88*, 70–75. [CrossRef]
7. Sensi, M.; del Barrio, M.; Baffert, C.; Fourmond, V.; Léger, C. New Perspectives in Hydrogenase Direct Electrochemistry. *Curr. Opin. Electrochem.* **2017**, *5*, 135–145. [CrossRef]

8. Kawai, S.; Yakushi, T.; Matsushita, K.; Kitazumi, Y.; Shirai, O.; Kano, K. The Electron Transfer Pathway in Direct Electrochemical Communication of Fructose Dehydrogenase with Electrodes. *Electrochem. Commun.* **2014**, *38*, 28–31. [CrossRef]
9. Kamitaka, Y.; Tsujimura, S.; Ikeda, T.; Kano, K. Electrochemical Quartz Crystal Microbalance Study of Direct Bioelectrocatalytic Reduction of Bilirubin Oxidase. *Electrochemistry* **2006**, *74*, 642–644. [CrossRef]
10. Bagdžiūnas, G.; Ramanavičius, A. Towards Direct Enzyme Wiring: A Theoretical Investigation of Charge Carrier Transfer Mechanisms between Glucose Oxidase and Organic Semiconductors. *Phys. Chem. Chem. Phys.* **2019**, *21*, 2968. [CrossRef]
11. Giese, B.; Graber, M.; Cordes, M. Electron Transfer in Peptides and Proteins. *Curr. Opin. Chem. Biol.* **2008**, *12*, 755. [CrossRef]
12. Sakai, K.; Xia, H.-Q.; Kitazumi, Y.; Shirai, O.; Kano, K. Assembly of Direct-Electron-Transfer-Type Bioelectrodes with High Performance. *Electrochim. Acta* **2018**, *271*, 305–311. [CrossRef]
13. Adachi, T.; Kitazumi, Y.; Shirai, O.; Kano, K. Direct Electron Transfer-Type Bioelectrocatalysis of Redox Enzymes at Nanostructured Electrodes. *Catalysts* **2020**, *10*, 236. [CrossRef]
14. Marcus, R.A. Chemical and Electrochemical Electron-Transfer Theory. *Annu. Rev. Phys. Chem.* **1964**, *15*, 155. [CrossRef]
15. Marcus, R.A. Electron Transfer Reactions in Chemistry: Theory and Experiment (Nobel Lecture). *Angew. Chem. Int. Ed.* **1993**, *32*, 1111. [CrossRef]
16. Sugimoto, Y.; So, K.; Xia, H.Q.; Kano, K. *Encyclopedia of Interfacial Chemistry: Surface Science and Electrochemistry*; Wandelt, K., Ed.; Elsevier: Amsterdam, The Netherlands, 2018; Volume 7.1.
17. Hitaishi, V.; Clement, R.; Bourassin, N.; Baaden, M.; de Poulpiquet, A.; Sacquin-Mora, S.; Ciaccafava, A.; Lojou, E. Controlling Redox Enzyme Orientation at Planar Electrodes. *Catalysts* **2018**, *8*, 192. [CrossRef]
18. Li, N.; Wang, J.; Li, M. Electrochemistry at Carbon Nanotube Electrodes. *Rev. Anal. Chem.* **2003**, *22*, 19. [CrossRef]
19. Gong, K.; Yan, Y.; Zhang, M.; Su, L.; Xiong, S.; Mao, L. Electrochemistry and Electroanalytical Application of Carbon Nanotubes: A Review. *Anal. Sci.* **2005**, *21*, 1383. [CrossRef]
20. Pumera, M. Graphene-based Nanomaterials and Their Electrochemistry. *Chem. Soc. Rev.* **2010**, *39*, 4146. [CrossRef]
21. Lawrence, K.; Baker, C.L.; James, T.D.; Bull, S.D.; Lawrence, R.; Mitchels, J.M.; Opallo, M.; Arotiba, O.A.; Ozoemena, K.I.; Marken, F. Functionalized Carbon Nanoparticles, Blacks and Soots as Electron-Transfer Building Blocks and Conduits. *Chem. Asian J.* **2014**, *9*, 1226. [CrossRef]
22. Flexer, V.; Durand, F.; Tsujimura, S.; Mano, N. Efficient Direct Electron Transfer of PQQ-Glucose Dehydrogenase on Carbon Cryogel Electrodes at Neutral pH. *Anal. Chem.* **2011**, *83*, 5721. [CrossRef]
23. Kamitaka, Y.; Tsujimura, S.; Kano, K. High Current Density Bioelectrolysis of D-Fructose at Fructose Dehydrogenase-Adsorbed and Ketjen Black-Modified Electrodes without a Mediator. *Chem. Lett.* **2007**, *36*, 218. [CrossRef]
24. Tasca, F.; Zafar, M.N.; Harreither, W.; Nöll, G.; Ludwig, R.; Gorton, L. A Third Generation Glucose Biosensor Based on Cellobiose Dehydrogenase from *Corynascus thermophilus* and Single-Walled Carbon Nanotubes. *Analyst* **2011**, *136*, 2033. [CrossRef] [PubMed]
25. Sakai, K.; Sugimoto, Y.; Kitazumi, Y.; Shirai, O.; Takagi, K.; Kano, K. Direct Electron Transfer-Type Bioelectrocatalytic Interconversion of Carbon Dioxide/Formate and NAD^+/NADH Redox Couples with Tungsten-Containing Formate Dehydrogenase. *Electrochim. Acta* **2017**, *228*, 537–544. [CrossRef]
26. Tsujimura, S.; Kamitaka, Y.; Kano, K. Diffusion-Controlled Oxygen Reduction on Multi-Copper Oxidase-Adsorbed Carbon Aerogel Electrodes without Mediator. *Fuel Cells* **2007**, *7*, 463–469. [CrossRef]
27. Tsujimura, S.; Miura, Y.; Kano, K. CueO-Immobilized Porous Carbon Electrode Exhibiting Improved Performance of Electrochemical Reduction of Dioxygen to Water. *Electrochim. Acta* **2008**, *53*, 5716. [CrossRef]
28. Xia, H.Q.; Kitazumi, Y.; Shirai, O.; Kano, K. Direct Electron Transfer-Type Bioelectrocatalysis of Peroxidase at Mesoporous Carbon Electrodes and Its Application for Glucose Determination Based on Bienzyme System. *Anal. Sci.* **2017**, *33*, 839–844. [CrossRef]

29. Yin, Z.; Ji, Z.; Zhang, W.; Taylor, E.W.; Zeng, X.; Wei, J. The Glucose Effect on Direct Electrochemistry and Electron Transfer Reaction of Glucose Oxidase Entrapped in a Carbon Nanotube-Polymer Matrix. *ChemistrySelect* **2020**, *5*, 12224. [CrossRef]
30. Wilson, G.S. Native Glucose Oxidase Does Not Undergo Direct Electron Transfer. *Biosens. Bioelectron.* **2016**, *82*, vii–viii. [CrossRef]
31. Bartlett, P.N.; Al-Lolage, F.A. There Is No Evidence to Support Literature Claims of Direct Electron Transfer (DET) for Native Glucose Oxidase (GOx) at Carbon Nanotubes or Graphene. *J. Electroanal. Chem.* **2018**, *819*, 26–37. [CrossRef]
32. Tsujimura, S.; Nishina, A.; Hamano, Y.; Kano, K.; Shiraishi, S. Electrochemical Reaction of Fructose Dehydrogenase on Carbon Cryogel Electrodes with Controlled Pore Sizes. *Electrochem. Commun.* **2010**, *12*, 446–449. [CrossRef]
33. Funabashi, H.; Murata, K.; Tsujimura, S. Effect of Pore Size of MgO-Templated Carbon on the Direct Electrochemistry of D-Fructose Dehydrogenase. *Electrochemistry* **2015**, *83*, 372–375. [CrossRef]
34. Tsujimura, S.; Oyama, M.; Funabashi, H.; Ishii, S. Effects of Pore Size and Surface Properties of MgO-Templated Carbon on the Performance of Bilirubin Oxidase-Modified Oxygen Reduction Reaction Cathode. *Electrochim. Acta* **2019**, *322*, 134744. [CrossRef]
35. Mazurenko, I.; Clément, R.; Byrne-Kodjabachian, D.; de Poulpiquet, A.; Tsujimura, S.; Lojou, E. Pore Size Effect of MgO-Templated Carbon on Enzymatic H_2 Oxidation by the Hyperthermophilic Hydrogenase from *Aquifex aeolicus*. *J. Electroanal. Chem.* **2018**, *812*, 221–226. [CrossRef]
36. Sugimoto, Y.; Kitazumi, Y.; Shirai, O.; Kano, K. Effects of Mesoporous Structures on Direct Electron Transfer-Type Bioelectrocatalysis: Facts and Simulation on a Three-Dimensional Model of Random Orientation of Enzymes. *Electrochemistry* **2017**, *85*, 82–87. [CrossRef]
37. Sugimoto, Y.; Takeuchi, R.; Kitazumi, Y.; Shirai, O.; Kano, K. Significance of Mesoporous Electrodes for Noncatalytic Faradaic Process of Randomly Oriented Redox Proteins. *J. Phys. Chem. C* **2016**, *120*, 26270. [CrossRef]
38. Wanibuchi, M.; Takahashi, Y.; Kitazumi, Y.; Shirai, O.; Kano, K. Significance of Nano-Structures of Carbon Materials for Direct-Electron-Transfer-Type Bioelectrocatalysis of Bilirubin Oxidase. *Electrochemistry* **2020**, *88*, 374. [CrossRef]
39. Funabashi, H.; Takeuchi, S.; Tsujimura, S. Hierarchical Meso/Macro-Porous Carbon Fabricated from Dual MgO Templates for Direct Electron Transfer Enzymatic Electrodes. *Sci. Rep.* **2017**, *7*, 45147. [CrossRef]
40. Hitaishi, V.P.; Clément, R.; Quattrocchi, L.; Parent, P.; Duché, D.; Zuily, L.; Ilbert, M.; Lojou, E.; Mazurenko, I. IInterplay between Orientation at Electrodes and Copper Activation of *Thermus thermophilus* Laccase for O_2 Reduction. *J. Am. Chem. Soc.* **2020**, *142*, 1394. [CrossRef]
41. Tominaga, M.; Otani, M.; Kishikawa, M.; Taniguchi, I. UV—Ozone Treatments Improved Carbon Black Surface for Direct Electron Transfer Reactions with Bilirubin Oxidase under Aerobic Conditions. *Chem. Lett.* **2006**, *35*, 1174. [CrossRef]
42. Xue, Q.; Kato, D.; Kamata, T.; Guo, Q.; You, T.; Niwa, O. Improved Direct Electrochemistry for Proteins Adsorbed on a UV/Ozone-Treated Carbon Nanofiber Electrode. *Anal. Sci.* **2013**, *29*, 611–618. [CrossRef] [PubMed]
43. Tominaga, M.; Togami, M.; Tsushida, M.; Kawai, D. Effect of N-Doping of Single-Walled Carbon Nanotubes on Bioelectrocatalysis of Laccase. *Anal. Chem.* **2014**, *86*, 5053. [CrossRef]
44. Filip, J.; Andicsová-Eckstein, A.; Vikartovská, A.; Tkac, J. Immobilization of Bilirubin Oxidase on Graphene Oxide Flakes with Different Negative Charge Density for Oxygen Reduction. The Effect of GO Charge Density on Enzyme Coverage, Electron Transfer Rate and Current Density. *Biosens. Bioelectron.* **2017**, *89*, 384–389. [CrossRef] [PubMed]
45. Xia, H.Q.; Kitazumi, Y.; Shirai, O.; Ozawa, H.; Onizuka, M.; Komukai, T.; Kano, K. Factors Affecting the Interaction between Carbon Nanotubes and Redox Enzymes in Direct Electron Transfer-Type Bioelectrocatalysis. *Bioelectrochemistry* **2017**, *118*, 70–74. [CrossRef] [PubMed]
46. Mohamed, A.A.; Salmi, Z.; Dahoumane, S.A.; Mekki, A.; Carbonnier, B.; Chehimi, M.M. Functionalization of Nanomaterials with Aryldiazonium Salts. *Adv. Colloid Interface Sci.* **2015**, *225*, 16–36. [CrossRef] [PubMed]

47. Xia, H.-Q.; Kitazumi, Y.; Shirai, O.; Kano, K. Enhanced Direct Electron Transfer-Type Bioelectrocatalysis of Bilirubin Oxidase on Negatively Charged Aromatic Compound-Modified Carbon Electrode. *J. Electroanal. Chem.* **2016**, *763*, 104–109. [CrossRef]
48. Tournus, F.; Latil, S.; Heggie, M.I.; Charlier, J.-C. π-Stacking Interaction between Carbon Nanotubes and Organic Molecules. *Phys. Rev. B* **2005**, *72*, 075431. [CrossRef]
49. Lalaoui, N.; Le Goff, A.; Holzinger, M.; Mermoux, M.; Cosnier, S. Wiring Laccase on Covalently Modified Graphene: Carbon Nanotube Assemblies for the Direct Bio-Electrocatalytic Reduction of Oxygen. *Chemistry* **2015**, *21*, 3198. [CrossRef]
50. Hoshikawa, Y.; Castro-Muñiz, A.; Tawata, H.; Nozaki, K.; Yamane, S.; Itoh, T.; Kyotani, T. Orientation Control of Trametes Laccases on a Carbon Electrode Surface to Understand the Orientation Effect on the Electrocatalytic Activity. *Bioconjug. Chem.* **2018**, *29*, 2927. [CrossRef]
51. Xia, H.-Q.; So, K.; Kitazumi, Y.; Shirai, O.; Nishikawa, K.; Higuchi, Y.; Kano, K. Dual Gas-Diffusion Membrane- and Mediatorless Dihydrogen/Air-Breathing Biofuel Cell Operating at Room Temperature. *J. Power Sources* **2016**, *335*, 105–112. [CrossRef]
52. Tang, J.; Yan, X.; Huang, W.; Engelbrekt, C.; Duus, J.Ø.; Ulstrup, J.; XIao, X.; Zhang, J. Bilirubin Oxidase Oriented on Novel Type Three-Dimensional Biocathodes with Reduced Graphene Aggregation for Biocathode. *Biosens. Bioelectron.* **2020**, *167*, 112500. [CrossRef] [PubMed]
53. Lalaoui, N.; Holzinger, M.; Le Goff, A.; Cosnier, S. Diazonium Functionalisation of Carbon Nanotubes for Specific Orientation of Multicopper Oxidases: Controlling Electron Entry Points and Oxygen Diffusion to the Enzyme. *Chemistry* **2016**, *22*, 10494. [CrossRef] [PubMed]
54. Gentil, S.; Carrière, M.; Cosnier, S.; Gounel, S.; Mano, N.; Le Goff, A. Direct Electrochemistry of Bilirubin Oxidase from Magnaporthe Orizae on Covalently-Functionalized MWCNT for the Design of High-Performance Oxygen-Reducing Biocathodes. *Chemistry* **2018**, *24*, 8404. [CrossRef] [PubMed]
55. Mazurenko, I.; Monsalve, K.; Rouhana, J.; Parent, P.; Laffon, C.; Goff, A.L.; Szunerits, S.; Boukherroub, R.; Giudici-Orticoni, M.T.; Mano, N.; et al. How the Intricate Interactions between Carbon Nanotubes and Two Bilirubin Oxidases Control Direct and Mediated O_2 Reduction. *ACS Appl. Mater. Interfaces* **2016**, *8*, 23074. [CrossRef] [PubMed]
56. Adachi, T.; Kitazumi, Y.; Shirai, O.; Kawano, T.; Kataoka, K.; Kano, K. Effects of Elimination of α Helix Regions on Direct Electron Transfer-Type Bioelectrocatalytic Properties of Copper Efflux Oxidase. *Electrochemistry* **2020**, *88*, 185. [CrossRef]
57. Blanford, C.F.; Heath, R.S.; Armstrong, F.A.; Stable, A. Electrode for High-Potential, Electrocatalytic O_2 Reduction Based on Rational Attachment of a Blue Copper Oxidase to a Graphite Surface. *Chem. Commun.* **2007**, *17*, 1710. [CrossRef]
58. Meredith, M.T.; Minson, M.; Hickey, D.; Artyushkova, K.; Glatzhofer, D.T.; Minteer, S.D. Anthracene-Modified Multi-Walled Carbon Nanotubes as Direct Electron Transfer Scaffolds for Enzymatic Oxygen Reduction. *ACS Catal.* **2011**, *1*, 1683. [CrossRef]
59. Giroud, F.; Minteer, S.D. Anthracene-Modified Pyrenes Immobilized on Carbon Nanotubes for Direct Electroreduction of O_2 by Laccase. *Electrochem. Commun.* **2013**, *34*, 157–160. [CrossRef]
60. Bourourou, M.; Elouarzaki, K.; Lalaoui, N.; Agnès, C.; Le Goff, A.; Holzinger, M.; Maaref, A.; Cosnier, S. Supramolecular Immobilization of Laccase on Carbon Nanotube Electrodes Functionalized with (Methylpyrenylaminomethyl) Anthraquinone for Direct Electron Reduction of Oxygen. *Chemistry* **2013**, *19*, 9371. [CrossRef]
61. Lalaoui, N.; Elouarzaki, K.; Le Goff, A.L.; Holzinger, M.; Cosnier, S. Efficient Direct Oxygen Reduction by Laccases Attached and Oriented on Pyrene-Functionalized Polypyrrole/Carbon Nanotube Electrodes. *Chem. Commun.* **2013**, *49*, 9281. [CrossRef]
62. Lalaoui, N.; David, R.; Jamet, H.; Holzinger, M.; Le Goff, A.; Cosnier, S. Hosting Adamantane in the Substrate Pocket of Laccase: Direct Bioelectrocatalytic Reduction of O_2 on Functionalized Carbon Nanotubes. *ACS Catal.* **2016**, *6*, 4259. [CrossRef]

63. Bollella, P.; Hibino, Y.; Kano, K.; Gorton, L.; Antiochia, R. Enhanced Direct Electron Transfer of Fructose Dehydrogenase Rationally Immobilized on a 2-Aminoanthracene Diazonium Cation Grafted Single-Walled Carbon Nanotube Based Electrode. *ACS Catal.* **2018**, *8*, 10279. [CrossRef]
64. Gentil, S.; Che Mansor, S.M.C.; Jamet, H.; Cosnier, S.; Cavazza, C.; Le Goff, A. Oriented Immobilization of [NiFeSe] Hydrogenases on Covalently and Noncovalently Functionalized Carbon Nanotubes for H_2/Air Enzymatic Fuel Cells. *ACS Catal.* **2018**, *8*, 3957. [CrossRef]
65. So, K.; Kawai, S.; Hamano, Y.; Kitazumi, Y.; Shirai, O.; Hibi, M.; Ogawa, J.; Kano, K. Improvement of a Direct Electron Transfer-Type Fructose/Dioxygen Biofuel Cell with a Substrate-Modified Biocathode. *Phys. Chem. Chem. Phys. PCCP* **2014**, *16*, 4823. [CrossRef] [PubMed]
66. Matanovic, I.; Babanova, S.; Chavez, M.S.; Atanassov, P. Protein-Support Interactions for Rationally Designed Bilirubin Oxidase Based Cathode: A Computational Study. *J. Phys. Chem. B* **2016**, *120*, 3634. [CrossRef]
67. Lopez, R.J.; Babanova, S.; Ulyanova, Y.; Singhal, S.; Atanassov, P. Improved Interfacial Electron Transfer in Modified Bilirubin Oxidase Biocathodes. *ChemElectroChem* **2014**, *1*, 241–248. [CrossRef]
68. Ulyanova, Y.; Babanova, S.; Pinchon, E.; Matanovic, I.; Singhal, S.; Atanassov, P. Effect of Enzymatic Orientation through the Use of Syringaldazine Molecules on Multiple Multi-Copper Oxidase Enzymes. *Phys. Chem. Chem. Phys. PCCP* **2014**, *16*, 13367. [CrossRef]
69. Lalaoui, N.; Le Goff, A.; Holzinger, M.; Cosnier, S. Fully Oriented Bilirubin Oxidase on Porphyrin-Functionalized Carbon Nanotube Electrodes for Electrocatalytic Oxygen Reduction. *Chemistry* **2015**, *21*, 16868. [CrossRef]
70. Xia, H.-Q.; Hibino, Y.; Kitazumi, Y.; Shirai, O.; Kano, K. Interaction between D-Fructose Dehydrogenase and Methoxy-Substituent-Functionalized Carbon Surface to Increase Productive Orientations. *Electrochim. Acta* **2016**, *218*, 41–46. [CrossRef]
71. Sakai, K.; Kitazumi, Y.; Shirai, O.; Takagi, K.; Kano, K. Direct Electron Transfer-Type Four-Way Bioelectrocatalysis of CO_2/Formate and NAD^+/NADH Redox Couples by Tungsten-Containing Formate Dehydrogenase Adsorbed on Gold Nanoparticle-Embedded Mesoporous Carbon Electrodes Modified with 4-Mercaptopyridine. *Electrochem. Commun.* **2017**, *84*, 75–79. [CrossRef]
72. Al-Lolage, F.A.; Bartlett, P.N.; Gounel, S.B.; Staigre, P.; Mano, N. Site-Directed Immobilization of Bilirubin Oxidase for Electrocatalytic Oxygen Reduction. *ACS Catal.* **2019**, *9*, 2068. [CrossRef]
73. Lalaoui, N.; Rousselot-Pailley, P.; Robert, V.; Mekmouche, Y.; Villalonga, R.; Holzinger, M.; Cosnier, S.; Tron, T.; Le Goff, A. Direct Electron Transfer between a Site-Specific Pyrene-Modified Laccase and Carbon Nanotube/Gold Nanoparticle Supramolecular Assemblies for Bioelectrocatalytic Dioxygen Reduction. *ACS Catal.* **2016**, *6*, 1894. [CrossRef]
74. Gentil, S.; Rousselot-Pailley, P.; Sancho, F.; Robert, V.; Mekmouche, Y.; Guallar, V.; Tron, T.; Le Goff, A. Efficiency of Site-Specific Clicked Laccase-Carbon Nanotubes Biocathodes towards O_2 Reduction. *Chemistry* **2020**, *26*, 4798. [CrossRef] [PubMed]
75. Dagys, M.; Laurynėnas, A.; Ratautas, D.; Kulys, J.; Vidžiūnaitė, R.; Talaikis, M.; Niaura, G.; Marcinkevičienė, L.; Meškys, R.; Shleev, S. Oxygen Electroreduction Catalysed by Laccase Wired to Gold Nanoparticles via the Trinuclear Copper Cluster. *Energy Environ. Sci.* **2017**, *10*, 498–502. [CrossRef]
76. Yang, S.; Liu, J.; Quan, X.; Zhou, J. Bilirubin Oxidase Adsorption onto Charged Self-Assembled Monolayers: Insights from Multiscale Simulations. *Langmuir* **2018**, *34*, 9818. [CrossRef]
77. Hexter, S.V.; Grey, F.; Happe, T.; Climent, V.; Armstrong, F.A. Electrocatalytic Mechanism of Reversible Hydrogen Cycling by Enzymes and Distinctions between the Major Classes of Hydrogenases. *Proc. Natl. Acad. Sci. USA* **2012**, *109*, 11516. [CrossRef]
78. Fourmond, V.; Léger, C. Modelling the Voltammetry of Adsorbed Enzymes and Molecular Catalysts. *Curr. Opin. Electrochem.* **2017**, *1*, 110. [CrossRef]
79. Kornienko, N.; Ly, K.H.; Robinson, W.E.; Heidary, N.; Zhang, J.Z.; Reisner, E. Advancing Techniques for Investigating the Enzyme—Electrode Interface. *Acc. Chem. Res.* **2019**, *52*, 1439. [CrossRef]
80. Sauerbrey, G. Verwendung von Schwingquarzen zur Wägung Dünner Schichten und zur Mikrowägung. *Z. Phys.* **1959**, *155*, 206. [CrossRef]
81. Singh, K.; Blanford, C.F. Electrochemical Quartz Crystal Microbalance with Dissipation Monitoring: A Technique to Optimize Enzyme Use in Bioelectrocatalysis. *ChemCatChem* **2014**, *6*, 921–929. [CrossRef]

82. Tahar, A.B.; Żelechowska, K.; Biernat, J.F.; Paluszkiewicz, E.; Cinquin, P.; Martin, D.; Zebda, A. High Catalytic Performance of Laccase Wired to Naphthylated Multiwall Carbon Nanotubes. *Biosens. Bioelectron.* **2020**, *151*, 111961. [CrossRef] [PubMed]
83. Kong, J.; Yu, S. Fourier Transform Infrared Spectroscopic Analysis of Protein Secondary Structures. *Acta Biochim. Biophys. Sin.* **2007**, *39*, 549–559. [CrossRef] [PubMed]
84. Wu, F.; Su, L.; Yu, P.; Mao, L. Role of Organic Solvents in Immobilizing Fungus Laccase on Single-Walled Carbon Nanotubes for Improved Current Response in Direct Bioelectrocatalysis. *J. Am. Chem. Soc.* **2017**, *139*, 1565. [CrossRef] [PubMed]
85. Han, Z.J.; Zhao, L.J.; Yu, P.; Chen, J.W.; Wu, F.; Mao, L.Q. Comparative Investigation of Small Laccase Immobilized on Carbon Nanomaterials for Direct Bioelectrocatalysis of Oxygen Reduction. *Electrochem. Commun.* **2019**, *101*, 82–87. [CrossRef]

Publisher's Note: MDPI stays neutral with regard to jurisdictional claims in published maps and institutional affiliations.

© 2020 by the authors. Licensee MDPI, Basel, Switzerland. This article is an open access article distributed under the terms and conditions of the Creative Commons Attribution (CC BY) license (http://creativecommons.org/licenses/by/4.0/).

Review

Direct Electrochemical Enzyme Electron Transfer on Electrodes Modified by Self-Assembled Molecular Monolayers

Xiaomei Yan, Jing Tang, David Tanner, Jens Ulstrup and Xinxin Xiao *

Department of Chemistry, Technical University of Denmark, 2800 Kongens Lyngby, Denmark; xiyan@kemi.dtu.dk (X.Y.); tangjing_12@163.com (J.T.); dt@kemi.dtu.dk (D.T.); ju@kemi.dtu.dk (J.U.)
* Correspondence: xixiao@kemi.dtu.dk

Received: 8 November 2020; Accepted: 10 December 2020; Published: 14 December 2020

Abstract: Self-assembled molecular monolayers (SAMs) have long been recognized as crucial "bridges" between redox enzymes and solid electrode surfaces, on which the enzymes undergo direct electron transfer (DET)—for example, in enzymatic biofuel cells (EBFCs) and biosensors. SAMs possess a wide range of terminal groups that enable productive enzyme adsorption and fine-tuning in favorable orientations on the electrode. The tunneling distance and SAM chain length, and the contacting terminal SAM groups, are the most significant controlling factors in DET-type bioelectrocatalysis. In particular, SAM-modified nanostructured electrode materials have recently been extensively explored to improve the catalytic activity and stability of redox proteins immobilized on electrochemical surfaces. In this report, we present an overview of recent investigations of electrochemical enzyme DET processes on SAMs with a focus on single-crystal and nanoporous gold electrodes. Specifically, we consider the preparation and characterization methods of SAMs, as well as SAM applications in promoting interfacial electrochemical electron transfer of redox proteins and enzymes. The strategic selection of SAMs to accord with the properties of the core redox protein/enzymes is also highlighted.

Keywords: self-assembled molecular monolayers; electron transfer; direct electron transfer; bioelectrocatalysis; oxidoreductase; gold electrode; metallic nanostructures

---

1. Introduction

Self-assembled molecular monolayers (SAMs) are surface monolayers that spontaneously bind to metal surfaces on which, for example, the unique metal-S bonding between metal and thiols offers a versatile pathway to tailor interfacial properties for electrochemical and bioelectrochemical applications [1–4]. Thiol SAMs on metal surfaces are core targets to provide an understanding of self-organization and interfacial interactions at the molecular level in biological systems [5,6]. Bigelow and associates were the first to demonstrate well-oriented SAMs adsorbed on a platinum wire [7]. However, SAMs did not attract much attention until Nuzzo and associates discovered disulfide monolayers on gold substrates in solution, as a phenomenon different from conventional Langmuir–Blodgett (LB) films [8]. Besides gold and platinum substrates, SAMs can also form on the surfaces of other metals including silver, copper, palladium, and mercury [9–14]. Gold is, however, the most extensively investigated because of its chemical inertness, relatively easy handling, and wide potential window, suitable for a range of electrochemical studies [1,15].

In parallel, the nature of the surface Au-S bond has been under intense focus over a number of years [1,2,4] and was recently overviewed [16,17]. In contrast to molecular Au(I)-S(-I) complexes, the "aurophilic" effect arising from collective interactions among surface Au atoms leads to displacement of the 6s Au electrons out of chemical reach and the filled 5d electrons taking over in Au-S bonding. The Au-S bond on Au surfaces is thus an intriguing example of (very) strong, aurophilically controlled van der Waals binding in an Au(0)-•S(0) gold(0)-thiyl bond.

Electron transfer (ET) reactions between an electrode surface and an oxidoreductase are one of the most important topics in bioelectrocatalysis [18–22]. For example, oxidoreductases possess redox/catalytic center(s) that catalyze the oxidation of fuels on a bioanode (e.g., glucose, fructose, lactate, and sulfite) [20,23–25], which can be assembled with a biocathode undergoing biocatalytic reduction reactions, typically dioxygen reduction, in enzymatic biofuel cells (EBFCs), allowing biopower generation [26–29]. The bioelectrochemical ET processes are classified into direct ET (DET) and mediated ET (MET) [19,20,30]. The MET-type system utilizes external and artificial redox mediators to shuttle the electrons between the electrode and the oxidoreductase, especially if the redox center(s) are buried deep inside the protein structure [31,32]. In DET-type systems, redox enzymes are able to communicate directly with the electrode surface if the redox cofactors/centers are spatially close to the electrode surfaces (generally less than 2 nm), facilitating electron tunneling [33]. DET is thus a simpler mechanism by eliminating the need for external redox mediators, and it is therefore amenable to more detailed mechanistic analysis.

Oxidoreductase immobilization is crucial to improving electrode reusability and stability [19]. To achieve efficient DET, it is important to consider the detailed surface characteristics of both enzyme and electrode for favorable enzyme orientation, leading to minimized electron tunneling distance. A wide range of carbon or metallic supports have been employed for effective enzyme immobilization [25,28,34–37]. Physical adsorption and covalent bonding are most commonly used. SAMs have been introduced into bioelectrocatalysis to serve as a bridge for gentle protein/enzyme immobilization on gold or other metal surfaces [36,38,39]. SAM structures are determined by the Au-S bond, the surface structure of the metal surfaces, lateral interactions, as well as the solvent and electrolytes [6]. Use of SAMs avoids the direct contact of enzyme and solid surfaces [40], mimicking the microenvironment in biological membranes. SAMs exhibit a variety of functional hydrophilic or hydrophobic terminal groups, such as carboxyl, hydroxyl, amino, and alkyl groups [41]. Frequently used alkanethiol and thiophenol molecules are summarized in Figure 1. In addition, the DET kinetics can also be governed by tuning the SAM molecular chain length, as the ET rate is strongly controlled by the tunneling distance [42–44]. As an emerging approach, protein engineering provides oxidoreductases directly with thiol residues, leading to controlled orientations either by direct protein thiol binding or by thioether bond formation with unsaturated maleimide [45].

Current density and enzyme loading can be promoted using nanostructured materials [3,44,46–48]. Among these, metallic nanomaterials exhibit excellent electronic conductivity and large surface area, with promising potential in improving the catalytic response and stability of redox enzymes [25]. Nanoporous gold (NPG), prepared via de-alloying Au alloys or electrodeposition, with three-dimensional porous architecture and a relatively uniform pore size, is a particular candidate for immobilizing enzymes in DET [25,28,32,49,50]. Moreover, gold nanoparticles (AuNPs) featuring a spherical nanostructure and large surface areas have been widely studied in bioelectrocatalysis [51–54]. The combination of SAMs and nanostructured gold offers new opportunities in bioelectrochemistry and has been extensively reviewed [21,32,55], but only a few reviews cover SAMs on planar or porous gold electrodes for controlling enzyme orientation [3,19]. DET based on carbon electrode materials (carbon nanotubes, graphene-based materials) is another major parallel sector, beyond the scope of the present focused review but recently reviewed elsewhere [18,56–58].

In this review, we review recent studies of SAMs in the DET-type bioelectrocatalysis of both atomically planar and nanostructured gold electrode surfaces. Preparations of SAMs coupled with characterization techniques, such as electrochemical, microscopic, and spectroscopic methods, are first overviewed. The use of structurally versatile SAMs and suitable electrode nanostructure supports achieving well-defined orientation for DET is highlighted next. The redox proteins and enzymes to be discussed are organized as (i) heme-containing proteins, i.e., cytochromes (cyts), fructose dehydrogenase

(FDH), cellobiose dehydrogenase (CDH), glucose dehydrogenase (GDH), and sulfite oxidase (SOx); (ii) blue copper proteins, i.e., azurin, copper nitrite reductase (CuNiR), bilirubin oxidase (BOD), and laccase (Lac); (iii) [FeS]-cluster hydrogenases, i.e., [FeFe]-, [NiFe]-, and [NiFeSe]-hydrogenase. This classification is warranted primarily by the different nature of the core ET cofactor, but also with the specific secondary and tertiary structures of the protein that envelopes the metallic or non-metallic catalytic sites. Conclusions and further perspectives are offered and discussed in the final section.

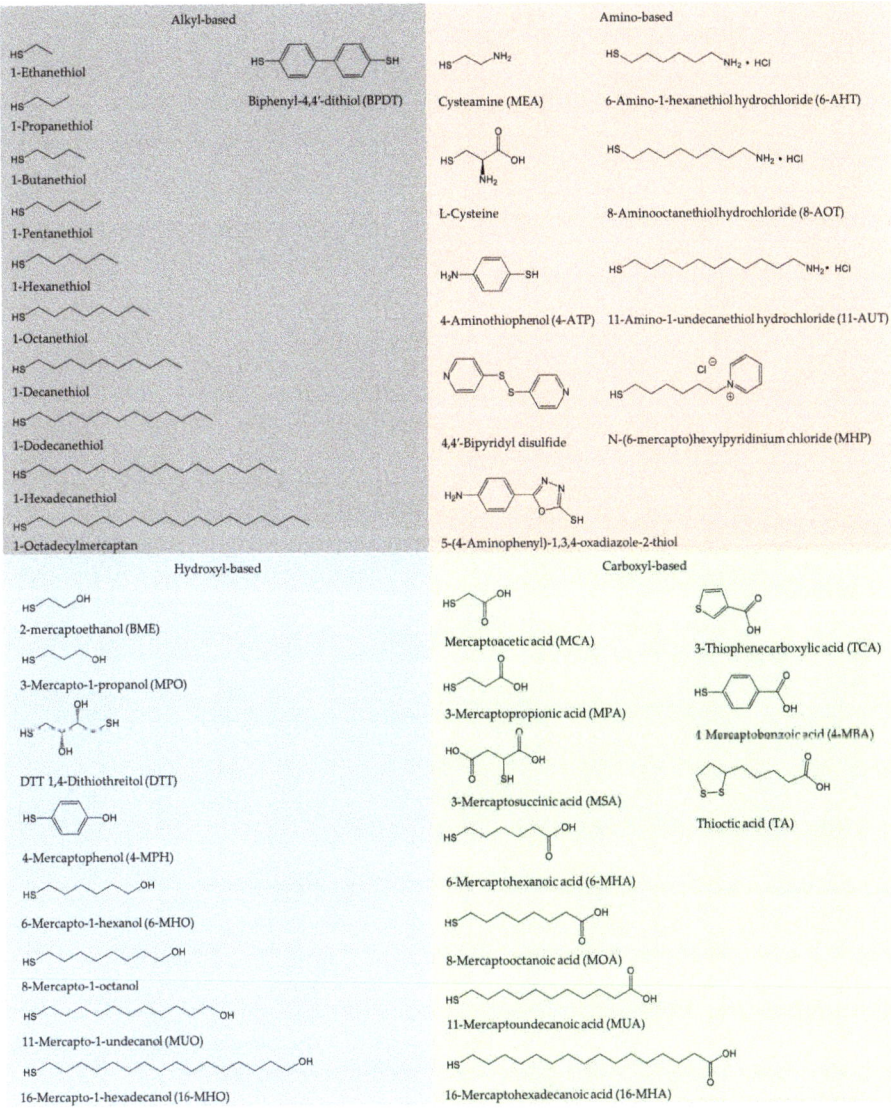

Figure 1. Frequently used alkanethiols and thiophenols in the formation of self-assembled molecular monolayers (SAMs) with alkyl, amino, hydroxyl, and carboxyl terminal groups.

2. Preparation and Characterization of SAM-Based Au- and Other Electrodes

2.1. Preparation of SAMs

A great merit of SAM-forming thiols is that they form spontaneously on electronically soft metal substrates such as gold surfaces from liquid or vapor media under mild conditions [5,59]. Well-defined SAMs are formed by immersing the clean metal surface into a thiol solution, commonly ethanol or aqueous solution, for a certain period of time (hours to days), followed by washing with the same solvent. The quality of the SAMs is determined by the variety and concentration of thiol solution, temperature, soaking time, and the metal surface structure. Vapor deposition is also used to prepare SAMs, but the SAM morphology is here hard to control. Single-crystal, atomically planar gold electrodes are the best substrates to investigate the SAM properties, which can then be characterized using several in situ techniques including electrochemical scanning tunneling microscopy (in situ STM). Nanomaterials with porous structure and high surface area are also used as substrates [40,60]. It is noteworthy that a potential-assisted method to accelerate the SAM formation process has been developed [9,61,62]. The potential range is then a key parameter that strongly affects the quality of the SAMs, which can be evaluated simply by the surface coverage.

2.2. Characterization Methods

2.2.1. Electrochemistry

SAMs can be characterized by a range of methods, among which electrochemical methods have been widely reported [59]. Electrochemical methods involve cyclic (CV) and linear sweep voltammetry (LSV), and electrochemical impedance spectroscopy (EIS). CV of non-redox active SAMs, i.e., capacitive voltammetry, generally shows decreased double-layer capacitance after SAM adsorption on a metal surface. LSV and CV records the irreversible reductive (and oxidative) desorption of SAMs at a negative (or positive) potential. Reductive desorption is well suited to evaluating the surface coverage and the thermodynamic stability of the SAM Au-S units. The shape and position of the desorption peaks are sensitive to the crystalline surface structure and the SAM species. EIS can disentangle the ET resistance from the mass transfer or diffusion resistance. As noted, the ET resistance increases with increasing chain length because of the increased tunneling distance [63]. Electrochemical quartz crystal microbalance (EQCM) is an in situ technique, allowing real-time monitoring of the SAM adsorption process with high sensitivity to the mass changes on the electrode [59].

2.2.2. Microscopy

Electrochemical scanning probe microscopies (SPMs), especially STM and atomic force microscopy (AFM), have been extensively used to provide structural SAM properties at the single-molecule level. The principle of STM is based on the quantum tunneling effect. A bias voltage is applied across the metal support and the extremely sharp STM probe, generating tunneling currents, which are transformed to high-resolution conductivity images when the STM tip is scanned across the SAM-modified metal support. Zhang and associates have reported extensive in situ STM investigations using a wide range of alkanethiols on single-crystal gold [6,64–66]. *In situ* STM combined with electrochemical control can also directly map real-time SAM dynamics and structural features. *In situ* AFM has also attracted attention in bioelectrochemistry [67]. AFM records complex forces, mapping the structural information by allowing the tip to directly contact the redox protein/enzyme molecules. High-resolution STM has been employed to record the SAM thickness and molecular orientations. Contact angle (CA) measurement is a simple and straightforward method for monitoring the hydrophobic/hydrophilic properties of SAMs [1,5].

2.2.3. Spectroscopy

Various spectroscopic techniques are needed to map the complex interactions between SAMs and the metal support [16,17]. X-ray photoelectron spectroscopy (XPS) records the element composition and the chemical state of the SAMs. The high-resolution XPS spectra of S 2p usually show a doublet at the binding energy ranging from 160 to 165 eV, attributed to the formation of the Au-S bond between alkanethiol molecules and the metal support. The functional SAM groups can be identified by Fourier-transform infrared spectroscopy (FTIR), and Raman spectroscopy can be used to detect structural changes in SAMs [68]. Surface plasmon resonance (SPR) spectroscopy is a powerful technique to measure the SAM thickness, showing a change in the tilt angle when the SAM thickness varies. Notably, the adsorption kinetics can be monitored online by SPR coupled with ellipsometry [62].

3. SAMs and Electrochemical DET of Redox Proteins/Enzymes

SAMs have received considerable early and recent attention as substrates for interfacial bioelectrochemical DET reactions, and a wide range of metalloproteins have been investigated in terms of catalytic mechanism and ET kinetics on various SAM-modified electrode surfaces. The redox active centers of protein/enzymes in DET can be roughly categorized into two major groups: metal-based (iron, copper and molybdenum centers etc.) [20,69–71] and non-metal-based (e.g., flavin adenine dinucleotide (FAD) and pyrrolo-quinoline quinone (PQQ)) centers [44,72,73]. Most, although not all, DET-capable enzymes harbor multiple redox centers, with internal ET relays via heme groups, copper clusters, or Fe-S clusters, which shuttle electrons between the electrode surface and the catalytic cofactors [45,74,75]. In this section, we overview recent studies where SAMs have been used in DET-type electrocatalysis and discuss how to obtain favorable orientations using versatile SAMs, as well as how to tune the interactions between the redox protein/enzyme and the SAM-modified electrodes. Redox proteins/enzymes which can be immobilized in well-defined orientations on SAM-modified supports are illustrated in Figure 2 [4,15,19,32,64,67,70,76–81]. We shall overview and discuss some selected examples from each of these enzyme classes in Sections 3.1–3.3.

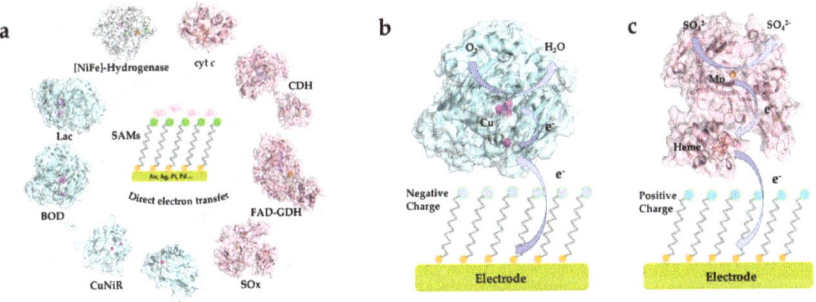

Figure 2. (**a**) Illustration of common proteins/enzymes capable of direct electron transfer (DET) on electrode modified by self-assembled molecular monolayers (SAMs). The proteins include: horse heart cytochrome c (cyt c), PDB 1HRC; cellobiose dehydrogenase (CDH) from *Neurospora crassa*, PDB 4QI7; gamma-alpha subunit of FAD-dependent glucose dehydrogenase (FAD-GDH) from *Burkholderia cepacia*, PDB 6A2U; chicken liver sulfite oxidase (SOx), PDB 1SOX; *Pseudomonas aeruginosa* azurin T30R1, PDB 5I28; monomer of *Achromobacter xylosoxidans* copper nitrite reductase (*Ax*CuNiR), PDB 1HAU; bilirubin oxidase from *Myrothecium verrucaria* (BOD), PDB 2XLL; laccase (Lac) from *Trametes versicolor*, PDB 1KYA; [NiFe]-hydrogenase from *Desulfov Vibrio Vulgaris* Miyazaki F (*Dv*MF), PDB 1UBU. Schematic view of representative DET processes of (**b**) *mv*BOD and (**c**) SOx on the electrode modified with negatively and positively charged SAMs, respectively.

3.1. Heme-Containing Proteins

3.1.1. Cytochrome c

As an electrochemical paradigm redox metalloprotein target, cytochrome c (cyt c) is a soluble heme protein extensively studied also as a model metalloprotein on thiol SAM surfaces [64,82–87]. Comprising 105 amino acid residues, cyt c (horse heart, ca. MW: 12.4 kDa) is an electron transport protein largely present in eukaryotic cells [88]. Cyt c is an ideal model system enabling an understanding of protein ET mechanisms in electrochemistry and homogeneous solution. The interfacial ET rate constant (k_{app}) can be obtained based on the Laviron equation [89]. Electroreflectance spectroscopy (ER) has been utilized to obtain more accurate k_{app} due to the elimination of the capacitive double-layer charging current [83]. Horse heart cyt c is the most studied cytochrome, containing a number of positively charged lysine residues around the heme edge. Cyt c docks electrostatically with natural partners including cyt c oxidases/peroxidases. To immobilize cyt c, SAMs with carboxyl terminal groups are suitable due to favorable electrostatic binding [84]. Collinson and associates demonstrated that horse heart cyt c shows similar orientations on the carboxyl terminated SAM-modified electrode for both covalent bonding and electrostatic adsorption, but covalent bonding led to more stable immobilization [85]. It was also noted that the formal redox potential (E°) of electrostatically adsorbed horse heart cyt c is shifted negatively due to the electrostatic interactions with the negatively charged SAM surface.

SAMs, consisting of a mixture of long-chain pyridine alkanethiols and short-chain alkanethiols, enhance the interfacial k_{app} because of more favorable electronic coupling between cyt c and the electrodes [84,86] than for pure SAMs. The effect of lysine residues on interfacial ET was explored by substituting lysine residues at specific positions. Niki and associates reported that replacement of lysine-13 with alanine in rat cyt c (RC9-K13A) showed a more than five-fold ET rate decrease compared with replacing lysine-72 and lysine-79 [90], which suggests that lysine-13 exhibits optimized coupling with the carboxyl SAM-modified electrode. Direct bonding to the heme group with axial pyridine or imidazole ligands onto the gold surfaces is another effective method for narrow orientation distribution of cyt c [84,87]. The tunneling distance-dependent ET was also investigated by surface-enhanced resonance Raman (SERR) spectroscopy, showing a declining signal with increasing SAM chain length from 2-mercaptoacetic acid to 16-mercaptohexadecanoic acid [91].

AuNPs enhance the interfacial ET rate of cyt c in bioelectrocatalysis. Insertion of 3–4 nm coated AuNPs between cyt c and the a SAM-modified Au(111)-electrode surfaces was shown to increase k_{app} by more than an order of magnitude [89] in spite of an ET distance increase exceeding 50 Å. This raises issues relating to the mechanism of the AuNP promotion even of simple ET processes, discussed in detail recently [92,93]. Engelbrekt and associates reported ultra-stable starch-coated AuNPs, enabling a clear redox signal of yeast cyt c on AuNP-modified basal plane graphite (BPG) electrodes but no signals on bare BPG and Au(111) electrode [94].

Other cytochromes, such as cyt b and cyt c_4, have also been investigated. Della Pia and associates reported that ET between the heme group in cyt b_{562} and the Au(111) electrode can be promoted by replacing the original aspartic acid residue with a cysteine residue, which provided specific protein orientation through a Au-S bond [95]. Chi and associates studied the interfacial and intramolecular ET kinetics of di-heme *Pseudomonas stutzeri* cyt c_4 compared with horse heart cyt c (Figure 3) [64]. *In situ* STM showed directly that the dipolar cyt c_4 is vertically oriented on the carboxyl SAM-modified Au(111) electrode (Figure 3c), resulting in intriguing asymmetric CVs. The authors could show that electrons were first transferred to the heme with the higher potential and then to the second, low-potential heme by fast intramolecular ET. Lisdat and coworkers reported extensive studies on a multilayered protein–enzyme system on SAM-modified gold electrodes [96–99]. For example, they described a sulfite oxidase/cyt c (SOx/cyt c) multilayer system without polyelectrolyte, repeatedly incubating the prepared cyt c-modified Au electrode into a mixture of SOx/cyt c solution and pure cyt c solution [97].

A notable current density was observed even up to eight SOx/cyt c layers, which could be explained by the direct electronic interactions between the two proteins.

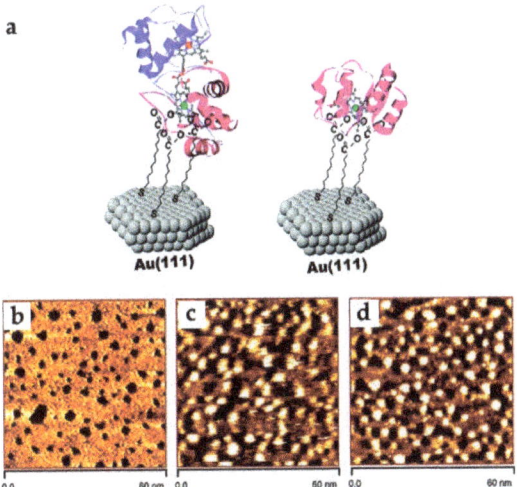

Figure 3. (a) Schematic illustration of *P. stutzeri* cyt c_4 (left) and horse heart cyt c (right) on SAM-modified Au(111); in situ STM images of a ω-mercapto-decanoic acid SAM-modified Au(111)-electrode surface (b) without protein as a reference, (c) with the two-domain *P. stutzeri* cyt c_4 vertically oriented (sharp roughly circular spots), and (d) horse heart cyt c (sharp roughly circular spots) in 5 mM pH 7.0 phosphate buffer under potential control in constant current mode; scan area, 60 × 60 nm² Reproduced with permission from [64]. Copyright 2010, American Chemical Society.

Overall, these reports highlight cyt c as a core electron carrier enabling efficient ET between redox enzymes and the electrode surface across suitably chosen SAMs and along with the blue ET protein azurin as a case for characterization in unique detail, right down to the level of the single molecule. Many oxidoreductases furthermore rely on cytochrome domains or subunits as "built-in" ET relays between the catalytically active cofactor and the electrode surfaces and will be discussed in the following sub-sections [100].

3.1.2. Fructose Dehydrogenase

Although not approaching the degree of detail associated with the simpler ET proteins cyt c and azurin, the molecular mechanistic mapping of several flavin dehydrogenases has reached an impressive level of detail over the last few years, represented by fructose dehydrogenase (FDH), CDH, and GDH in particular. The three enzymes display a common pattern, with a FAD catalytic center where fructose, cellobiose, and glucose, respectively, is oxidized, and an ET relays temporarily populated ET sites through which the liberated electrons are transmitted to the electrode surface. As the first FAD enzyme in our overview, *Gluconobacter* sp. FDH is a membrane-bound FAD-dependent oxidoreductase with a molecular mass around 140 kDa [25,30]. The protein holds three subunits: subunit I (67 kDa) contains a FAD cofactor, serving as the catalytic center for the two-electron oxidation of D-fructose to keto-D fructose. Subunit II (51 kDa) has three heme groups with the formal potentials of 0.15, 0.06, and −0.01 V vs. Ag/AgCl electrode (sat. KCl), respectively. Only two heme groups with the relatively lower redox potentials are proposed to participate in DET [30]. Subunit III (20 kDa) plays an important role in maintaining the structural integrity of the enzyme complex. A number of recent studies illustrate the employment of FDH for the development of biosensors and biofuel cells with high current density and operational stability [30,81,101,102].

The three-dimensional crystal structure, especially the enzyme surface properties, is essential for rational tuning of the redox enzyme immobilization. The detailed crystallography of FDH is still unclear, but homology models have helped to provide a clearer picture of intramolecular electron transfer (IET) [103]. Kano and associates constructed FDH variants with glutamine instead of the axial methionine ligands (M301, M450, or M578) of heme 1c, heme 2c, and heme 3c, respectively, illustrating that the ET pathway leads from M578 to M450 bypassing M301 [104]. Heme 1c with the highest formal potential of 0.15 V vs. Ag/AgCl electrode was evaluated not to be involved in DET, whereas heme 2c was identified as the ET bridge between FDH and the electrode surface. Heme 3c with the lowest formal potential of −0.01 V vs. Ag/AgCl (sat. KCl) was suggested as a bridge between FAD and heme 2c in the IET process (Figure 4) [30]. In addition, the catalytic current density of FDH was dramatically increased by deleting the amino acid residues on the N- or C-terminus of subunit II [30,105,106]. The deletion not only promotes enzyme loading but also provides more opportunities for favorable orientations on the electrode surface. Some researchers reported that hydrophobic anthracene groups anchored on single-walled carbon nanotubes are favorable for enhanced catalytic activity and stability via the hydrophobic C-terminal region of subunit II [101,107]. These studies suggest that orientation and enzyme loading are crucial for controlling the catalytic activity in DET-type bioelectrocatalysis.

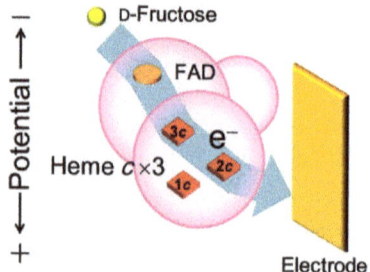

Figure 4. Proposed ET pathway from D-fructose to the electrode surface in the DET of FDH. The ET route involves FAD, heme 3c, heme 2c, but not heme 1c. Reproduced with permission from [30]. Copyright 2019, Elsevier.

Favorable FDH SAM immobilization rests on the following consideration: the isoelectric point (IEP) of FDH is 6.59, which means that FDH is overall positively charged in slightly acidic electrolyte [81]. Recent studies to improve the catalytic performance of FDH on various SAM-modified electrodes have been reported [25,41,44,81]. Bollella and associates reported extensive research on the catalytic activity of FDH on highly porous gold (h-PG) electrodes modified with 4-mercaptobenzoic acid (4-MBA), 4-mercaptophenol (4-MPh), and 4-aminothiophenol (4-APh) SAMs. The data showed that high bioelectrocatalytic activity and stability of FDH was only observed with -OH terminated SAMs [81], suggesting optimized enzyme orientation on this particular SAM. Negatively charged SAMs may help to favor FDH coverage due to preferred electrostatic interaction, but this is not necessarily the most favorable FDH orientation for DET. Murata and associates reached similar conclusions for FDH immobilized on 2-mercaptoethanol (MET) SAM-modified AuNPs [41]. Considering the importance of the gold nanostructure for the ET rate, the AuNP size is furthermore a crucial parameter. Kizling and associates synthesized 1.0 to 3.5 nm AuNP clusters functionalized with 1,6-hexanedithiol and 1-butanethiol for investigating the ET mechanism. A channel of "mediated" catalysis, i.e., electron "hopping", was observed for the smallest AuNP clusters around 1 nm with the half-wave potential close to the first oxidation potential of the AuNP at the edge of the HOMO-LUMO gap [44]. Such a mode accords with theoretical notions recently reported [92]. Siepenkoetter and associates reported the catalytic performance of covalently bonded FDH on NPG electrodes with varying pore sizes [25]. A large number of findings thus show that FDH has high affinity for well-defined SAM-modified electrodes with the polar but electrostatically neutral hydroxyl terminal group as the most efficient, indicating the importance of hydrophilicity of the electrode surface towards the mixed surface charge distribution of the FDH target enzyme. However, details of the underlying mechanism remain unknown.

3.1.3. Cellobiose Dehydrogenase

The second FAD enzyme, cellobiose dehydrogenase (CDH), is a versatile oxidoreductase for direct bioelectrocatalysis. CDH contains a catalytic dehydrogenase domain (DH) harboring a FAD as the redox center and an ET cytochrome b (CYT) domain to shuttle electrons from the FAD to the electrode surface [23,45,108]. The two domains are separated in the crystalline structure, but an integrated IET pathway can be opened through a flexible and hydrophilic amino acid linker between the catalytic and CYT domains [80]. This motif is encountered also for the molybdenum sulfite oxidases, cf. Section 3.1.4. CDH is extracted from the phyla *Basidiomycota* and *Ascomycota*, divided into class I, class II, and class III, respectively. Class III CDH from *Ascomycota* remains uncertain compared to class I and class II CDH [23].

The natural substrates of CDH mainly include cellulose, lactose, and glucose [108]. Class I CDH with short amino acid sequences shows direct, strongly pH-dependent catalytic activity only in solutions with pH below 5.5 [23]. Harreither and associates reported extensive studies on class II CDH from *Chaetomium attrobrunneum* (*Ca*CDH), *Corynascus thermophiles* (*Ct*CDH), *Dichomera saubinetii* (*Ds*CDH), *Hypoxylon haematostroma* (*Hh*CDH), *Neurospora crassa* (*Nc*CDH), and *Stachybotrys bisbyi* (*Sb*CDH) [109]. pH-dependent catalytic activity was observed for neutral and slightly alkaline electrolytes with cyt c and 2,6-dichloroindophenol (DCIP) as electron acceptors. The different electrostatic environment in class I and II CDH reveals different optimal IET processes, with optimum IET in acidic electrolyte for class I CDH and neutral or slightly alkaline electrolyte for class II CDH [110]. Schulz and associates demonstrated turnover and non-turnover DET performance of CDH on polycrystalline gold modified with MUO or 6-MHO [23]. A clear catalytic current between the FAD and the electrode surface was reported. The midpoint potential of bound FAD cofactor was −163 mV vs. SCE at pH 3.0, approximately 130 mV less than that of the heme b relay, leading to a much lower onset potential for lactose oxidation. The authors concluded that the tunneling distance between the FAD cofactor and the electrode was around 12–15 Å, thereby allowing direct electrochemical communication. The authors also demonstrated that only class I CDH from *Trametes villosa* (*Tv*CDH) and *Phanerochaete sordida* (*Ps*CDH) displayed DET activity, whereas no DET signal was observed for class II CDH from reconstructed *Myriococcum thermophilum* (rec*Mt*CDH) and reconstructed *Corynascus thermophilus* (rec*Ct*CDH) at low pH.

The IEP of CDH (DH domain ~5, CYT domain ~3) gives a negatively charged surface around the ET path exit, indicating that positively charged SAMs are suitable for favorable DET [110]. Lamberg and associates anchored *Humicola insolens* CDH on various SAMs with different terminal groups for investigating the effects of charge and hydrophobicity [111] and found that hydrophilic SAMs were favorable for high catalytic activities, with lower enzyme orientation variations than for hydrophobic SAMs. Tavahodi and associates reported a DET-type lactose biosensor of *Ps*CDH on polyethyleneimine (PEI)-coated AuNP electrodes (PEI@AuNP) [112]. PEI with positively charged amino groups not only optimized the orientation of *Ps*CDH on the electrode to increase the IET rate but also gave high enzyme stability and sensitivity. Bollella and associates reported a lactose biosensor based on DET of *Ct*CDH on gold electrodes [113]. The authors demonstrated that BPDT SAMs with two thiol groups can be used to anchor metal NPs on a gold electrode surface by covalent bonding, showing the best ET rate on AuNPs/BPDT/Au electrode. A mediator-free *Hi*CDH/*Mv*BOD BFC using a positively charged MHP SAM for immobilization on AuNP-modified gold electrodes was also reported [114]. The half-life time, i.e., the time duration over which the activity decreased to half the initial value, was 13 and 44 h with 5 mM glucose and 10 mM lactose in neutral buffer solution, respectively. Hirotoshi and associates found an efficient ET process of *Phanerochaete chrysosporium* (*Pc*CDH) on a mixed 11-AUT/MUO SAM-modified AuNP electrode [46]. Al-Lolage and coworkers addressed the catalytic performance and stability of *Mt*CDH by introducing cysteine mutants (E522 and T701) on the protein surface (Figure 5) [45]. The cysteine mutant with a surface thiol group made it possible to form stable thioether bonds with maleimide groups via click-chemistry, thereby controlling the orientations of the *Mt*CDH mutant on the electrode surface and revealing efficient electrochemical glucose oxidation. Another example reported by Meneghello and associates clearly indicated that the DET-type bioelectrocatalysis of CDH is highly sensitive to the cysteine residues introduced at particular positions [115]. Experimental results also showed that divalent cations (i.e., Ca^{2+} and Mg^{2+}) affect the IET kinetics [23,115].

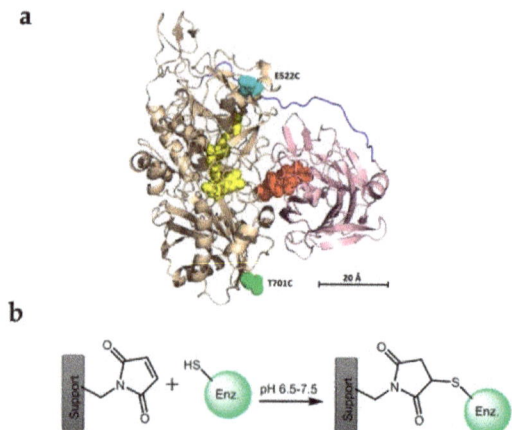

Figure 5. (a) Structure of *Mt*CDH with cysteine E522 and T701 mutations shown in blue and green, respectively; FAD and the heme group are highlighted in yellow and red, respectively. The flexible chain linking the two domains is in blue; (b) the maleimide group anchored on an electrode surface can react with the mutant cysteine residue in the enzyme. Reproduced with permission from [45]. Copyright 2017, John Wiley and Sons.

3.1.3. FAD-Dependent Glucose Dehydrogenase

Our final FAD enzyme target, glucose dehydrogenase (FAD-GDH), is one of the most widely known dehydrogenases with a tightly bound cofactor, making it different from NAD^+-dependent GDH. FAD-GDH has been extensively studied as an emerging alternative to glucose oxidase (GOx) due to its favorable DET capability, insensitivity to dioxygen, and the fact that no hydrogen peroxide is generated [116–119]. The structure of GDH is analogous to that of FDH, comprising three subunits: an FAD-dependent catalytic subunit, an ET subunit with three heme groups, and a small "hitch-hiker" protein used for the flexibility of the catalytic subunit into the periplasm [116,120]. The catalytic subunit harbors a 3Fe-4S cluster close to the ET subunit protein surface, allowing efficient DET on the electrode without the need for mediators. Lee and associates demonstrated controlled DET of bacterial FAD-GDH from *Burkholderia cepacia* on three different SAM-modified electrodes, on which catalytic current density decreases with increasing SAM chain length [118]. A glucose biosensor based on FAD-GDH was reported recently by introducing a gold-binding peptide (GBP) for enzyme immobilization on a screen-printed electrode (SPE) [24]. GBP composed of 12 amino acids exhibiting a strong binding affinity to the gold electrode surface was fused to the enzyme terminus, thereby determining the enzyme orientation on the electrode. Around 10 times higher catalytic response toward 100 mM glucose was observed with FAD-GDH-GBP/Au compared with normal GDH/Au.

A DET-type glucose biosensor could also be fabricated coupled with electrochemical impedance spectroscopy (EIS) [120]. Three variable-length thiols, dithiobis(succinimidyl hexanoate) (DSH), dithiobis(succinimidyl octanoate) (DSO), and dithiobis(succinimidyl undecanoate) (DSU), were employed to modify the electrode surface. Charge transfer resistance (R_{ct}) was a key parameter reflecting the DET efficiency, with the lowest resistance when FAD-GDH was immobilized on DSH SAMs because of the shortest tunneling distance. In addition, the steady-state catalytic current density of FAD-GDH was dramatically increased on AuNPs assembled on the gold electrode [73,121]. Ratautas and associates reported high glucose oxidation activities of FAD-GDH extracted from

Ewingella Americana without mediator [73,121]. They demonstrated that FAD-GDH immobilized on 4-ATP-modified AuNPs displayed higher catalytic activity and lower overpotential than on 4-MBA-modified electrodes. Notably, 4-ATP can be oxidized in neutral media and further converted to 4-mercapto-N-phenylquinone monoimine (MPQM), the quinone groups of which could bind covalently with primary amino groups of the enzyme, thereby enhancing the ET rate. The catalytic response was found to be positively correlated to the ratio of 4-ATP/4-MBA for electrodes modified by mixed thiol SAMs [121]. PQQ-dependent GDH is still another promising GOx candidate, widely utilized for glucose biosensors and biofuel cells [35,95,96]. Kim and coworkers prepared a glucose biosensor by immobilizing positively charged PQQ-GDH on MUA SAM-modified gold electrodes [122]. In this case, PQQ-GDH exhibited higher current density and detection sensitivity via electrostatic adsorption than via covalent bonding, due to less enzyme inactivation. Covalent bonding could thus enhance the interactions between enzyme and electrode surface, but the surface enzyme characteristics are the more critical factors that determine the electrochemical behavior.

3.1.4. Sulfite Oxidase

Sulfite oxidase (SOx) is another heme-containing redox enzyme which has been studied for a long time. Two kinds of SOx are mainly employed in DET-type bioelectrocatalysis: chicken liver, *c*SOx [123], and human, *h*SOx [124]. SOx acts by a principle resembling that of the FAD enzymes, with a catalytic center and an electronic relay, but also with some important differences. As a metalloprotein, the catalytic domain is composed of a pyranopterin molybdenum (Mo) cofactor effecting two-electron oxidation of sulfite to sulfate and connected to a N-terminal cytochrome b_5 (cyt b_5) domain by a flexible linker [123,125]. Although the crystallographic structure of *c*SOx shows that the distance between the catalytic Mo domain and the cyt b_5 domain is more than 32 Å, rapid internal ET in DET-type bioelectrocatalysis is still observed [126], effected by a conformational change on the electrode surface via the flexible tether. This "on-off" switch enables cyt b_5 to be either adjacent or remote from the Mo cofactor [70,127] in a gated ET mode, which is quite different from the rigidly bound ET relays in the FDH and GDH enzymes, but similar to the CDH operational mode.

The electrostatic surface charge distributions around both domains are complex and highlight the importance of subtle tuning of the electrode surfaces [124]. A positively charged SAM surface is favorable for immobilization of *h*SOx via the Mo domain, directing the smaller heme domain towards the electrode surface. As noted, a similar pattern with a flexible linker connecting the catalytic and ET domain applies to CDH [80]; cf. Section 3.1.3. Sezer and associates investigated the catalytic activity of *h*SOx on a mixed SAM-modified silver electrode, showing a significantly increased SERR signal when increasing the ionic strength. High ionic strength is favored to shorten the distance between the Mo cofactor and the heme domain, thereby facilitating both intramolecular and interfacial ET. Wollenberger and associates reported other studies of the electrochemical behavior of *h*SOx [127–130]. In situ scanning tunneling microscopy and spectroscopy to single-molecule resolution has, finally been reported quite recently and disclosed intriguing patterns of tunneling via the Mo- and heme group redox centers [130]. AuNPs with a diameter less than 10 nm were covalently bonded on the MUA/MUO SAM-modified gold electrode, giving a significant enhancement of the interfacial ET rate of *h*SOx [129]. The interfacial ET rate could be further increased by using SAMs with 3,3'-dithiodipropionic acid di(N-hydroxysuccinimide ester) (DTSP) thinner than MUA/MUO SAMs [128]. They also reported that the introduction of $BaSO_4$ nanoparticles played a significant role in the DET-type bioelectrocatalysis of *h*SOx. The electrochemical communication between the active sites of *h*SOx and the electrode surface could be further enhanced by using a positively charged biopolymer. Kalimuthu and coworkers thus reported direct catalytic activity of *h*SOx on a chitosan-covered gold electrode [131]. Both non-catalytic redox signals corresponding to the heme group and catalytic signals in the presence of 4 mM sulfite, on chitosan-covered MPA-, MSA-, and 4-MBA-modified electrodes were observed. However, there were no catalytic signals on MUA-based electrodes despite an observed non-turnover feature, highlighting the importance of rational SAM selection.

3.2. Blue Copper Proteins

3.2.1. Azurin

Pseudomonas aeruginosa azurin is a simple blue copper protein that undergoes single-ET between the Type 1 copper atom and the electrode, now developed as a single-molecule "core" target [6,132–136], as for cyt *c* characterized and mapped in unique detail. A hydrophobic patch and the disulfide group at opposite ends of the azurin molecule are both critical for well-defined orientations on the electrode. Chi, Ulstrup, Zhang and associates conducted extensive investigations of the electrochemical behavior of azurin [39,43,51,137]. *In situ* STM disclosed arrays of well-organized self-assembled azurin monolayers on single-crystal Au(111)-electrodes mapped to single-molecule resolution [51]. Hydrophobic alkanethiol monolayers were employed for gentle immobilization by hydrophobic interactions with the hydrophobic patch of azurin [43]. Notably, exponential decay of the ET rate constant with increasing chain length was observed for chain lengths longer than six carbon atoms, reflecting a dual mechanism, with tunneling dominating for the longer chains [39]. The ET kinetics and redox mechanism of azurin have been analyzed theoretically at different levels, as discussed in detail elsewhere [6,138]. Inserting 3–4 nm coated AuNPs as for horse heart cyt *c* [83] results in a 20-fold enhancement of k_{app} (220 ± 16 s^{-1}) for azurin compared with the AuNP-free system (10.2 ± 0.4 s^{-1}) [51]. Armstrong and associates compared the ET kinetics of azurin on the electrode modified with a synthetic {3,5-diethoxy-4-[(E)-2-(4-ethylphenyl)vinyl]-phenyl} methanethiol or commercial-CH$_3$ terminal alkanethiol [139]. The front hydrophobic ethyl group of SAMs served as the protein-binding bridge on the surface, yielding a very fast ET rate with a k_{app} over 1600 s^{-1}.

3.2.2. Copper Nitrite Reductase

A common feature of the blue multi-copper oxidases is a blue T$_1$ center for the electron inlet and a T$_{II}$ or combined T$_{II}$/T$_{III}$ catalytic center for the electron "outlet" in the catalytic process (nitrite, or, dioxygen reduction). Such an electrochemical mode of action is only feasible via an efficient (short) intramolecular ET channel. This feature has been mapped in considerable detail for three blue copper enzyme classes, the copper nitrite reductases, bilirubin oxidases, and the laccases. The trimeric blue copper nitrite reductase (CuNiR) is crucial in the global nitrogen cycle, catalyzing the single-electron reduction of nitrite to nitrogen monoxide [140]. CuNiR effects direct bioelectrocatalysis, including an ET relay (Cu$_I$) and a catalytic site (Cu$_{II}$) in each monomer. Ulstrup and associates reported DET-based electrocatalysis of CuNiR from *Achromobacter xylosoxidans* (*Ax*CuNiR) on a cysteamine SAM-modified Au(111) electrode [141]. *In situ* STM displayed single *Ax*CuNiR molecules but, intriguingly, only in the presence of nitrite substrate. Further, the combination of varying alkanethiols with charged, neutral, hydrophilic, and hydrophobic properties showed that mixed hydrophilic/hydrophobic SAMs were the most favorable for facile *Ax*CuNiR electrocatalysis [142]. *In situ* AFM is also reported and disclosed *Ax*CuNiR conformational changes during catalytic reaction [67], with the apparent height of *Ax*CuNiR rising from 4.5 nm in the resting state to 5.5 nm in the nitrite reduction state in the presence of nitrite.

3.2.3. Bilirubin Oxidase

Bilirubin oxidase (BOD) is another well-known blue multicopper oxidase class containing four copper centers (T$_1$, T$_2$/T$_3$), catalyzing the dioxygen reduction reaction (ORR) into water in four-electron direct bioelectrocatalysis [75,96]. The blue T$_1$ center has a copper atom located closed to the protein surface for accepting electrons from the natural electron donor (bilirubin) or an electrode. The external electrons are relayed via a short ligand-bound intramolecular peptide bridge to the T$_2$/T$_3$ center, where ORR takes place. Target BODs are mainly from *Myrothecium verrucaria* (*Mv*BOD) [26,29,75,143–145], *Trachyderma tsunodae* (*Tt*BOD) [146], *Bacillus pumilus* (*Bp*BOD) [147], and *Magnaporthe oryzae* (*Mo*BOD) [148]. Single-crystal gold electrodes modified with -NH$_2$, -COOH, -OH, and -CH$_3$ terminated SAMs have been studied and show that negatively charged SAMs are favorable for proper *Mv*BOD orientations due to the highly positively charged region close to the T$_1$

center [42]. E° values of T_1 (0.69 V vs. NHE) and T_2/T_3 (0.39 V vs. NHE) in *Tt*BOD have been observed under aerobic conditions [146], but E° of the T_2/T_3 centers drops to 0.36 V vs. NHE in the resting state of the enzyme, indicative of an IET pathway from T_1 to T_2/T_3 triggered by the ORR process.

Characterization of enzyme loading and conformation on the electrode is crucial [149]. Lojou and associates studied the adsorption of *Mv*BOD on both positively (NH_3^+) and negatively (COO^-) charged SAM-modified electrodes with no significant difference in enzyme loading as disclosed by surface plasmon resonance (SPR) spectroscopy [26]. DET and MET processes were found to dominate on the COO^- and NH_3^+ surfaces, respectively, consistent with the positively charged surroundings of the T_1 Cu [26]. Polarization-modulated infrared reflection absorption spectroscopy (PMIRRAS) showed strong electrostatic interactions between the negatively charged SAMs and the positively charged *Mv*BOD surface near the T_1 Cu [26]. PMIRRAS and SPR ellipsometry were jointly used to demonstrate the effect of electrostatic interactions on the enzyme adsorption, catalytic performance, and stability of *Mv*BOD at four different pH values (Figure 6) [75]. The dipole moment of the enzyme is distinct at different pH levels, showing a direction towards the T_1 center in neutral or slightly acidic electrolytes but significantly shifted in the strongly acid electrolyte (Figure 6a). This accords with the different charge distributions at the T_1 Cu center. PMIRRAS spectra showed the amide I (ca. 1680 cm^{-1}) and amide II (ca. 1550 cm^{-1}) peaks, related to vibrational C=O and N-H modes in *Mv*BOD, respectively, see Figure 6b. The wavelengths of the two peaks remained unchanged upon enzyme immobilization, which suggests that the secondary structure of the *Mv*BOD is retained. In addition, amide I/amide II ratios were similar at different pH levels, further indicating that the orientations of *Mv*BOD were independent of pH.

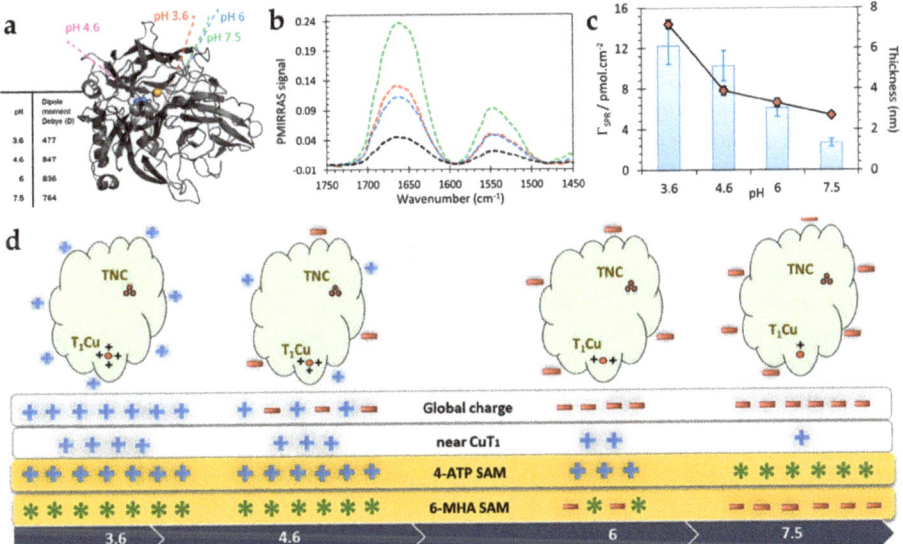

Figure 6. (a) *Mv*BOD structure and dipole moments in the pH range 3.6 to 7.5. CuT_2/T_3 contains three copper atoms marked in blue and CuT_1 with a single copper atom marked in gold; (b) PMIRRAS signals of *Mv*BOD-modified bioelectrodes with 6-MHA SAM; (c) Enzyme coverage (Γ) and enzyme layer thickness at different pH recorded by SPR ellipsometry; (d) Cartoon illustration of the charge distributions of *Mv*BOD at different pH levels, with the CuT_1 center, as well the 6-MHA and 4-ATP SAM-modified electrode surfaces. The neutral electrode surface is indicated with green stars. Reproduced with permission from [75]. Copyright 2018, American Chemical Society.

Enzyme loading, critical to catalytic activity, declines as pH increases (Figure 6c [75]). The authors demonstrated that *Mv*BOD on a negatively charged 6-MHA SAM-modified electrode does not form a saturated monolayer at pH 7.5 but possibly more than a single monolayer at pH 3.6. The electrostatic interactions between enzyme and SAMs thus not only strongly affect the enzyme adsorption, the dipole moment of the enzyme, and the charge around the T_1 center, but they also determine the enzyme orientation and catalytic rate (Figure 6d). Gholami and associates immobilized *Mv*BOD on a gold microfilm by electropolymerization of TCA [150]. Molecular dynamics simulation has provided a more comprehensive understanding of *Mv*BOD in DET-type bioelectrocatalysis [151]. In particular, *Mv*BOD showed various orientations reflecting wide charge distributions, representing a "back-on" and "lying-on" state on positively and negatively charged electrodes, respectively.

Nanostructured gold surfaces are expected to improve the DET current density of BOD [29]. However, the apparent and real current densities normalized to the geometric and real gold surface area, respectively, should be distinguished. In comparison to FDH on 1 nm AuNPs [44], this is particularly important when the gold nanostructures are of larger size than BOD. Pankratov and associates investigated the size effect of sub-monolayer AuNPs (diameter: 20, 40, 60, and 80 nm, significantly larger than *Mv*BOD) on the bioelectrocatalytic performance of *Mv*BOD [152]. Although the apparent ORR current density increased with increasing size from 20 to 80 nm (proportionally to the real surface area), similar values (15 ± 3 uA cm^{-2}) for the real current density were obtained. This can be explained by the fact that the ET rate constant is independent of the AuNP size in this case, with similar values of 10.3 ± 0.5 s^{-1} and 10.7 ± 0.3 s^{-1} for the *Mv*BOD-based bioelectrode with and without AuNPs, respectively. Siepenkoetter and associates immobilized *Mv*BOD on NPG using diazonium grafting coupled with an MPA SAM [28]. NPG electrodes with average pore sizes between 9 and 62 nm were finally investigated. The maximum apparent ORR current density was achieved with 10 and 25 nm pores, slightly larger than the enzyme and likely due to a compromise between the real gold surface area and enzyme loading, but similar real current densities were registered for these pore sizes.

3.2.4. Laccase

The laccases (Lac) constitute a third important member class of the multicopper oxidases, containing T_1 and T_2/T_3 centers [27,153,154]. Similar to BOD, T_1 and T_2/T_3 Cu centers act as electron acceptance and ORR centers, respectively. E° of the T_1 center from tree Lac is lower than those from fungal Lac, varying in the range 300 to 800 mV vs. NHE [155]. Bioelectrocatalysis of fungal Lac has been extensively investigated [156]. SAMs with specific functional groups favor productive orientations of Lac at the electrode surface. Thorum and associates reported that the overpotential of ORR of Lac from *Trametes versicolor* (*Tv*Lac) could be decreased by employing an anthracene-2-methanethiol SAM on gold. The aromatic anthracene presumably penetrates into the hydrophobic pocket close to the T_1 center, facilitating DET [157]. Climent and associates investigated the DET-type catalytic activity of three different Lacs (*Coprinus cinereus* (*Cc*Lac), *Myceliophthora thermophila* (*Mt*Lac), and *Streptomyces coelicolor* (*Sc*Lac)) [38]. *In situ* STM enabled single-molecule understanding of enzyme–electrode electronic interactions and the IET process on well-defined Au(111) surfaces with various SAMs. MPA SAMs with the carboxyl terminal group was best for *Cc*Lac, while alkyl and amino SAMs were most suitable for *Sc*Lac. No catalytic signal was found for *Mt*Lac on any SAMs. As for *Ax*CuNiR, single-molecule in situ STM contrasts were observed only in the presence of enzyme substrate, nitrite, and dioxygen, respectively. Traunsteiner and coworkers exploited DET-type bioelectrocatalysis of *Tv*Lac using diluted MPA SAMs and a linker molecule of thiolated veratric acid (tVA) that could approach the T_1 site [158]. Optimal catalytic activity was shown when the tVA and MPA SAMs were mixed uniformly, whereas the catalytic activity decreased dramatically due to the aggregation of tVA. Molecular dynamics simulations also showed that the positively charged SAMs were more favorable for the DET of *Tv*Lac, with a narrow orientation distribution.

A recent study evaluated *Thermus thermophilus* (*Tt*Lac), exhibiting a methionine rich domain. Figure 7 shows the detailed three-dimensional structure and electrostatic charge distributions of the surface amino acid residues of *Tt*Lac [69]. The T_1 center, used for transferring electrons to an external electrode, showed a negatively charged zone. Gold electrodes modified with negatively (-COO$^-$), uncharged (-OH), and positively charged groups (-NH$_3^+$) were adopted, with the highest catalytic current on the positively charged electrode and with no or only weak catalytic signals on the negatively charged and uncharged electrodes, respectively [69]. Nanostructured materials have been developed recently to optimize the Lac orientation. NPG modified with 4-ATP SAM could increase the catalytic performance and stability at high temperatures [159]. Cristina and associates reported that AuNPs (particle size: 5 nm, comparable to the size of the enzyme) served as electronic bridges, thus promoting DET with a heterogeneous ET rate constant over 400 s^{-1} [54]. Nanostructured electrodes consisting of low-density graphite (LDG) and gold nanorods (AuNRs, average length: 31 ± 6 nm, width: 5 ± 1 nm) have, finally, been prepared to orient Lac for improved ORR [27].

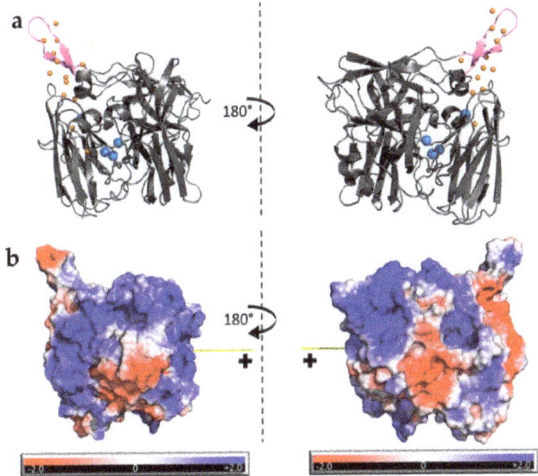

Figure 7. (a) Model structure of *Tt*Lac (PDB 2XU9) showing the hairpin domain (magenta), the T_1 and T_2/T_3 Cu centers (blue spheres) and all Met sulfurs (yellow spheres); (b) Electrostatic potentials at the surface of *Tt*Lac in the same orientation as in the top panel at pH 5: positive charges in blue, negative charges in red, and neutral in white. The positive end of the dipole moment vector is shown as a yellow stick. Reproduced with permission from [69]. Copyright 2020, American Chemical Society.

3.3. [FeS]-Cluster Hydrogenases

Hydrogenases are a class of [FeS] cluster-based metalloproteins which reversibly catalyze the two-electron reactions of dihydrogen oxidation and evolution [71,74]. The commonly reported membrane-bound hydrogenases can be classified based on their intrinsic catalytic cofactors ([FeFe], [NiFe], and [NiFeSe]), where hydrogen conversion is combined with [FeS] electron relay to accomplish the entire ET process. It has been reported that the smallest [FeFe]-hydrogenase *Cr*HydA1 isolated from *Chlamydomonas reinhardtii* exhibits dioxygen insensitivity and shows high catalytic activity in dihydrogen evolution [160]. The conditions of immobilized [FeFe]-hydrogenase *Cr*HydA1 on a SAM-modified gold electrode were characterized by in situ surface-enhanced infrared absorption spectroscopy (SEIRAS) and SPR spectroscopy [160]. Madden and associates investigated the catalytic activity of [FeFe]-hydrogenase CaHydA from *Clostridium acetobutylicum* immobilized on negatively charged MHA-modified Au(111) electrodes [74]. Electrochemical STM showed that the apparent height of [FeFe]-hydrogenase CaHydA continuously increased when the potential of the STM substrate was shifted from −0.4 to −0.6 V (vs. Ag/AgCl electrode), in which the hydrogen evolution response was observed by CV. Notably, the catalysis of hydrogenase is easily quenched by carbon monoxide and cyanide binding to the Fe active sites.

Membrane-bound [NiFe]-hydrogenase and [NiFeSe]-hydrogenase have attracted considerable attention due to their resistance to dioxygen, carbon monoxide, as well as to high temperature [161]. Armstrong and associates reported a number of studies into the oxygen tolerance of [NiFe]-hydrogenase [162,163]. Typically, the catalytic sites of well-known hydrogenases are quenched by dissolved oxygen gas due to the high oxygen sensitivity of the internal peptide chains. Modification of Fe-S clusters by changing two oxygen-sensitive cysteine residues into glycines showed significant improvements in the long-term dihydrogen oxidation performance [162]. The authors demonstrated that the mechanism of oxygen tolerance mainly relates to removal of oxide species rather than preventing oxygen access into the protein. Lojou and associates demonstrated that carboxyl-terminated SAMs were favorable for optimizing the orientations of [NiFe]-hydrogenase from *Aquifex aeolicus* for efficient DET-type and MET-type bioelectrocatalysis of hydrogen oxidation, whereas hydrophobic SAMs only resulted in a MET process with the need for methylene blue mediator [71]. The electrochemical behavior of [NiFe]-hydrogenase from *Allochromatium vinosum*, *DesulfoVibrio Vulgaris* Miyazaki F (*DvMF*), and *Ralstonia eutropha* H16 has also been reported [164–166].

Immobilization of membrane-bound hydrogenase on the SAM-modified electrode surface appears promising, but the ET progress of membrane-bound hydrogenase is challenging due to the complex enzyme structure compared with soluble redox enzymes. Gutiérrez-Sánchez and coworkers demonstrated that the introduction of a phospholipidic bilayer on positively charged 4-ATP modified Au electrode surfaces effectively controls the orientation of membrane-bound [NiFeSe]-hydrogenase for hydrogen oxidation (Figure 8) [167]. The hydrophobic lipid tail of hydrogenase can be embedded into the phospholipidic bilayer, thereby reducing the orientation distribution and promoting the ET process. All these results suggest that SAM-modified electrodes are paramount to provide a versatile platform for understanding how to tune the right enzyme orientation for DET.

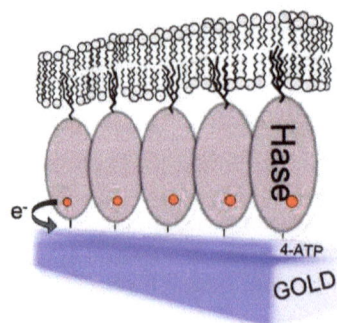

Figure 8. Schematic illustration of [NiFeSe]-hydrogenase covalently inserted into a phospholipidic bilayer. Reproduced with permission from [167]. Copyright 2011, American Chemical Society.

4. Conclusions and Perspectives

Redox proteins and enzymes immobilized on solid surfaces are fragile biomolecular entities. With a few exceptions, they retain ET of enzyme function only in prepared microenvironments that somehow emulate their natural aqueous/membrane reaction media. Self-assembled molecular monolayers with multifariously functionalized thiols are close to ideal SAM building blocks which have emerged as the core class over the last couple of decades. The -SH thiol end ascertains robust Au-S linking which is now increasingly well understood as dominated by the Au(0)-•S(0) gold-thiyl and strong van der Waals Au-S interaction. The opposite functionalized end of the SAM thiol molecules offers the choice of hydrophilic or hydrophobic, electrostatically charged or neutral, and structurally large or small terminal groups, well suited for designed gentle protein/enzyme linking to the solid

SAM-modified electrochemical surface. The option of varying the length of the SAM-forming molecules offers additional control of the electron tunneling process as well as of the dielectric and other local environmental properties crucial in the overall control of the electrochemical activity of the immobilized proteins or enzymes.

We have first overviewed the preparation and comprehensive characterization of thiol-based SAMs on Au surfaces in particular, both *per se* and as developed in redox protein/enzyme electrochemical research over the last couple of decades. Preparation is, in principle, straightforward, although the ultimate SAM properties depend in subtle ways on thiol exposure time, temperature, and other external controlling factors. A variety of sophisticated SAM surface techniques have brought the understanding of the fundamental molecular and electronic SAM structure to a high level, ranging all the way from the ordered domain right down to the single molecule. The techniques include spectroscopy (XPS, FTIR, SPR, Raman, NEXUS, and others), microscopy (AFM, STM in the electrochemical in situ/*operando* modes), and mass balance techniques (QCM), supported by large-scale electronic structure calculations [4,6,16,17,66].

Well-defined solid SAM-modified Au- and other electrode surfaces are a pre-requisite for the productive immobilization of bioelectrochemically active DET enzymes. An outstanding challenge is that polycrystalline Au- and other metallic electrodes are nearly always used, with single-crystal, atomically planar, e.g. Au(111), electrodes only relatively recently introduced as electrochemical biomolecular target surfaces. A variety of local low- and higher-index surface structures are distributed over the polycrystalline Au- and other surfaces, expectedly giving quite different surface ET activities [92]. This presents a challenge to the microscopic characterization of the pure and SAM-modified electrode surfaces but at the same time also provides new openings in the way of more robust protein/enzyme monolayers and higher enzyme activity, if the polycrystallinity can be controlled such as for NPG and other nanoporous metallic electrodes.

We have next overviewed and discussed adsorption and controlled protein and enzyme orientation on strategically chosen SAMs, of both simple ET metalloproteins (cyt. c and c_4, azurin) and a variety of much more complex redox metallo- and nonmetalloenzymes (blue copper oxidases, FAD-based enzymes, cellobiose dehydrogenase, [FeS]-cluster dehydrogenases). We have shown that properly chosen SAMs can be brought to control efficiently the enzyme surface orientation in the ways most favorable both for productive bioelectrocatalysis and for detailed mapping of the molecular mechanisms involved. We have also shown that, although of more complex molecular/atomic surface structure, SAM-modified NPG and other nanoporous metallic electrochemical surfaces, structurally characterized to intermediate levels of resolution, may offer other advantages. These extend to increased enzyme stability and even enhanced catalytic efficiency compared to atomically planar electrode surfaces.

Overall, the present state of detailed structural and mechanistic protein and enzyme bioelectrochemical mapping has now advanced in impressive detail, approaching the true level of the single molecule [3,4,6,168,169] and supported by large-scale electronic structure computations [16,17,92,138]. With new and rapidly increasing understanding of the complex, heterogeneous, and anisotropic electrode/SAM/protein/enzyme/aqueous interface, its real exploitation in the strategic design and development of enzyme biofuel cells, next-generation bioelectrochemical sensors, and other high-technology applications is rapidly coming close.

Author Contributions: Writing—original draft, X.Y.; Supervision, D.T., J.U., X.X.; Writing—review and editing, J.T., D.T., J.U., X.X. All authors have read and agreed to the published version of the manuscript.

Funding: This project has received funding from the European Union's Horizon 2020 research and innovation program under the Marie Skłodowska-Curie grant agreement No. 713683. Financial support was received also from the Danish Council for Independent Research for the YDUN project (DFF 4093-00297), the Russian Science Foundation (project No. 17-13-01274), and from Villum Experiment (grant No. 35844).

Acknowledgments: X.Y. acknowledges support from the China Scholarship Council (No. 201806650009).

Conflicts of Interest: The authors declare no conflict of interest.

References

1. Love, J.C.; Estroff, L.A.; Kriebel, J.K.; Nuzzo, R.G.; Whitesides, G.M. Self-assembled monolayers of thiolates on metals as a form of nanotechnology. *Chem. Rev.* **2005**, *105*, 1103–1170. [CrossRef] [PubMed]
2. Nöll, T.; Nöll, G. Strategies for "wiring" redox-active proteins to electrodes and applications in biosensors, biofuel cells, and nanotechnology. *Chem. Soc. Rev.* **2011**, *40*, 3564–3576. [CrossRef] [PubMed]
3. Bollella, P. Porous Gold: A New Frontier for Enzyme-Based Electrodes. *Nanomaterials* **2020**, *10*, 722. [CrossRef] [PubMed]
4. Chi, Q.; Ford, M.J.; Halder, A.; Hush, N.S.; Reimers, J.R.; Ulstrup, J. Sulfur ligand mediated electrochemistry of gold surfaces and nanoparticles: What, how, and why. *Curr. Opin. Electrochem.* **2017**, *1*, 7–15. [CrossRef]
5. Ulman, A. Formation and structure of self-assembled monolayers. *Chem. Rev.* **1996**, *96*, 1533–1554. [CrossRef]
6. Zhang, J.; Kuznetsov, A.M.; Medvedev, I.G.; Chi, Q.; Albrecht, T.; Jensen, P.S.; Ulstrup, J. Single-molecule electron transfer in electrochemical environments. *Chem. Rev.* **2008**, *108*, 2737–2791. [CrossRef]
7. Bigelow, W.C.; Pickett, D.L.; Zisman, W.A. Oleophobic monolayers. I. Films adsorbed from solution in non-polar liquids. *J. Colloid Sci.* **1946**, *1*, 513–538. [CrossRef]
8. Nuzzo, R.G.; Allara, D.L. Adsorption of Bifunctional Organic Disulfides on Gold Surfaces. *J. Am. Chem. Soc.* **1983**, *105*, 4481–4483. [CrossRef]
9. Muskal, N.; Mandler, D. Thiol self-assembled monolayers on mercury surfaces: The adsorption and electrochemistry of v-mercaptoalkanoic acids. *Electrochim. Acta* **1999**, *45*, 537–548. [CrossRef]
10. Williams, J.A.; Gorman, C.B. Alkanethiol Reductive Desorption from Self-Assembled Monolayers on Gold, Platinum, and Palladium Substrates. *J. Phys. Chem. C* **2007**, *111*, 12804–12810. [CrossRef]
11. Hoque, E.; Derose, J.A.; Houriet, R.; Hoffmann, P.; Mathieu, H.J.; Fe, E.P. Stable Perfluorosilane Self-Assembled Monolayers on Copper Oxide Surfaces: Evidence of Siloxy-Copper Bond Formation. *Chem. Mater.* **2007**, *19*, 798–804. [CrossRef]
12. Fischer, S.; Papageorgiou, A.C.; Marschall, M.; Reichert, J.; Diller, K.; Klappenberger, F.; Allegretti, F.; Nefedov, A.; Wöll, C.; Barth, J.V. L-Cysteine on Ag(111): A combined STM and X-ray spectroscopy study of anchorage and deprotonation. *J. Phys. Chem. C* **2012**, *116*, 20356–20362. [CrossRef]
13. Alejandra, M.; Addato, F.; Rubert, A.A.; Benítez, G.A.; Fonticelli, M.H.; Carrasco, J.; Carro, P.; Salvarezza, R.C. Alkanethiol Adsorption on Platinum: Chain Length Effects on the Quality of Self-Assembled Monolayers. *J. Phys. Chem. C* **2011**, *115*, 17788–17798. [CrossRef]
14. Feliciano-ramos, I.; Caban-acevedo, M.; Scibioh, M.A.; Cabrera, C.R. Self-assembled monolayers of L-cysteine on palladium electrodes. *J. Electroanal. Chem.* **2010**, *650*, 98–104. [CrossRef]
15. Gooding, J.J.; Ciampi, S. The molecular level modification of surfaces: From self-assembled monolayers to complex molecular assemblies. *Chem. Soc. Rev.* **2011**, *40*, 2704–2718. [CrossRef]
16. Reimers, J.R.; Ford, M.J.; Marcuccio, S.M.; Ulstrup, J.; Hush, N.S. Competition of van der Waals and chemical forces on gold-sulfur surfaces and nanoparticles. *Nat. Rev. Chem.* **2017**, *1*, 1–13. [CrossRef]
17. Reimers, J.R.; Ford, M.J.; Halder, A.; Ulstrup, J.; Hush, N.S. Gold surfaces and nanoparticles are protected by Au(0)-thiyl species and are destroyed when Au(I)-thiolates form. *Proc. Natl. Acad. Sci. USA* **2016**, *113*, E1424–E1433. [CrossRef]
18. Tang, J.; Yan, X.; Engelbrekt, C.; Ulstrup, J.; Magner, E.; Xiao, X.; Zhang, J. Development of graphene-based enzymatic biofuel cells: A minireview. *Bioelectrochemistry* **2020**, *134*, 107537. [CrossRef]
19. Hitaishi, V.P.; Clement, R.; Bourassin, N.; Baaden, M.; de Poulpiquet, A.; Sacquin-Mora, S.; Ciaccafava, A.; Lojou, E. Controlling redox enzyme orientation at planar electrodes. *Catalysts* **2018**, *8*, 192. [CrossRef]
20. Milton, R.D.; Minteer, S.D. Direct enzymatic bioelectrocatalysis: Differentiating between myth and reality. *J. R. Soc. Interface* **2017**, *14*, 20170253. [CrossRef]
21. Mano, N.; De Poulpiquet, A. O_2 Reduction in Enzymatic Biofuel Cells. *Chem. Rev.* **2018**, *118*, 2392–2468. [CrossRef] [PubMed]
22. Moser, C.C.; Keske, J.M.; Warncke, K.; Farid, R.S.; Dutton, P.L. Nature of biological electron transfer. *Nature* **1992**, *355*, 796–802. [CrossRef] [PubMed]
23. Schulz, C.; Kittl, R.; Ludwig, R.; Gorton, L. Direct Electron Transfer from the FAD Cofactor of Cellobiose Dehydrogenase to Electrodes. *ACS Catal.* **2016**, *6*, 555–563. [CrossRef]

24. Lee, H.; Lee, Y.S.; Reginald, S.S.; Baek, S.; Lee, E.M.; Choi, I.G.; Chang, I.S. Biosensing and electrochemical properties of flavin adenine dinucleotide (FAD)-Dependent glucose dehydrogenase (GDH) fused to a gold binding peptide. *Biosens. Bioelectron.* **2020**, *165*, 112427. [CrossRef]
25. Siepenkoetter, T.; Salaj-Kosla, U.; Magner, E. The Immobilization of Fructose Dehydrogenase on Nanoporous Gold Electrodes for the Detection of Fructose. *ChemElectroChem* **2017**, *4*, 905–912. [CrossRef]
26. Gutierrez-Sanchez, C.; Ciaccafava, A.; Blanchard, P.Y.; Monsalve, K.; Giudici-Orticoni, M.T.; Lecomte, S.; Lojou, E. Efficiency of Enzymatic O_2 Reduction by Myrothecium verrucaria Bilirubin Oxidase Probed by Surface Plasmon Resonance, PMIRRAS, and Electrochemistry. *ACS Catal.* **2016**, *6*, 5482–5492. [CrossRef]
27. Di Bari, C.; Shleev, S.; De Lacey, A.L.; Pita, M. Laccase-modified gold nanorods for electrocatalytic reduction of oxygen. *Bioelectrochemistry* **2016**, *107*, 30–36. [CrossRef]
28. Siepenkoetter, T.; Salaj-Kosla, U.; Xiao, X.; Conghaile, P.; Pita, M.; Ludwig, R.; Magner, E. Immobilization of Redox Enzymes on Nanoporous Gold Electrodes: Applications in Biofuel Cells. *ChemPlusChem* **2017**, *82*, 553–560. [CrossRef]
29. Murata, K.; Kajiya, K.; Nakamura, N.; Ohno, H. Direct electrochemistry of bilirubin oxidase on three-dimensional gold nanoparticle electrodes and its application in a biofuel cell. *Energy Environ. Sci.* **2009**, *2*, 1280–1285. [CrossRef]
30. Adachi, T.; Kaida, Y.; Kitazumi, Y.; Shirai, O.; Kano, K. Bioelectrocatalytic performance of D-fructose dehydrogenase. *Bioelectrochemistry* **2019**, *129*, 1–9. [CrossRef]
31. Xiao, X.; Si, P.; Magner, E. An overview of dealloyed nanoporous gold in bioelectrochemistry. *Bioelectrochemistry* **2016**, *109*, 117–126. [CrossRef] [PubMed]
32. Xiao, X.; Xia, H.Q.; Wu, R.; Bai, L.; Yan, L.; Magner, E.; Cosnier, S.; Lojou, E.; Zhu, Z.; Liu, A. Tackling the Challenges of Enzymatic (Bio)Fuel Cells. *Chem. Rev.* **2019**, *119*, 9509–9558. [CrossRef] [PubMed]
33. Adachi, T.; Kitazumi, Y.; Shirai, O.; Kano, K. Direct Electron Transfer-Type Bioelectrocatalysis of Redox Enzymes at Nanostructured Electrodes. *Catalysts* **2020**, *10*, 236. [CrossRef]
34. Vaz-Dominguez, C.; Campuzano, S.; Rüdiger, O.; Pita, M.; Gorbacheva, M.; Shleev, S.; Fernandez, V.M.; De Lacey, A.L. Laccase electrode for direct electrocatalytic reduction of O_2 to H_2O with high-operational stability and resistance to chloride inhibition. *Biosens. Bioelectron.* **2008**, *24*, 531–537. [CrossRef]
35. Lindgren, A.; Gorton, L.; Ruzgas, T.; Baminger, U.; Haltrich, D.; Schülein, M. Direct electron transfer of cellobiose dehydrogenase from various biological origins at gold and graphite electrodes. *J. Electroanal. Chem.* **2001**, *496*, 76–81. [CrossRef]
36. Yan, Y.M.; Baravik, I.; Tel-Vered, R.; Willner, I. An ethanol/O_2 biofuel cell based on an electropolymerized bilirubin oxidase/Pt nanoparticle bioelectrocatalytic O_2-reduction cathode. *Adv. Mater.* **2009**, *21*, 4275–4279. [CrossRef]
37. Schubert, K.; Goebel, G.; Lisdat, F. Bilirubin oxidase bound to multi-walled carbon nanotube-modified gold. *Electrochim. Acta* **2009**, *54*, 3033–3038. [CrossRef]
38. Climent, V.; Zhang, J.; Friis, E.P.; Østergaard, L.H.; Ulstrup, J. Voltammetry and single-molecule in situ scanning tunneling microscopy of laccases and bilirubin oxidase in electrocatalytic dioxygen reduction on Au(111) single-crystal electrodes. *J. Phys. Chem. C* **2012**, *116*, 1232–1243. [CrossRef]
39. Chi, Q.; Farver, O.; Ulstrup, J. Long-range protein electron transfer observed at the single-molecule level: In situ mapping of redox-gated tunneling resonance. *Proc. Natl. Acad. Sci. USA* **2005**, *102*, 16203–16208. [CrossRef]
40. Xiao, X.; Li, H.; Wang, M.; Zhang, K.; Si, P. Examining the effects of self-assembled monolayers on nanoporous gold based amperometric glucose biosensors. *Analyst* **2013**, *139*, 488–494. [CrossRef]
41. Murata, K.; Suzuki, M.; Kajiya, K.; Nakamura, N.; Ohno, H. High performance bioanode based on direct electron transfer of fructose dehydrogenase at gold nanoparticle-modified electrodes. *Electrochem. Commun.* **2009**, *11*, 668–671. [CrossRef]
42. Tominaga, M.; Ohtani, M.; Taniguchi, I. Gold single-crystal electrode surface modified with self-assembled monolayers for electron tunneling with bilirubin oxidase. *Phys. Chem. Chem. Phys.* **2008**, *10*, 6928–6934. [CrossRef] [PubMed]
43. Chi, Q.; Zhang, J.; Andersen, J.E.T.; Ulstrup, J. Ordered assembly and controlled electron transfer of the blue copper protein azurin at gold (111) single-crystal substrates. *J. Phys. Chem. B* **2001**, *105*, 4669–4679. [CrossRef]

44. Kizling, M.; Dzwonek, M.; Więckowska, A.; Bilewicz, R. Size Does Matter—Mediation of Electron Transfer by Gold Clusters in Bioelectrocatalysis. *ChemCatChem* **2018**, *10*, 1988–1992. [CrossRef]
45. Al-Lolage, F.A.; Meneghello, M.; Ma, S.; Ludwig, R.; Bartlett, P.N. A Flexible Method for the Stable, Covalent Immobilization of Enzymes at Electrode Surfaces. *ChemElectroChem* **2017**, *4*, 1528–1534. [CrossRef]
46. Matsumura, H.; Ortiz, R.; Ludwig, R.; Igarashi, K.; Samejima, M.; Gorton, L. Direct electrochemistry of phanerochaete chrysosporium cellobiose dehydrogenase covalently attached onto gold nanoparticle modified solid gold electrodes. *Langmuir* **2012**, *28*, 10925–10933. [CrossRef]
47. Wang, X.; Falk, M.; Ortiz, R.; Matsumura, H.; Bobacka, J.; Ludwig, R.; Bergelin, M.; Gorton, L.; Shleev, S. Mediatorless sugar/oxygen enzymatic fuel cells based on gold nanoparticle-modified electrodes. *Biosens. Bioelectron.* **2012**, *31*, 219–225. [CrossRef]
48. Rauf, S.; Zhou, D.; Abell, C.; Klenerman, D.; Kang, D.J. Building three-dimensional nanostructures with active enzymes by surface templated layer-by-layer assembly. *Chem. Commun.* **2006**, 1721–1723. [CrossRef]
49. Xiao, X.; Siepenkoetter, T.; Conghaile, P.; Leech, D.; Magner, E. Nanoporous Gold-Based Biofuel Cells on Contact Lenses. *ACS Appl. Mater. Interfaces* **2018**, *10*, 7107–7116. [CrossRef]
50. Siepenkoetter, T.; Salaj-Kosla, U.; Xiao, X.; Belochapkine, S.; Magner, E. Nanoporous Gold Electrodes with Tuneable Pore Sizes for Bioelectrochemical Applications. *Electroanalysis* **2016**, *28*, 2415–2423. [CrossRef]
51. Jensen, P.S.; Chi, Q.; Zhang, J.; Ulstrup, J. Long-Range interfacial electrochemical electron transfer of Pseudomonas aeruginosa azurin-gold nanoparticle hybrid systems. *J. Phys. Chem. C* **2009**, *113*, 13993–14000. [CrossRef]
52. Xu, S.; Han, X. A novel method to construct a third-generation biosensor: Self-assembling gold nanoparticles on thiol-functionalized poly(styrene-co-acrylic acid) nanospheres. *Biosens. Bioelectron.* **2004**, *19*, 1117–1120. [CrossRef] [PubMed]
53. Murata, K.; Suzuki, M.; Nakamura, N.; Ohno, H. Direct evidence of electron flow via the heme c group for the direct electron transfer reaction of fructose dehydrogenase using a silver nanoparticle-modified electrode. *Electrochem. Commun.* **2009**, *11*, 1623–1626. [CrossRef]
54. Gutiérrez-Sánchez, C.; Pita, M.; Vaz-Domínguez, C.; Shleev, S.; De Lacey, A.L. Gold nanoparticles as electronic bridges for laccase-based biocathodes. *J. Am. Chem. Soc.* **2012**, *134*, 17212–17220. [CrossRef]
55. Meredith, M.T.; Minteer, S.D. Biofuel Cells: Enhanced Enzymatic Bioelectrocatalysis. *Annu. Rev. Anal. Chem.* **2012**, *5*, 157–179. [CrossRef]
56. Gross, A.J.; Holzinger, M.; Cosnier, S. Buckypaper bioelectrodes: Emerging materials for implantable and wearable biofuel cells. *Energy Environ. Sci.* **2018**, *11*, 1670–1687. [CrossRef]
57. Mazurenko, I.; de Poulpiquet, A.; Lojou, E. Recent developments in high surface area bioelectrodes for enzymatic fuel cells. *Curr. Opin. Electrochem.* **2017**, *5*, 74–84. [CrossRef]
58. Mazurenko, I.; Hitaishi, V.P.; Lojou, E. Recent advances in surface chemistry of electrodes to promote direct enzymatic bioelectrocatalysis. *Curr. Opin. Electrochem.* **2020**, *19*, 113–121. [CrossRef]
59. Chaki, N.K.; Vijayamohanan, K. Self-assembled monolayers as a tunable platform for biosensor applications. *Biosens. Bioelectron.* **2002**, *17*, 1–12. [CrossRef]
60. Xiao, X.; Ulstrup, J.; Li, H.; Wang, M.; Zhang, J.; Si, P. Nanoporous gold assembly of glucose oxidase for electrochemical biosensing. *Electrochim. Acta* **2014**, *130*, 559–567. [CrossRef]
61. Serleti, A.; Salaj-Kosla, U.; Magner, E. The spatial and sequential immobilisation of cytochrome c at adjacent electrodes. *Chem. Commun.* **2013**, *49*, 8395–8397. [CrossRef] [PubMed]
62. Meunier-Prest, R.; Legay, G.; Raveau, S.; Chiffot, N.; Finot, E. Potential-assisted deposition of mixed alkanethiol self-assembled monolayers. *Electrochim. Acta* **2010**, *55*, 2712–2720. [CrossRef]
63. Bradbury, C.R.; Zhao, J.; Fermín, D.J. Distance-independent charge-transfer resistance at gold electrodes modified by thiol monolayers and metal nanoparticles. *J. Phys. Chem. C* **2008**, *112*, 10153–10160. [CrossRef]
64. Chi, Q.; Zhang, J.; Arslan, T.; Borg, L.; Pedersen, G.W.; Christensen, H.E.M.; Nazmudtinov, R.R.; Ulstrup, J. Approach to interfacial and intramolecular electron transfer of the diheme protein cytochrome c_4 assembled on Au(111) surfaces. *J. Phys. Chem. B* **2010**, *114*, 5617–5624. [CrossRef]
65. Engelbrekt, C.; Nazmutdinov, R.R.; Zinkicheva, T.T.; Glukhov, D.V.; Yan, J.; Mao, B.; Ulstrup, J.; Zhang, J. Chemistry of cysteine assembly on Au(100): Electrochemistry, in situ STM and molecular modeling. *Nanoscale* **2019**, *11*, 17235–17251. [CrossRef]

66. Zhang, J.; Welinder, A.C.; Chi, Q.; Ulstrup, J. Electrochemically controlled self-assembled monolayers characterized with molecular and sub-molecular resolution. *Phys. Chem. Chem. Phys.* **2011**, *13*, 5526–5545. [CrossRef]
67. Hao, X.; Zhang, J.; Christensen, H.E.M.; Wang, H.; Ulstrup, J. Electrochemical single-molecule AFM of the redox metalloenzyme copper nitrite reductase in action. *ChemPhysChem* **2012**, *13*, 2919–2924. [CrossRef]
68. Marmisollé, W.A.; Capdevila, D.A.; De La Llave, E.; Williams, F.J.; Murgida, D.H. Self-assembled monolayers of NH_2-terminated thiolates: Order, pKa, and specific adsorption. *Langmuir* **2013**, *29*, 5351–5359. [CrossRef]
69. Hitaishi, V.P.; Clément, R.; Quattrocchi, L.; Parent, P.; Duché, D.; Zuily, L.; Ilbert, M.; Lojou, E.; Mazurenko, I. Interplay between Orientation at Electrodes and Copper Activation of Thermus thermophilus Laccase for O_2 Reduction. *J. Am. Chem. Soc.* **2020**, *142*, 1394–1405. [CrossRef]
70. Tang, J.; Werchmeister, R.M.L.; Preda, L.; Huang, W.; Zheng, Z.; Leimkühler, S.; Wollenberger, U.; Xiao, X.; Engelbrekt, C.; Ulstrup, J.; et al. Three-Dimensional Sulfite Oxidase Bioanodes Based on Graphene Functionalized Carbon Paper for Sulfite/O_2 Biofuel Cells. *ACS Catal.* **2019**, *9*, 6543–6554. [CrossRef]
71. Ciaccafava, A.; Infossi, P.; Ilbert, M.; Guiral, M.; Lecomte, S.; Giudici-Orticoni, M.T.; Lojou, E. Electrochemistry, AFM, and PM-IRRA spectroscopy of immobilized hydrogenase: Role of a hydrophobic helix in enzyme orientation for efficient H_2 oxidation. *Angew. Chem. Int. Ed.* **2012**, *51*, 953–956. [CrossRef] [PubMed]
72. Loew, N.; Scheller, F.W.; Wollenberger, U. Characterization of self-assembling of glucose dehydrogenase in mono- and multilayers on gold electrodes. *Electroanalysis* **2004**, *16*, 1149–1154. [CrossRef]
73. Yehezkeli, O.; Tel-Vered, R.; Raichlin, S.; Willner, I. Nano-engineered flavin-dependent glucose dehydrogenase/gold nanoparticle-modified electrodes for glucose sensing and biofuel cell applications. *ACS Nano* **2011**, *5*, 2385–2391. [CrossRef] [PubMed]
74. Madden, C.; Vaughn, M.D.; Díez-Pérez, I.; Brown, K.A.; King, P.W.; Gust, D.; Moore, A.L.; Moore, T.A. Catalytic turnover of [FeFe]-hydrogenase based on single-molecule imaging. *J. Am. Chem. Soc.* **2012**, *134*, 1577–1582. [CrossRef]
75. Hitaishi, V.P.; Mazurenko, I.; Harb, M.; Clément, R.; Taris, M.; Castano, S.; Duché, D.; Lecomte, S.; Ilbert, M.; De Poulpiquet, A.; et al. Electrostatic-Driven Activity, Loading, Dynamics, and Stability of a Redox Enzyme on Functionalized-Gold Electrodes for Bioelectrocatalysis. *ACS Catal.* **2018**, *8*, 12004–12014. [CrossRef]
76. Guo, L.H.; Allen, H.; Hill, O. Direct electrochemistry of proteins and enzymes. *Adv. Inorg. Chem.* **1991**, *36*, 341–375. [CrossRef]
77. Albery, W.J.; Eddowes, M.J.; Allen, H.; Hill, O.; Hillman, A.R. Mechanism of the Reduction and Oxidation Reaction of Cytochrome c at a Modified Gold Electrode. *J. Am. Chem. Soc.* **1981**, *103*, 3904–3910. [CrossRef]
78. Armstrong, F.A.; Hill, H.A.O.; Walton, N.J. Direct Electrochemistry of Redox Proteins. *Acc. Chem. Res.* **1988**, *21*, 407–413. [CrossRef]
79. Armstrong, F.A.; Heering, H.A.; Hirst, J. Reaction of complex metalloproteins studied by protein-film voltammetry. *Chem. Soc. Rev.* **1997**, *26*, 169–179. [CrossRef]
80. Ma, S.; Laurent, C.V.F.P.; Meneghello, M.; Tuoriniemi, J.; Oostenbrink, C.; Gorton, L.; Bartlett, P.N.; Ludwig, R. Direct electron-transfer anisotropy of a site-specifically immobilized cellobiose dehydrogenase. *ACS Catal.* **2019**, *9*, 7607–7615. [CrossRef]
81. Bollella, P.; Hibino, Y.; Kano, K.; Gorton, L.; Antiochia, R. Highly Sensitive Membraneless Fructose Biosensor Based on Fructose Dehydrogenase Immobilized onto Aryl Thiol Modified Highly Porous Gold Electrode: Characterization and Application in Food Samples. *Anal. Chem.* **2018**, *90*, 12131–12136. [CrossRef] [PubMed]
82. Karlsson, J.J.; Nielsen, M.F.; Thuesen, M.H.; Ulstrup, J. Electrochemistry of cytochrome c_4 from Pseudomonas stutzeri. *J. Phys. Chem. B* **1997**, *101*, 2430–2436. [CrossRef]
83. Avila, A.; Gregory, B.W.; Niki, K.; Cotton, T.M. An electrochemical approach to investigate gated electron transfer using a physiological model system: Cytochrome C immobilized on carboxylic acid-terminated alkanethiol self-assembled monolayers on gold electrodes. *J. Phys. Chem. B* **2000**, *104*, 2759–2766. [CrossRef]
84. Wei, J.; Liu, H.; Dick, A.R.; Yamamoto, H.; He, Y.; Waldeck, D.H. Direct wiring of cytochrome c's heme unit to an electrode: Electrochemical studies. *J. Am. Chem. Soc.* **2002**, *124*, 9591–9599. [CrossRef] [PubMed]
85. Collinson, M.; Bowden, E.F.; Tarlov, M.J. Voltammetry of Covalently Immobilized Cytochrome c on Self-Assembled Monolayer Electrodes. *Langmuir* **1992**, *8*, 1247–1250. [CrossRef]
86. El Kasmi, A.; Wallace, J.M.; Bowden, E.F.; Binet, S.M.; Linderman, R.J. Controlling interfacial electron-transfer kinetics of cytochrome c with mixed self-assembled monolayers. *J. Am. Chem. Soc.* **1998**, *120*, 225–226. [CrossRef]

87. Yue, H.; Waldeck, D.H.; Schrock, K.; Kirby, D.; Knorr, K.; Switzer, S.; Rosmus, J.; Clark, R.A. Multiple sites for electron tunneling between cytochrome c and mixed self-assembled monolayers. *J. Phys. Chem. C* **2008**, *112*, 2514–2521. [CrossRef]
88. Scott, R.A.; Mauk, A.G. *Cytochrome c: A Multidisciplinary Approach*; Univ Science Books: Sausalito, CA, USA, 1996.
89. Jensen, P.S.; Chi, Q.; Grumsen, F.B.; Abad, J.M.; Horsewell, A.; Schiffrin, D.J.; Ulstrup, J. Gold nanoparticle assisted assembly of a heme protein for enhancement of long-range interfacial electron transfer. *J. Phys. Chem. C* **2007**, *111*, 6124–6132. [CrossRef]
90. Niki, K.; Hardy, W.R.; Hill, M.G.; Li, H.; Sprinkle, J.R.; Margoliash, E.; Fujita, K.; Tanimura, R.; Nakamura, N.; Ohno, H.; et al. Coupling to lysine-13 promotes electron tunneling through carboxylate-terminated alkanethiol self-assembled monolayers to cytochrome c. *J. Phys. Chem. B* **2003**, *107*, 9947–9949. [CrossRef]
91. Murgida, D.H.; Hildebrandt, P. Heterogeneous electron transfer of cytochrome c on coated silver electrodes. Electric field effects on structure and redox potential. *J. Phys. Chem. B* **2001**, *105*, 1578–1586. [CrossRef]
92. Shermukhamedov, A.S.; Nazmutdinov, R.R.; Zinkicheva, T.T.; Bronshtein, D.M.; Zhang, J.; Mao, B.; Tian, Z.; Yan, J.; Wu, D.-Y.; Ulstrup, J. Electronic Spillover from a Metallic Nanoparticle: Can Simple Electrochemical Electron Transfer Processes Be Catalyzed by Electronic Coupling of a Molecular Scale Gold Nanoparticle Simultaneously to the Redox Molecule and the Electrode? *J. Am. Chem. Soc.* **2020**, *142*, 10646–10658. [CrossRef] [PubMed]
93. Chazalviel, J.-N.; Allongue, P. On the Origin of the Efficient Nanoparticle Mediated Electron Transfer across a Self-Assembled Monolayer. *J. Am. Chem. Soc.* **2010**, *133*, 762–764. [CrossRef] [PubMed]
94. Engelbrekt, C.; Sørensen, K.H.; Zhang, J.; Welinder, A.C.; Jensen, P.S.; Ulstrup, J. Green synthesis of gold nanoparticles with starch-glucose and application in bioelectrochemistry. *J. Mater. Chem.* **2009**, *19*, 7839–7847. [CrossRef]
95. Della Pia, E.A.; Chi, Q.; Jones, D.D.; MacDonald, J.E.; Ulstrup, J.; Elliott, M. Single-molecule mapping of long-range electron transport for a cytochrome b_{562} variant. *Nano Lett.* **2011**, *11*, 176–182. [CrossRef]
96. Dronov, R.; Kurth, D.G.; Möhwald, H.; Scheller, F.W.; Lisdat, F. Communication in a protein stack: Electron transfer between cytochrome c and bilirubin oxidase within a polyelectrolyte multilayer. *Angew. Chem. Int. Ed.* **2008**, *47*, 3000–3003. [CrossRef]
97. Dronov, R.; Kurth, D.G.; Möhwald, H.; Spricigo, R.; Leimkühler, S.; Wollenberger, U.; Rajagopalan, K.V.; Scheller, F.W.; Lisdat, F. Layer-by-layer arrangement by protein-protein interaction of sulfite oxidase and cytochrome c catalyzing oxidation of sulfite. *J. Am. Chem. Soc.* **2008**, *130*, 1122–1123. [CrossRef]
98. Feifel, S.C.; Kapp, A.; Lisdat, F. Electroactive nanobiomolecular architectures of laccase and cytochrome c on electrodes: Applying silica nanoparticles as artificial matrix. *Langmuir* **2014**, *30*, 5363–5367. [CrossRef]
99. Beissenhirtz, M.K.; Scheller, F.W.; Stöcklein, W.F.M.; Kurth, D.G.; Möhwald, H.; Lisdat, F. Electroactive cytochrome c multilayers within a polyelectrolyte assembly. *Angew. Chem. Int. Ed.* **2004**, *43*, 4357–4360. [CrossRef]
100. Ma, S.; Ludwig, R. Direct Electron Transfer of Enzymes Facilitated by Cytochromes. *ChemElectroChem* **2019**, *6*, 958–975. [CrossRef]
101. Bollella, P.; Hibino, Y.; Kano, K.; Gorton, L.; Antiochia, R. Enhanced Direct Electron Transfer of Fructose Dehydrogenase Rationally Immobilized on a 2-Aminoanthracene Diazonium Cation Grafted Single-Walled Carbon Nanotube Based Electrode. *ACS Catal.* **2018**, *8*, 10279–10289. [CrossRef]
102. Bollella, P.; Gorton, L.; Antiochia, R. Direct electron transfer of dehydrogenases for development of 3rd generation biosensors and enzymatic fuel cells. *Sensors* **2018**, *18*, 1319. [CrossRef] [PubMed]
103. Bollella, P.; Hibino, Y.; Kano, K.; Gorton, L.; Antiochia, R. The influence of pH and divalent/monovalent cations on the internal electron transfer (IET), enzymatic activity, and structure of fructose dehydrogenase. *Anal. Bioanal. Chem.* **2018**, *410*, 3253–3264. [CrossRef] [PubMed]
104. Hibino, Y.; Kawai, S.; Kitazumi, Y.; Shirai, O.; Kano, K. Mutation of Heme c Axial Ligands in D-Fructose Dehydrogenase For Investigation of Electron Transfer Pathways and Reduction of Overpotential in Direct Electron Transfer-type Bioelectrocatalysis. *Electrochem. Commun.* **2016**, *67*, 43–46. [CrossRef]
105. Hibino, Y.; Kawai, S.; Kitazumi, Y.; Shirai, O.; Kano, K. Construction of a protein-engineered variant of D-fructose dehydrogenase for direct electron transfer-type bioelectrocatalysis. *Electrochem. Commun.* **2017**, *77*, 112–115. [CrossRef]

106. Sugimoto, Y.; Kawai, S.; Kitazumi, Y.; Shirai, O.; Kano, K. Function of C-terminal hydrophobic region in fructose dehydrogenase. *Electrochim. Acta* **2015**, *176*, 976–981. [CrossRef]
107. Meredith, M.T.; Minson, M.; Hickey, D.; Artyushkova, K.; Glatzhofer, D.T.; Minteer, S.D. Anthracene-modified multi-walled carbon nanotubes as direct electron transfer scaffolds for enzymatic oxygen reduction. *ACS Catal.* **2011**, *1*, 1683–1690. [CrossRef]
108. Ludwig, R.; Harreither, W.; Tasca, F.; Gorton, L. Cellobiose dehydrogenase: A versatile catalyst for electrochemical applications. *ChemPhysChem* **2010**, *11*, 2674–2697. [CrossRef]
109. Harreither, W.; Nicholls, P.; Sygmund, C.; Gorton, L.; Ludwig, R. Investigation of the pH-dependent electron transfer mechanism of ascomycetous class II cellobiose dehydrogenases on electrodes. *Langmuir* **2012**, *28*, 6714–6723. [CrossRef]
110. Scheiblbrandner, S.; Ludwig, R. Cellobiose dehydrogenase: Bioelectrochemical insights and applications. *Bioelectrochemistry* **2020**, *131*, 107345. [CrossRef]
111. Lamberg, P.; Hamit-Eminovski, J.; Toscano, M.D.; Eicher-Lorka, O.; Niaura, G.; Arnebrant, T.; Shleev, S.; Ruzgas, T. Electrical activity of cellobiose dehydrogenase adsorbed on thiols: Influence of charge and hydrophobicity. *Bioelectrochemistry* **2017**, *115*, 26–32. [CrossRef]
112. Tavahodi, M.; Ortiz, R.; Schulz, C.; Ekhtiari, A.; Ludwig, R.; Haghighi, B.; Gorton, L. Direct Electron Transfer of Cellobiose Dehydrogenase on Positively Charged Polyethyleneimine Gold Nanoparticles. *ChemPlusChem* **2017**, *82*, 546–552. [CrossRef] [PubMed]
113. Bollella, P.; Mazzei, F.; Favero, G.; Fusco, G.; Ludwig, R.; Gorton, L.; Antiochia, R. Improved DET communication between cellobiose dehydrogenase and a gold electrode modified with a rigid self-assembled monolayer and green metal nanoparticles: The role of an ordered nanostructuration. *Biosens. Bioelectron.* **2017**, *88*, 196–203. [CrossRef] [PubMed]
114. Krikstolaityte, V.; Lamberg, P.; Toscano, M.D.; Silow, M.; Eicher-Lorka, O.; Ramanavicius, A.; Niaura, G.; Abariute, L.; Ruzgas, T.; Shleev, S. Mediatorless carbohydrate/oxygen biofuel cells with improved cellobiose dehydrogenase based bioanode. *Fuel Cells* **2014**, *14*, 792–800. [CrossRef]
115. Meneghello, M.; Al-Lolage, F.A.; Ma, S.; Ludwig, R.; Bartlett, P.N. Studying Direct Electron Transfer by Site-Directed Immobilization of Cellobiose Dehydrogenase. *ChemElectroChem* **2019**, *6*, 700–713. [CrossRef]
116. Yamashita, Y.; Lee, I.; Loew, N.; Sode, K. Direct electron transfer (DET) mechanism of FAD dependent dehydrogenase complexes ~from the elucidation of intra- and inter-molecular electron transfer pathway to the construction of engineered DET enzyme complexes~. *Curr. Opin. Electrochem.* **2018**, *12*, 92–100. [CrossRef]
117. Korani, A.; Salimi, A.; Karimi, B. Guanine/Ionic Liquid Derived Ordered Mesoporous Carbon Decorated with AuNPs as Efficient NADH Biosensor and Suitable Platform for Enzymes Immobilization and Biofuel Cell Design. *Electroanalysis* **2017**, *29*, 2646–2655. [CrossRef]
118. Lee, I.; Loew, N.; Tsugawa, W.; Lin, C.E.; Probst, D.; La Belle, J.T.; Sode, K. The electrochemical behavior of a FAD dependent glucose dehydrogenase with direct electron transfer subunit by immobilization on self-assembled monolayers. *Bioelectrochemistry* **2018**, *121*, 1–6. [CrossRef]
119. Xiao, X.; Conghaile, P.; Leech, D.; Ludwig, R.; Magner, E. An oxygen-independent and membrane-less glucose biobattery/supercapacitor hybrid device. *Biosens. Bioelectron.* **2017**, *98*, 421–427. [CrossRef]
120. Ito, Y.; Okuda-Shimazaki, J.; Tsugawa, W.; Loew, N.; Shitanda, I.; Lin, C.E.; La Belle, J.; Sode, K. Third generation impedimetric sensor employing direct electron transfer type glucose dehydrogenase. *Biosens. Bioelectron.* **2019**, *129*, 189–197. [CrossRef]
121. Ratautas, D.; Laurynenas, A.; Dagys, M.; Marcinkevičiene, L.; Meškys, R.; Kulys, J. High current, low redox potential mediatorless bioanode based on gold nanoparticles and glucose dehydrogenase from Ewingella americana. *Electrochim. Acta* **2016**, *199*, 254–260. [CrossRef]
122. Kim, Y.P.; Park, S.J.; Lee, D.; Kim, H.S. Electrochemical glucose biosensor by electrostatic binding of PQQ-glucose dehydrogenase onto self-assembled monolayers on gold. *J. Appl. Electrochem.* **2012**, *42*, 383–390. [CrossRef]
123. Ferapontova, E.E.; Ruzgas, T.; Gorton, L. Direct electron transfer of heme- and molybdopterin cofactor-containing chicken liver sulfite oxidase on alkanethiol-modified gold electrodes. *Anal. Chem.* **2003**, *75*, 4841–4850. [CrossRef] [PubMed]

124. Sezer, M.; Spricigo, R.; Utesch, T.; Millo, D.; Leimkuehler, S.; Mroginski, M.A.; Wollenberger, U.; Hildebrandt, P.; Weidinger, I.M. Redox properties and catalytic activity of surface-bound human sulfite oxidase studied by a combined surface enhanced resonance Raman spectroscopic and electrochemical approach. *Phys. Chem. Chem. Phys.* **2010**, *12*, 7894–7903. [CrossRef] [PubMed]
125. Spricigo, R.; Dronov, R.; Lisdat, F.; Leimkühler, S.; Scheller, F.W.; Wollenberger, U. Electrocatalytic sulfite biosensor with human sulfite oxidase co-immobilized with cytochrome c in a polyelectrolyte-containing multilayer. *Anal. Bioanal. Chem.* **2009**, *393*, 225–233. [CrossRef]
126. Utesch, T.; Sezer, M.; Weidinger, I.M.; Mroginski, M.A. Adsorption of sulfite oxidase on self-assembled monolayers from molecular dynamics simulations. *Langmuir* **2012**, *28*, 5761–5769. [CrossRef]
127. Zeng, T.; Leimkühler, S.; Wollenberger, U.; Fourmond, V. Transient Catalytic Voltammetry of Sulfite Oxidase Reveals Rate Limiting Conformational Changes. *J. Am. Chem. Soc.* **2017**, *139*, 11559–11567. [CrossRef]
128. Zeng, T.; Frasca, S.; Rumschöttel, J.; Koetz, J.; Leimkühler, S.; Wollenberger, U. Role of Conductive Nanoparticles in the Direct Unmediated Bioelectrocatalysis of Immobilized Sulfite Oxidase. *Electroanalysis* **2016**, *28*, 2303–2310. [CrossRef]
129. Frasca, S.; Rojas, O.; Salewski, J.; Neumann, B.; Stiba, K.; Weidinger, I.M.; Tiersch, B.; Leimkühler, S.; Koetz, J.; Wollenberger, U. Human sulfite oxidase electrochemistry on gold nanoparticles modified electrode. *Bioelectrochemistry* **2012**, *87*, 33–41. [CrossRef]
130. Yan, J.; Frøkjær, E.; Engebrekt, C.; Leimkühler, S.; Ulstrup, J.; Wollenberger, U.; Xiao, X.; Zhang, J. Voltammetry and Single-molecule in Situ Scanning Tunnelling Microscopy of the Redox Metalloenzyme Human Sulfite Oxidase. *ChemElectroChem* **2020**. [CrossRef]
131. Kalimuthu, P.; Belaidi, A.A.; Schwarz, G.; Bernhardt, P.V. Chitosan-Promoted Direct Electrochemistry of Human Sulfite Oxidase. *J. Phys. Chem. B* **2017**, *121*, 9149–9159. [CrossRef]
132. Khoshtariya, D.E.; Dolidze, T.D.; Shushanyan, M.; Davis, K.L.; Waldeck, D.H.; Van Eldik, R. Fundamental signatures of short- And long-range electron transfer for the blue copper protein azurin at Au/SAM junctions. *Proc. Natl. Acad. Sci. USA* **2010**, *107*, 2757–2762. [CrossRef] [PubMed]
133. Vargo, M.L.; Gulka, C.P.; Gerig, J.K.; Manieri, C.M.; Dattelbaum, J.D.; Marks, C.B.; Lawrence, N.T.; Trawick, M.L.; Leopold, M.C. Distance dependence of electron transfer kinetics for azurin protein adsorbed to monolayer protected nanoparticle film assemblies. *Langmuir* **2010**, *26*, 560–569. [CrossRef] [PubMed]
134. Zhang, J.; Chi, Q.; Hansen, A.G.; Jensen, P.S.; Salvatore, P.; Ulstrup, J. Interfacial electrochemical electron transfer in biology—Towards the level of the single molecule. *FEBS Lett.* **2012**, *586*, 526–535. [CrossRef] [PubMed]
135. Alessandrini, A.; Facci, P. Electron transfer in nanobiodevices. *Eur. Polym. J.* **2016**, *83*, 450–466. [CrossRef]
136. López-Martínez, M.; Artés, J.M.; Sarasso, V.; Carminati, M.; Díez-Pérez, I.; Sanz, F.; Gorostiza, P. Differential Electrochemical Conductance Imaging at the Nanoscale. *Small* **2017**, *13*, 1–7. [CrossRef]
137. Chin, Q.; Zhang, J.; Nielsen, J.U.; Friis, E.P.; Chorkendorff, I.; Canters, G.W.; Andersen, J.E.T.; Ulstrup, J. Molecular monolayers and interfacial electron transfer of Pseudomonas aeruginosa azurin on Au(111). *J. Am. Chem. Soc.* **2000**, *122*, 4047–4055. [CrossRef]
138. Romero-Muñiz, C.; Ortega, M.; Vilhena, J.G.; Díez-Pérez, I.; Cuevas, J.C.; Pérez, R.; Zotti, L.A. Mechanical deformation and electronic structure of a blue copper azurin in a solid-state junction. *Biomolecules* **2019**, *9*, 506. [CrossRef]
139. Armstrong, F.A.; Barlow, N.L.; Burn, P.L.; Hoke, K.R.; Jeuken, L.J.C.; Shenton, C.; Webster, G.R. Fast, long-range electron-transfer reactions of a 'blue' copper protein coupled non-covalently to an electrode through a stilbenyl thiolate monolayer. *Chem. Commun.* **2004**, *4*, 316–317. [CrossRef]
140. Chi, Q.; Zhang, J.; Jensen, P.S.; Christensen, H.E.M.; Ulstrup, J. Long-range interfacial electron transfer of metalloproteins based on molecular wiring assemblies. *Faraday Discuss.* **2006**, *131*, 181–195. [CrossRef]
141. Zhang, J.; Welinder, A.C.; Hansen, A.G.; Christensen, H.E.M.; Ulstrup, J. Catalytic Monolayer Voltammetry and In Situ Scanning Tunneling Microscopy of Copper Nitrite Reductase on Cysteamine-Modified Au(111) Electrodes. *J. Phys. Chem. B* **2003**, *107*, 12480–12484. [CrossRef]
142. Welinder, A.C.; Zhang, J.; Hansen, A.G.; Moth-Poulsen, K.; Christensen, H.E.M.; Kuznetsov, A.M.; Bjørnholm, T.; Ulstrup, J. Voltammetry and Electrocatalysis of Achromobacter Xylosoxidans Copper Nitrite Reductase on Functionalized Au(111)-Electrode Surfaces. *Z. Phys. Chem.* **2007**, *221*, 1343–1378. [CrossRef]

143. Tang, J.; Yan, X.; Huang, W.; Engelbrekt, C.; Duus, J.Ø.; Ulstrup, J.; Xiao, X.; Zhang, J. Bilirubin oxidase oriented on novel type three-dimensional biocathodes with reduced graphene aggregation for biocathode. *Biosens. Bioelectron.* **2020**, *167*. [CrossRef] [PubMed]
144. Xiao, X.; Leech, D.; Zhang, J. An oxygen-reducing biocathode with "oxygen tanks. " *Chem. Commun.* **2020**, *56*, 9767–9770. [CrossRef] [PubMed]
145. Lopez, F.; Siepenkoetter, T.; Xiao, X.; Magner, E.; Schuhmann, W.; Salaj-Kosla, U. Potential pulse-assisted immobilization of Myrothecium verrucaria bilirubin oxidase at planar and nanoporous gold electrodes. *J. Electroanal. Chem.* **2018**, *812*, 194–198. [CrossRef]
146. Ramírez, P.; Mano, N.; Andreu, R.; Ruzgas, T.; Heller, A.; Gorton, L.; Shleev, S. Direct electron transfer from graphite and functionalized gold electrodes to T1 and T2/T3 copper centers of bilirubin oxidase. *Biochim. Biophys. Acta Bioenerg.* **2008**, *1777*, 1364–1369. [CrossRef]
147. Lalaoui, N.; De Poulpiquet, A.; Haddad, R.; Le Goff, A.; Holzinger, M.; Gounel, S.; Mermoux, M.; Infossi, P.; Mano, N.; Lojou, E.; et al. A membraneless air-breathing hydrogen biofuel cell based on direct wiring of thermostable enzymes on carbon nanotube electrodes. *Chem. Commun.* **2015**, *51*, 7447–7450. [CrossRef]
148. Al-Lolage, F.A.; Bartlett, P.N.; Gounel, S.; Staigre, P.; Mano, N. Site-Directed Immobilization of Bilirubin Oxidase for Electrocatalytic Oxygen Reduction. *ACS Catal.* **2019**, *9*, 2068–2078. [CrossRef]
149. McArdle, T.; McNamara, T.P.; Fei, F.; Singh, K.; Blanford, C.F. Optimizing the Mass-Specific Activity of Bilirubin Oxidase Adlayers through Combined Electrochemical Quartz Crystal Microbalance and Dual Polarization Interferometry Analyses. *ACS Appl. Mater. Interfaces* **2015**, *7*, 25270–25280. [CrossRef]
150. Gholami, F.; Navaee, A.; Salimi, A.; Ahmadi, R.; Korani, A.; Hallaj, R. Direct Enzymatic Glucose/O_2 Biofuel Cell based on Poly-Thiophene Carboxylic Acid alongside Gold Nanostructures Substrates Derived through Bipolar Electrochemistry. *Sci. Rep.* **2018**, *8*, 1–14. [CrossRef]
151. Yang, S.; Liu, J.; Quan, X.; Zhou, J. Bilirubin Oxidase Adsorption onto Charged Self-Assembled Monolayers: Insights from Multiscale Simulations. *Langmuir* **2018**, *34*, 9818–9828. [CrossRef]
152. Pankratov, D.; Sundberg, R.; Suyatin, D.B.; Sotres, J.; Barrantes, A.; Ruzgas, T.; Maximov, I.; Shleev, S. The influence of nanoparticles on enzymatic bioelectrocatalysis. *RSC Adv.* **2014**, *4*, 38164–38168. [CrossRef]
153. Cabrita, J.F.; Abrantes, L.M.; Viana, A.S. N-Hydroxysuccinimide-terminated self-assembled monolayers on gold for biomolecules immobilisation. *Electrochim. Acta* **2005**, *50*, 2117–2124. [CrossRef]
154. Balland, V.; Hureau, C.; Cusano, A.M.; Liu, Y.; Tron, T.; Limoges, B. Oriented immobilization of a fully active monolayer of histidine-tagged recombinant laccase on modified gold electrodes. *Chem. A Eur. J.* **2008**, *14*, 7186–7192. [CrossRef] [PubMed]
155. Pita, M.; Gutierrez-Sanchez, C.; Olea, D.; Velez, M.; Garcia-Diego, C.; Shleev, S.; Fernandez, V.M.; Lacey, A.L.D. High redox potential cathode based on laccase covalently attached to gold electrode. *J. Phys. Chem. C* **2011**, *115*, 13420–13428. [CrossRef]
156. Shleev, S.; Christenson, A.; Serezhenkov, V.; Burbaev, D.; Yaropolov, A.; Gorton, L.; Ruzgas, T. Electrochemical redox transformations of T1 and T2 copper sites in native Trametes hirsuta laccase at gold electrode. *Biochem. J.* **2005**, *385*, 745–754. [CrossRef]
157. Thorum, M.S.; Anderson, C.A.; Hatch, J.J.; Campbell, A.S.; Marshall, N.M.; Zimmerman, S.C.; Lu, Y.; Gewirth, A.A. Direct, electrocatalytic oxygen reduction by laccase on anthracene-2-methanethiol-modified gold. *J. Phys. Chem. Lett.* **2010**, *1*, 2251–2254. [CrossRef]
158. Traunsteiner, C.; Sek, S.; Huber, V.; Valero-Vidal, C.; Kunze-Liebhäuser, J. Laccase immobilized on a mixed thiol monolayer on Au(111)—Structure-dependent activity towards oxygen reduction. *Electrochim. Acta* **2016**, *213*, 761–770. [CrossRef]
159. Hakamada, M.; Takahashi, M.; Mabuchi, M. Enhanced thermal stability of laccase immobilized on monolayer-modified nanoporous Au. *Mater. Lett.* **2012**, *66*, 4–6. [CrossRef]
160. Krassen, H.; Stripp, S.T.; Böhm, N.; Berkessel, A.; Happe, T.; Ataka, K.; Heberle, J. Tailor-made modification of a gold surface for the chemical binding of a high-activity [FeFe] hydrogenase. *Eur. J. Inorg. Chem.* **2011**, 1138–1146. [CrossRef]
161. Gutiérrez-Sanz, Ó.; Tapia, C.; Marques, M.C.; Zacarias, S.; Vélez, M.; Pereira, I.A.C.; De Lacey, A.L. Induction of a Proton Gradient across a Gold-Supported Biomimetic Membrane by Electroenzymatic H_2 Oxidation. *Angew. Chem.* **2015**, *127*, 2722–2725. [CrossRef]

162. Goris, T.; Wait, A.F.; Saggu, M.; Fritsch, J.; Heidary, N.; Stein, M.; Zebger, I.; Lendzian, F.; Armstrong, F.A.; Friedrich, B.; et al. A unique iron-sulfur cluster is crucial for oxygen tolerance of a [NiFe]-hydrogenase. *Nat. Chem. Biol.* **2011**, *7*, 310–318. [CrossRef] [PubMed]
163. Cracknell, J.A.; Wait, A.F.; Lenz, O.; Friedrich, B.; Armstrong, F.A. A kinetic and thermodynamic understanding of O_2 tolerance in [NiFe]-hydrogenases. *Proc. Natl. Acad. Sci. USA* **2009**, *106*, 20681–20686. [CrossRef] [PubMed]
164. Sezer, M.; Frielingsdorf, S.; Millo, D.; Heidary, N.; Utesch, T.; Mroginski, M.A.; Friedrich, B.; Hildebrandt, P.; Zebger, I.; Weidinger, I.M. Role of the HoxZ subunit in the electron transfer pathway of the membrane-bound [NiFe]-hydrogenase from Ralstonia eutropha immobilized on electrodes. *J. Phys. Chem. B* **2011**, *115*, 10368–10374. [CrossRef] [PubMed]
165. Millo, D.; Pandelia, M.E.; Utesch, T.; Wisitruangsakul, N.; Mroginski, M.A.; Lubitz, W.; Hildebrandt, P.; Zebger, I. Spectroelectrochemical study of the [NiFe] hydrogenase from Desulfovibrio vulgaris miyazaki F in solution and immobilized on biocompatible gold surfaces. *J. Phys. Chem. B* **2009**, *113*, 15344–15351. [CrossRef]
166. Léger, C.; Jones, A.K.; Albracht, S.P.J.; Armstrong, F.A. Effect of a dispersion of interfacial electron transfer rates on steady state catalytic electron transport in [NiFe]-hydrogenase and other enzymes. *J. Phys. Chem. B* **2002**, *106*, 13058–13063. [CrossRef]
167. Gutiérrez-Sánchez, C.; Olea, D.; Marques, M.; Fernández, V.M.; Pereira, I.A.C.; Vélez, M.; Lacey, A.L.D. Oriented immobilization of a membrane-bound hydrogenase onto an electrode for direct electron transfer. *Langmuir* **2011**, *27*, 6449–6457. [CrossRef]
168. Zhang, B.; Song, W.; Pang, P.; Lai, H.; Chen, Q.; Zhang, P.; Lindsay, S. Role of contacts in long-range protein conductance. *Proc. Natl. Acad. Sci. USA* **2019**, *116*, 5886–5891. [CrossRef]
169. Lagunas, A.; Guerra-Castellano, A.; Nin-Hill, A.; Díaz-Moreno, I.; De la Rosa, M.A.; Samitier, J.; Rovira, C.; Gorostiza, P. Long distance electron transfer through the aqueous solution between redox partner proteins. *Nat. Commun.* **2018**, *9*, 3–9. [CrossRef]

Publisher's Note: MDPI stays neutral with regard to jurisdictional claims in published maps and institutional affiliations.

© 2020 by the authors. Licensee MDPI, Basel, Switzerland. This article is an open access article distributed under the terms and conditions of the Creative Commons Attribution (CC BY) license (http://creativecommons.org/licenses/by/4.0/).

Review

Membrane Protein Modified Electrodes in Bioelectrocatalysis

Huijie Zhang [†], Rosa Catania [†] and Lars J. C. Jeuken *

School of Biomedical Sciences and the Astbury Centre for Structural Molecular Biology, University of Leeds, Leeds LS2 9JT, UK; H.Zhang6@leeds.ac.uk (H.Z.); R.Catania@leeds.ac.uk (R.C.)
* Correspondence: L.J.C.Jeuken@leeds.ac.uk; Tel.: +44-113-343-3829
† These authors have contributed equally.

Received: 18 November 2020; Accepted: 3 December 2020; Published: 6 December 2020

Abstract: Transmembrane proteins involved in metabolic redox reactions and photosynthesis catalyse a plethora of key energy-conversion processes and are thus of great interest for bioelectrocatalysis-based applications. The development of membrane protein modified electrodes has made it possible to efficiently exchange electrons between proteins and electrodes, allowing mechanistic studies and potentially applications in biofuels generation and energy conversion. Here, we summarise the most common electrode modification and their characterisation techniques for membrane proteins involved in biofuels conversion and semi-artificial photosynthesis. We discuss the challenges of applications of membrane protein modified electrodes for bioelectrocatalysis and comment on emerging methods and future directions, including recent advances in membrane protein reconstitution strategies and the development of microbial electrosynthesis and whole-cell semi-artificial photosynthesis.

Keywords: membrane protein; bioelectrocatalysis; electrode modification; biofuel cells; photosynthesis; liposomes; hybrid vesicles; microbial electrosynthesis

1. Introduction

Membrane proteins constitute 20–30% of all proteins encoded by both prokaryotic and eukaryotic cells. They perform a wide variety of functions, including material transport, signal transduction, catalysis, proton and electron transport (Figure 1) [1]. They are also key to a number of earth's most fundamental reactions, such as respiration and photosynthesis [2,3]. Redox enzymes in the respiratory chain catalyse a variety of fundamental processes for energy conversion and fuel production, including H_2 oxidation, O_2 reduction, and carbon and nitrogen cycling. Membrane proteins that are involved in the light reaction of photosynthesis harvest light and facilitate electron transfer essential for solar energy conversion. The amphiphilic nature of membrane proteins makes them difficult to isolate, study and manipulate. Despite these challenges, membrane proteins have been widely advocated and studied for applications in bioelectrocatalysis, such as biofuel cells [4] and semi-artificial photosynthesis [5]. Here, we will review electrochemical studies of membrane proteins with the view to using these systems for bioelectrocatalysis. To aid discussion later on, we will briefly introduce a small selection of membrane enzymes active in bioenergy conversion, although this is far from a comprehensive list. We will then summarise the main strategies to immobilise membrane proteins on electrodes and discuss common techniques used to characterise these electrodes, including electrochemistry, spectroscopy, spectroelectrochemistry, microscopy and quartz crystal microbalance. Finally, some critical application challenges and potential future research directions will be highlighted that might find application in bioelectrocatalysis. Specifically, we will focus on two emerging directions. One is the reconstitution of membrane proteins into hybrid vesicles to extend their functional lifetime. The other is the use of microorganisms for microbial electrosynthesis and semi-artificial photosynthesis.

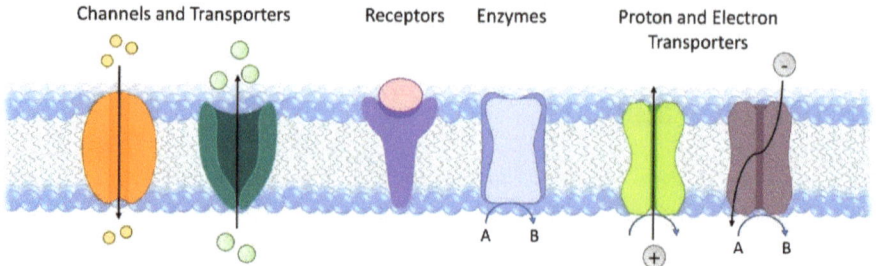

Figure 1. Representative functions of transmembrane proteins.

2. Membrane Proteins in Biofuel Conversion and Photosynthesis

2.1. Membrane Enzymes in Biofuel Conversion

Nature offers highly specialised enzyme machineries that can be exploited for (bio)fuel conversion. Among them, membrane-bound hydrogenases, found in many bacteria, archaea and lower eukaryotes, are metalloenzymes capable of catalysing the reversible oxidation of H_2 to protons and electrons [6]. Depending on the metal located in the active site, three phylogenetically unrelated classes can be identified: [NiFe], [FeFe] and, less common, [Fe] hydrogenases. Most hydrogenases are rapidly and almost completely inactivated by O_2 [7]. However, aerobic or facultative aerobic H_2-oxidising bacteria have a particular subtype of O_2-tolerant [NiFe] hydrogenase, which withstands the presence of O_2 and are membrane-bound hydrogenases (MBHs) [8]. The most studied O_2-tolerant [NiFe] hydrogenases are found in *Ralstonia eutropha*, *Ralstonia metallidurans*, *Aquifex aeolicus*, *Hydrogenovibrio marinus* and *Escherichia coli*. MBHs are multimeric proteins with one large subunit and one small subunit, bound to the periplasmic side of the cytoplasmic membrane through a transmembrane protein (*b*-type cytochrome) [8]. The [NiFe] active site, with the same configuration as other [NiFe] hydrogenases, is deeply buried in the large subunit. MBHs couple oxidation of hydrogen in the large subunit to the reduction of either menaquinone-7 or ubiquinone-8 in the *b*-type cytochrome in the membrane.

In the respiratory chain, terminal oxidases catalyse the reduction of molecular oxygen to water without formation of reactive oxygen species (ROS). They oxidise quinones (ubiquinone and menaquinone oxidases) or cytochromes (cytochrome *c* oxidases) and can be classed into haem-copper oxidases [9], *bd* oxidases [10] or alternative oxidases [11]. For instance, cytochrome bo_3 enzymes are haem-copper oxidases in bacteria such as *E. coli* and couple oxidation of ubiquinol-8 to the reduction of oxygen [12]. Cytochrome *bd* is a quinol-dependent terminal oxidase found exclusively in prokaryotes and is structurally unrelated to haem-copper oxidases [13].

Nitrate reductases are molybdoenzymes capable of reducing nitrate (NO_3^-) to nitrite (NO_2^-). This enzyme family can be found in eukaryotic and prokaryotic cells. Eukaryotic nitrate reductases are present in plants, algae and fungi, and are involved in the assimilation of nitrate. Prokaryotic nitrate reductases are classified into three classes: assimilatory nitrate reductases (Nas), periplasmic nitrate reductases (Nap) and respiratory nitrate reductase (Nar). The latter are transmembrane enzymes which use nitrate as the electron acceptor of an anaerobic respiratory chain [14]. Facultative anaerobic bacteria use these alternative respiratory enzymes in oxygen depleted environments, replacing oxygen with different electron acceptors [15]. For instance, under anaerobic conditions, *E. coli* expresses a respiratory membrane-bound NarGHI coupled with a membrane-bound formate dehydrogenases (FDH-N). These two membrane enzymes form the supermolecular formate:nitrate oxidoreductase system. FDH-N oxidises formate to CO_2, after which electrons are transferred to nitrate with the formation of a proton-motive force across the cytoplasmic membrane [16].

2.2. Membrane Proteins in Photosynthesis

Photosynthesis is the natural process by which photosynthetic organisms convert light energy into chemical forms. The key steps, which include light harvesting, charge separation and electron transport, occur in the membrane. For oxygenic photosynthesis in algae, higher plants and cyanobacteria, two photosynthetic complexes are involved: photosystem I (PSI) and photosystem II (PSII) [17], each with a peripheral antenna system: light harvesting complex I (LH I) for PSI and light harvesting complex II (LH II) for PSII [18]. Anoxygenic photosynthesis in bacteria, such as purple bacteria, is conducted by just one type of photosystem [19]. There are two types of light-harvesting complexes in most purple bacteria. The light-harvesting complex I and reaction centre forms the RC-LH I complex which is surrounded by multiple light-harvesting complex II [20]. In both oxygenic and anoxygenic photosynthesis, light-harvesting complexes exert the vital function to effectively absorb light and transfer the energy to the reaction centres [21]. Light-harvesting complexes absorb a limited spectral range depending on their natural pigment. The light-induced charge separation occurs at the reaction centres: PSII catalyses the light-driven water oxidation to reduce quinones, while the reaction centre of purple bacteria only catalyses the reduction of quinones (part of the cyclic electron transfer process). PSI is not catalytically active but instead serves as an "electron pump" which can be potentially coupled to the other redox catalysts. Electrodes interfaced with photosynthetic proteins have found broad applications in biosensors, biophotovoltaic cells and solar fuels generation [5,22,23].

3. Membrane Protein Electrode Design

Redox enzymes are molecular electrocatalysts that can function either in solution or while immobilised on an electrode surface. Immobilising enzymes on the electrodes has several advantages for electrocatalysis and this review will consider only immobilised systems. It is important that the enzymes retain their structural integrity and catalytic activity upon immobilisation [24]. Unlike soluble proteins, transmembrane proteins exist within lipid membranes and, consequently, are less stable in an aqueous environment where the amphiphilic properties of membrane proteins can lead to aggregation and denaturation. Therefore, it is challenging to retain the stability and function of membrane proteins in in vitro studies [25]. Suitable detergents or mixed lipid/detergent systems are needed to maintain an amphiphilic environment surrounding the membrane protein and mimic the original membrane conditions [26,27]. Alternatively, for some multisubunit, heteromeric membrane proteins the solubility problem can be circumvented by either purifying only the soluble subunits or engineering the protein to express only the soluble subunits [28]. For instance, the large and small subunits of membrane-bound hydrogenases can be separated from the transmembrane "anchor" subunit (b-type cytochrome). More sophisticated approaches have been developed for the mammalian membrane-bound cytochrome P450 that has been bioengineered without the hydrophobic membrane anchor domain [29]. However, similar methods are not always applicable for other proteins, either because the catalytic centre is located in a polytopic membrane subunit or because the water-soluble subunits on their own are not stable. It is worth mentioning that the activity of certain membrane proteins can be dependent on the presence of particular annular lipids [30] and therefore the strategies that improve the solubility by removing the need of the membrane environment can have some potential downsides.

Various immobilisation methods have been developed to achieve efficient electron transfer between a membrane protein and an electrode. There are several aspects to consider when designing and assembling (membrane) protein modified electrodes for bioelectrocatalysis: (1) orientation of the protein on the electrode surface, (2) preservation of the protein structural integrity and functionality, (3) low overpotential to minimise the energy loss, (4) protein loading of the electrode [24,31,32]. Some methods developed for electrocatalysis using soluble proteins can be adapted to detergent solubilised membrane proteins. An alternative strategy to the use of detergent is represented by reconstitution of membrane proteins within a lipid membrane on the electrode surface and several strategies have been developed to achieve this. We will discuss the most commonly used immobilisation methods.

3.1. Unmodified Electrode

Detergent solubilised membrane proteins can be absorbed directly on carbon, metal or semi-conductor electrodes in an approach known as protein film electrochemistry (Figure 2a) [33]. A frequently used electrode material is pyrolytic graphite "edge" (PGE) which has a rough, negatively charged surface, which can be tailored with polycations such as polymixin B or poly-L-lysine if required. Respiratory nitrate reductase (NarGHI) isolated from *Paracoccus pantotrophus* in *n*-dodecyl β-D-maltoside (DDM) buffer was adsorbed onto PGE electrodes and showed direct electron transfer (DET) with the electrode [34]. The catalytic activity of NarGHI from *E. coli* was studied over a wide pH range (5 < pH < 9), nitrate concentrations and in the presence of inhibitor [35]. Similarly, purple bacteria RC-LH I was solubilised in 0.1% lauryldimethylamine N-oxide (LDAO) buffer and immobilised directly on a bare Au electrode [36]. RC-LH I did not exhibit DET, but showed photocurrents (~64 nA/cm^2) when ubiquinone-0 and cytochrome *c* were present in solution, indicating that electron transfer needs to be mediated by small redox compounds in this system (mediated electron transfer, MET). Like water-soluble proteins, membrane proteins on bare electrodes often lack a specific orientation and this can impede efficient electron transfer between electrode and proteins. Making use of the amphiphilic nature of membrane proteins, a densely packed protein monolayer with defined orientation can be formed on electrode surfaces by Langmuir–Blodgett deposition and this was shown to increase the electron transfer efficiency [37–41]. An example of this strategy can be found in Kamran et al. where detergent solubilised-RC-LH I was first pre-assembled in a monolayer at a water-air interface and then transferred onto a gold-coated electrode, reaching photocurrent values of ~45 μA/cm^2 [37].

3.2. SAM Modified Electrode

An electrode surface can be modified to promote protein immobilisation and control the orientation of the protein on the surface. A common method to functionalise metallic electrode surfaces, in particular gold electrodes, is to form a self-assembly monolayer (SAM) of thiols (e.g., alkanethiols) (Figure 2b). By changing the length of the thiol compound (e.g., the alkyl chain), the distance between the protein and electrode can be controlled. More importantly, by changing the terminal group of the SAM (e.g., *n*-hydroxy-alkanethiol or *n*-amino-alkanethiol), the surface chemistry of the electrode can be controlled. As protein binding to surfaces is typically governed by van-der-Waals and electrostatic interactions, the surface chemistry strongly influences the orientation of proteins on the surface [42,43].

Besides tuning the chemistry of the electrode surface, proteins can be genetically engineered to control their orientation on the electrode through affinity interactions. RCs from *Rhodobacter sphaeroides* in LDAO detergent have been immobilised on SAM-modified electrodes terminated by Ni–NTA groups. By genetically engineering a poly-histidine tag (His7) at the C-terminus of the M-subunit of the RC, the primary donor of RC was positioned to face the electrode [44]. The histidine tag can also be engineered on H subunit of RCs to achieve the opposite orientation [45]. The Ni-NTA modified gold electrodes have also been used to immobilise His-tagged PSII solubilised in DDM buffer [46]. A different strategy that has been explored is to modify the gold electrode surface with an amine terminated SAM for further reaction with terephthaldialdehyde (TPDA). The TPDA modified SAM reacts with lysine residues from PSI to form covalent imine bonds [47]. Such an approach does not create the same orientational control compared to the His-tag/NTA coupling. Irrespective of selective orientation on the surface, mediators are often required for efficient electron exchange with membrane proteins. Small proteins like cytochrome *c* are widely used as mediators for PSI and purple bacteria RC [48]. Water-soluble redox mediators such as 2,6-dichloro-1,4-benzoquinone (PSII), 2,6-dichlorophenolindophenol (PSI), sodium ascorbate (PSI) and ferricyanide (PSI) are also commonly used to facilitate efficient electron transfer between the protein and electrode [49].

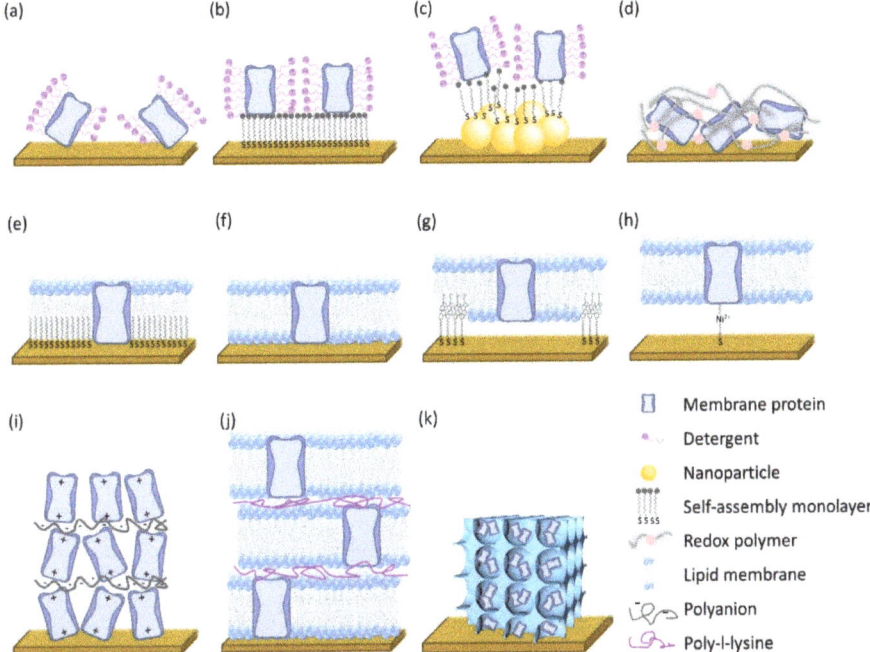

Figure 2. Overview of the strategies for membrane protein modified electrodes (not to scale). Proteins absorbed on (**a**) an unmodified electrode, (**b**) a SAM modified electrode and (**c**) a nanoparticle modified electrode. (**d**) Immobilisation of membrane proteins within a redox polymer. Examples of lipid membrane-modified electrodes: (**e**) a hybrid bilayer lipid membrane (hBLM), (**f**) a solid supported bilayer lipid membrane (sBLM), (**g**) a tethered bilayer lipid membrane (tBLM) and (**h**) a protein tethered bilayer lipid membrane. (ptBLM). (**i**) Layer-by-layer deposition of alternating charged films. (**j**) Multilayered lipid membrane stacks. (**k**) 3D structure electrode for protein immobilisation.

3.3. Nanoparticle Modified Electrode

Nanoparticles (NPs) have been shown to enhance interfacial electron transfer between a variety of electrodes and proteins [50] and gold NP-modified electrodes have been successfully used to study membrane proteins (Figure 2c), such as terminal oxidases: *E. coli* cytochrome *bd* (a quinol oxidase) [51], *Paracoccus denitrificans* cytochrome aa_3 [52] and *Thermus thermophilus* cytochrome ba_3 [53] (two cytochrome *c* oxidases). The use of NPs facilitated DET between the electrode and protein, enabling more in-depth studies on enzyme activity. The size of the NPs significantly affects the electron transfer rates and smaller particles reduced the requirement of overpotential for O_2 reduction activity by *E. coli* cytochrome bo_3 [54]. The group of Hellwig has recently studied the role of the surface charge of thiol-modified gold NPs, the length of the thiols and the effect of phospholipid composition on the interaction and DET between NP and the membrane enzyme, cytochrome *bd* [55]. Both cytochromes bo_3 and *bd* used in the aforementioned studies were isolated and stabilised in DDM buffer. Besides gold NP, various other particles have been used for protein immobilisation [56]. The Armstrong group used graphite microparticles to pair electron donor and acceptor membrane enzymes [57]. [NiFe] membrane-bound hydrogenase was reconstituted with either *E. coli* nitrate reductase (NarGHI) or *E. coli* fumarate reductase (FrdAB; not a membrane protein) on single microparticles; both systems catalysed the reductions of nitrate or fumarate, respectively, by hydrogen. In 2015, Duca et al. [58] demonstrated a cascade electrochemical reduction of nitrate to ammonia by immobilising *E. coli* respiratory nitrate reductase (NarGHI) on an electrode (in polyoxyethylene 9-dodecyl ether detergent) with Pt or Rh nanoparticles.

3.4. Redox Polymers

Redox polymers are widely used for wiring redox proteins on electrode surface (Figure 2d) [59]. Redox polymers act simultaneously as immobilisation matrix and as redox mediators. High coverages of proteins can be achieved and the ubiquitous localisation of redox mediators bound to the polymer matrix negates the need to control protein orientation. These polymer-bound redox mediators overcome mass transport limitations, typically observed with freely diffusing redox mediators. The chemical and physical properties of the redox polymers can be tailored by tuning the polymer backbone and the redox mediators. The Schuhmann group has extensively explored Os redox polymers to "wire" DDM-solubilised PSI [60] and PSII [61] on electrode surfaces for photocurrent generation. A two-compartment cell with a PSII photoanode and a PSI photocathode was constructed to mimic the Z-scheme of natural photosynthesis and an open-circuit voltage (OCV) of 90 mV was achieved [62]. By tuning the redox potential of Os complexes in the redox polymers to match the redox sites of the proteins (PSII and PSI), the OCV could be increased from 90 mV to 372 mV [63]. Interestingly, by exploiting the pH-dependent properties of Os-modified polymer, the group of Plumeré improved the interfacial electron transfer rates for the PSI photocathode, even exceeding rates observed in natural photosynthesis [64].

3.5. Membrane Modified Electrode

Because of the nature and functions of the membrane proteins, it is important that their assembly on electrodes preserves their structural integrity and functionality. In the examples provided so far, detergents are used to retain stability of the immobilised membrane enzymes. Although detergents are required to accommodate the amphiphilic nature of membrane proteins, they also have adverse effects to membrane protein stability and are likely to influence the electrode-protein interaction. To mimic the native environment of membrane proteins, electrodes have been modified with model membranes, as reviewed previously [65–67]. These membrane-modified electrodes can be categorised into hybrid bilayer lipid membrane (hBLM) system (Figure 2e), solid supported bilayer lipid membrane (sBLM) system (Figure 2f), tethered bilayer lipid membrane (tBLM) system (Figure 2g), and protein tethered bilayer lipid membrane (ptBLM) system (Figure 2h).

In hBLM systems a phospholipid layer is absorbed onto a self-assembled monolayer (SAM) of alkylthiols (Figure 2e). This strategy was used by the Hawkridge group to study cytochrome c oxidase immobilised on gold electrode [68,69]. In sBLMs, a lipid bilayer is non-covalently bound to the electrode surface (Figure 2f). Photosynthetic reaction centres including *Rhodobacter sphaeroides* RC [70,71], spinach photosystem I [72] and photosystem II [73] have been integrated into sBLMs on pyrolytic graphite. Electron transfer between electrode and the aforementioned proteins was achieved and showed well-defined peaks in voltammetry which corresponded to the redox sites of the proteins. Noji et al. incorporated *Rhodopseudomonas palustris* RC-LH I into sBLMs on an ITO electrode. Anionic phospholipid like phosphatidylglycerol (PG) were shown to stabilise the charge-separated state of RC-LH I and enhance the photocurrent [74]. In tBLM systems, the membrane is 'tethered' to the lipid-modified electrode surface via a linker (Figure 2g). Cytochrome bo_3 has been incorporated into tBLMs and was shown to retain its catalytic activity [75]. We previously included the membrane-bound [NiFe]-hydrogenases (MBH) from *R. eutropha* into a tBLM approach and used electrochemistry to study the activity of the entire heterotrimeric membrane-bound form of the enzyme [76]. In 2016, Pelster and Minteer isolated mitochondrial electron transport chain (ECT) enzymes, reconstituted them into liposomes and immobilised them onto a gold electrode in a tBLM. The authors used this reconstructed mitochondrial inner membrane biomimic to show the interdependence of the different complexes on bioelectrocatalytic activity [77]. In ptBLMs, the membrane protein is first anchored to an electrode via a His-tag/NTA interaction and subsequently reconstituted into a lipid membrane (Figure 2h). Ataka et al. have immobilised the *Rhodobacter sphaeroides* cytochrome aa_3 (a cytochrome c oxidase) by using a ptBLM system [78–80]. Two opposite protein orientations were compared, by varying the position of the His-tag. Cytochrome c bound and exchanged electrons with cytochrome c oxidase only when the latter was orientated with its subunit II (the binding domain for cytochrome *c*) facing the bulk aqueous medium.

3.6. Multi-Layer Assembly, Multilayered Lipid Membrane Stacks, 3D Structure Electrode

An important aspect to improve the performance of bioelectrocatalysis is catalyst loading on the electrode surface. For membrane proteins, this can be achieved via two alternative strategies: multi-layer (membrane) assembly (Figure 2d,i,j) and/or 3D electrode structure (Figure 2k). For a more detailed discussion on layer-by-layer assembly approach for redox proteins, we also refer to [65,81]. Multilayer immobilisation within a redox polymer matrix was already discussed in Section 3.4. As an example of a layer-by-layer approach (Figure 2i), PSI has been co-assembled with cytochrome *c* as mediator using DNA as an anionic polymer [82]. The purple bacteria RC was also co-assembled into a multilayer architecture by alternating layers of RC with a cationic polymer poly(dimethyldiallylammonium chloride) (PDDA) [83,84]. Our group used poly-L-lysine (PLL) as an electrostatic polymer to construct multilayered lipid membrane stacks (Figure 2j) [85,86]. Two membrane proteins, *E. coli* cytochrome bo_3 and *R. eutropha* MBH, were incorporated into these lipid membranes stacks. Lipophilic quinones, the natural substrates of cytochrome bo_3 and MBH, diffuse freely between the multilayered membranes and mediated electron transfer between the proteins and the electrode. Catalytic activity was shown to increase linearly with the number of membrane layers for at least up to 5 layers [86].

The high surface area of so-called 3D electrodes can be used to immobilise proteins at high loading [87]. Mesoporous metal electrodes have been shown to increase protein loading compared to planar electrodes, e.g., rough silver electrode for bacteria RC [88] and nanoporous gold electrode for PSI [89]. A mesoporous WO_3-TiO_2 film electrode has been reported for the entrapment for bacterial RCs [90] and mesoporous indium tin oxide (ITO) electrodes have been used to immobilise PSII [91]. The ITO electrode can be modified with SAM to covalently bind and orientate PSII with the electron acceptor side of towards the electrode surface, enhancing electron transfer kinetics and electrode stability [92]. A hierarchically structured, inverse opal, mesoporous (IO-meso) ITO electrode was later developed to provide even larger surface areas [93]. These electrodes were combined with redox polymers to electrically wire PSII with the 3D structure of IO-meso ITO, yielding photocurrent densities of up to ~410 $\mu A \cdot cm^{-2}$ [94]. Mesoporous ITO electrodes were also applied to co-immobilise cytochrome *c* and PSI and photocurrent densities >150 $\mu A \cdot cm^{-2}$ were achieved [95]. The effect of pore size was studied by comparing photocurrents between mesoporous (20–100 nm) and macroporous (5 μm) electrodes using 2,6 dichlorophenolindophenol (DCPIP) and ascorbate as redox mediator for PSI. The macroporous electrode showed three times higher photocurrent than the mesoporous electrode [96]. The authors observed that the macroporous electrode increased the active surface area twice compared to the mesoporous electrode with the same PSI mass loading. They concluded that the increase in photocurrent was due to multilayers of PSI deposited along pore walls and the macropores enhanced the MET within a single pore.

4. Methods to Characterise Membrane Protein Modified Electrode

The characterisation of electrodes and protein films is very often achieved through a combination of powerful and advanced techniques and their combination with electrochemical tools. In this section, we will focus on the most commonly used techniques to study membrane proteins on electrode surfaces and we will report some representative works.

4.1. Electrochemical Methods

Protein film electrochemistry (PFE) is a well-established and important technique to study protein modified electrodes and has proven powerful to probe the thermodynamic, kinetic and catalytic properties of redox proteins, typically with voltammetry [97–99]. PFE is also compatible with membrane-modified electrode and allows to probe the membrane proteins within detergent solutions, polymer matrices or model-membrane environment [100,101]. We refer to the following reviews for a detailed description of PFE methods [102–106].

Besides voltammetry, impedance spectroscopy has been used to study the electrode-membrane protein interface. The quality and the structure of electrode surfaces modified with planar lipid membranes (Figure 2e–h) are particularly suitable for investigation by electrochemical impedance spectroscopy (EIS). EIS can monitor the resistance and the capacitance of a planar membrane covering the electrode. For ideal planar lipid bilayers on the electrode, the capacitance value should be in the range of ~0.5 $\mu F \cdot cm^{-2}$ with a resistance of >MΩ cm^2 [107]. Disorders and defects of these bilayers will result in lower resistance and/or higher capacitance values. For instance, we monitored the formation of multilayer membrane stacks (Figure 2j) by EIS [86] and observed only small reductions in capacitance upon the formation of each additional bilayer, indicating that the additional lipid bilayers permeable to ions and thus contain large or many defects. EIS can also provide information on whether protein incorporation in the membranes affects the electrode structure. For instance, incorporation of cytochrome bo_3 into tethered membranes (tBLM, Figure 2g) had almost no effect on the capacitance, indicating that cytochrome bo_3 did not induce large defects in the tBLM [75]. Similarly, only small changes in capacitance were observed of a hBLM (Figure 2e) after integrating human enzyme cytochrome P450, indicating that the hBLM retains its integrity upon protein adsorption [108]. Besides the membrane, the quality of a SAM on metal electrodes is also often evaluated with EIS. Compact, well-formed SAMs act as ideal insulating layers with high resistance values. The compactness of the SAM will influence subsequent protein immobilisation steps and can determine the thickness of the SAM, impacting on interfacial electron transfer kinetics.

Protein-film photoelectrochemistry (PF-PEC) is a PFE technique which has been developed for photosynthetic proteins [22,109–111]. PF-PEC combines illumination with PFE to investigate photoactive proteins. Voltammetry and chronoamperometry can be used to illustrate the charge transfer processes and the kinetics of the light driven reactions. Recently, other electrochemical setups have been used to investigate photosynthetic proteins, including rotating ring disk electrode (RRDE) and scanning electrochemical microscopy (SECM). The set-up of RRDE includes a central rotating disc electrode and a ring electrode surrounding it. The potential on these two electrodes can be controlled independently and products generated by a protein film at the central disk electrode are transferred to the ring electrode for electrochemical analysis. For instance, oxygenic photoreactivity of PSII was studied with RRDE [112] and products generated by PSII at the central disc, e.g., oxygen and radical species, were detected by the ring electrode. This study revealed ET pathways that generate reactive oxygen species and O_2 by PSII. SECM employs a microelectrode as a tip above the protein modified electrode to detect the products generated locally. It was used to monitor H_2 evolution of a PSI-Pt complex within redox polymer under illumination [113]. SECM has also been used to quantify the reduction of charge carriers (methyl viologen) by PSI and compared this to the photocurrent [114]. Methyl viologen is often used as a charge carrier to collect electron from PSI in biophotovoltaic systems, but reoxidation on the electrode leads to charge recombination. The authors compared PSI on gold and silicon surfaces, both with Os-based redox polymers. SECM showed that gold and silicon exhibits different photocurrents due to different charge recombination kinetics, i.e., electron transfer kinetics for the reoxidation of the methyl viologen radical cation. Finally, by monitoring the photocurrent and H_2O_2 generation of a PSI photocathode with SECM, it was shown that light-induced formation of reactive oxygen species caused degradation of PSI modified electrodes [115].

4.2. Spectroscopic Methods

While electrochemical methods provide information about redox reactions, in situ spectroscopic methods can be applied to characterise the structure of the protein and the electrode assembly. Surface-enhanced Raman spectroscopy (SERS) is a surface-sensitive technique for molecules and proteins immobilised on roughened metallic surfaces, in particular Ag [116]. The SERS effect is due to a large increase in the Raman cross-section of molecules in contact with the metal surface, which is enhanced at metallic nanostructured surfaces due to local surface plasmon resonance (LSPR) effects. SERS has been used to successfully characterise oxidised and reduced forms of adsorbed cytochromes (cytochrome c, haemoglobin and myoglobin) under precise potential control with low laser power [117]. In 2005, Hrabakova et al. showed that no structural changes occurred to the haem sites (e.g., a and a_3) of cytochrome c oxidase embedded in a phospholipid bilayer tethered to a functionalised silver electrode via a histidine-tag [118]. In 2011, Weidinger group used SERS to study the electron transfer of MBH from R. eutropha H16 [119]. Interestingly, the authors compared the behaviour of the entire enzyme with just the transmembrane "anchor" subunit (b-type cytochrome) and determined two independent pathways for the electrons from the active site of the enzyme to the electrode. A slow rate pathway crosses all the three subunits of the enzyme, whilst a faster pathway only crosses two subunits and leaves out the transmembrane anchor subunit which, however, contributes to the stabilisation of the enzyme on the electrode.

A different variation of conventional infrared spectroscopy is represented by surface-enhanced infrared absorption spectroscopy (SEIRAS) in which the signal enhancement is due to plasmon resonance from a nanostructured metal thin film [120]. The Hellwig group used SEIRAS to characterise the deposition of gold NPs on gold electrodes, the modification of gold NPs with thiols and, finally, the absorption of cytochrome bo_3 on these modified electrodes [51,54]. Wiebalck et al. characterised the formation of a tBLM and incorporation of functional cytochrome bo_3 from E. coli by SEIRA spectroscopy [121].

4.3. Spectroelectrochemistry

Spectroelectrochemistry combines electrochemistry and spectroscopic approaches to monitor a specific molecular response while varying the electrode potential [122,123]. This can be done with proteins in solution [124] or immobilised on the electrode surface [125,126]. Insights into reaction mechanisms of membrane enzymes could be gained by methods such as UV-vis spectroelectrochemistry and infrared spectroelectrochemistry. UV-vis spectra of a protein are often specific for the redox state of a cofactor, e.g., haems, FeS clusters or flavins. A common application is a spectroelectrochemical titration to determine the reduction potential of a cofactor in an enzyme in solution, where results are usually evaluated by fitting the titration data to the Nernst equation. An example is a study of the cytochrome bc_1 complex where the midpoint potentials of the cofactors were determined by UV-vis and IR spectroelectrochemical titrations [127,128]. The redox potentials of the primary electron acceptor pheophytin a in photosystem II [129] and the primary electron donor P700 in photosystem I [130] were also revealed by UV-vis spectroelectrochemistry with proteins in solution. UV-vis spectroelectrochemistry can also be used to study proteins on electrode surfaces. For instance, Haas et al. used UV-vis spectroelectrochemical titrations to determine the midpoint potentials for cytochrome c oxidase either in solution or in a Langmuir-Blodgett monolayer film [131]. A multi-protein system was studied with UV-vis spectroelectrochemistry in which an outer-membrane cytochrome, MtrC from Shewanella oneidensis MR-1 (a peripheral membrane protein), was used as an electron conduit between an electrode and other redox enzymes. MtrC was shown to be an effectively transfer-electron conduit by monitoring the absorbance of reduced Fe^{II}-haems in MtrC [132].

Infrared (IR) spectroscopy is a powerful technique to detect structural changes of proteins during or in response to an electrochemical reaction although few studies have been reported for membrane proteins. Reactions can be triggered either by light or redox potential and infrared difference spectroscopy of proteins is typically measured to investigate the reaction mechanisms. This approach has been developed for membrane proteins involved in photosynthesis, respiration and metabolic pathways [100]. Electrochemical SEIRA was used to study the conformational changes of the R. sphaeroides cytochrome c oxidase induced by direct electron transfer in ptBLM system [133]. The cytochrome c oxidase was shown to change from a non-activated to an activated state after it involving enzymatic reaction. By applying periodic potential pulses switching between −800 mV and open circuit potential to control the state of cytochrome c oxidase, the kinetics of the conformational changes was monitored by time-resolved SEIRA spectroelectrochemistry [134]. The bacterial respiratory ubiquinol/cytochrome bo_3 couple was incorporated into a tethered bilayer lipid membrane (tBLM) on SAM modified electrode. The transmembrane proton gradient was successfully monitored by spectroelectrochemical SEIRA [121].

4.4. Microscopy

4.4.1. Electron Microscopy

Electron microscopy (EM) techniques are available to investigate membrane proteins (typically by cryo transmission electron microscopy, TEM or cryo-EM) and electrode structures (typically by scanning electron microscopy, SEM) at various scales. However, EM has not been widely used to study membrane proteins immobilised on electrodes as TEM and SEM are based on different modes of observation [135]. In 2015, Monsalve et al. [136] used SEM and TEM to characterise the morphology and size distribution of gold nanoparticles on a gold electrode used for direct absorption of *Aquifex aeolicus* [NiFe] membrane-bound hydrogenase surrounded by DDM detergent.

4.4.2. Atomic Force Microscopy

Atomic force microscopy (AFM) is a scanning probe microscopic technique based on a nanoprobe, a tip placed at the end of a long and narrow cantilever, that interacts with the sample surface to measure the topography of a surface [137]. Although AFM has been widely used to study lipid membranes [138–140], membrane proteins [141–144] and electrochemical systems with soluble proteins [145–147], AFM is less often used to study membrane proteins on electrode surfaces. In 2006, we used tapping-mode AFM to analyse the distribution of cytochrome bo_3 on a tethered bilayer lipid membrane (tBLM) on stripped gold electrode [75]. AFM combined with Polarisation Modulation Infrared Reflection-Absorption Spectroscopy (PM-IRRAS) was used to study the different orientation of membrane-bound *Aquifex aeolicus* (Aa) [NiFe] hydrogenase immobilised on hydrophilic and hydrophobic SAM on gold electrodes. This work highlighted that on charged or hydrophilic interfaces, H_2 oxidation proceeds through both direct and mediated electron transfer processes, while on hydrophobic surfaces, a mediator is required [148]. In 2014, Gutiérrez-Sanz et al. [149] characterised the functional reconstitution of respiratory complex I on SAM on gold electrode. Their AFM study showed the formation of a phospholipid bilayer on SAM modified gold electrode and protrusions of 6–8 nm height were observed which were ascribed to the hydrophilic arm of complex I as this arm extends outside the membrane.

4.5. Quartz Crystal Microbalance

Quartz crystal microbalance (QCM) is a sensitive mass sensor which utilises acoustic waves generated by oscillating a piezoelectric, single crystal quartz plate to measure mass changes in the order of nanograms. The association of QCM with dissipation monitoring (QCM-D) allows to also measure the energy loss or dissipation (ΔD) of the system [150]. QCM was used to monitor the different stages of the ptBLM formation with cytochrome *c* oxidase from *Rhodobacter sphaeroides* as model protein [151]. The adsorption of cytochrome *c* oxidase at the surface decreased the resonance frequency while increased the dissipation. The Hawkridge group investigated the electrostatic association between cytochrome *c* and cytochrome *c* oxidase immobilised in hBLM with QCM [68,69]. The QCM data revealed the binding of cytochrome *c* and cytochrome *c* oxidase at different ionic strength which was related to the mediated electron transfer. QCM-D was also used to characterise the formation of the cystamine–pyrroloquinoline quinone–thylakoids layers onto SAMs [152].

5. Membrane Protein Modified Electrodes for Bioelectrocatalysis

The increased demand to produce energy and value-added chemicals from cheap and environmentally friendly renewable resources has driven the recent advances in bioelectrocatalysis research towards the development of alternative systems. The applications of bioelectrocatalysis range from biosensors, energy conversion devices and bioelectrosynthesis [153–156]. The majority of the studies employ soluble proteins; however, some membrane proteins with suitable catalytic or electron transfer properties also find applications in bioelectrocatalysis. Some of them show advantages over soluble proteins or have unique functions. Here, we will discuss the membrane protein electrode applications in the field of bioelectrocatalysis.

5.1. Enzymatic Biofuel Cells (EBCs)

One of the most studied energy conversion devices is the enzymatic biofuel cell (EBC) which uses oxidoreductase enzymes as biocatalysts to convert chemical energy into electrical energy [157–159]. A typical EBC consists of a two-electrode cell in which the biofuels (such as H_2, formate) are oxidised at the bioanode and the oxidants are reduced at the biocathode (usually O_2 is reduced to water) [160]. Among these, H_2/O_2 biofuel cells are one of the most investigated enzymatic systems [161]. Hydrogenases are promising biocatalysts to fabricate high performance H_2-oxidation bioanodes. However, the extreme oxygen sensitivity of highly active hydrogenases is one of the main limitations of hydrogenase based H_2 bioanodes [162]. The O_2 sensitive hydrogenase can be protected by a low-potential viologen redox polymer matrix [163] or enzymatic O_2 scavenger [164]. However, these protection mechanisms either consume the electron from H_2 oxidation or add chemicals for protection. O_2-tolerant MBH therefore shows great advantage for H_2/O_2 biofuel cell. The first membrane-less H_2/O_2 cell was assembled by Armstrong group. Two pyrolytic graphite electrodes, coated respectively with MBH from *R. eutropha* and laccase from *Trametes versicolor* (*Tv*), were immersed in a H_2/air flushed solution. The system reached an OCV of ~970 mV and a maximum power output of ~5 µW [165]. High OCVs were achieved because MBH directly exchanged electrons with the electrode and no mediators were required. The *Re* MBH was solubilised from the membrane extracts by 2% Triton-X 114 and isolated via a *Strep*-tag sequence on the small subunit. It is possible that the MBH was devoid of the transmembrane cytochrome *b* anchor and this could explain why there was no need to use mediators. The *Re* MBH also showed CO-tolerance [165]. The same group further improved the performance of the H_2/O_2 cell by replacing the *Re* MBH with MBH (similarly *Strep*-tagged on the small subunit) from *R. metallidurans*, which is more active and stable to O_2 exposure (Figure 3a). The authors emphasised that such a system would have the advantage to perform H_2 conversion even in H_2-poor mixtures. They also showed that three cells in series provided a total OCV of 2.7 V which was sufficient to power a wristwatch for 24 h [166].

As described earlier, we have presented a strategy for using the full heterotrimeric MBH as a biocatalyst (including the cytochrome *b* anchor) and have used multi-layer membrane stacks on gold electrode to increase MBH loading [86]. In our approach, we used cytoplasmic membrane extracts of *R. eutropha* and created interconnected layers of membranes with each layer containing anchored MBHs. Lipophilic quinones were used as mediator, shuttling electron between electrode and protein. However, the irreversible electrochemical behaviour of the quinone redox reaction increases the required overpotential for H_2 oxidation, which will limit the power output of the devices and more research is needed to resolve this.

To optimise EBCs, gas diffusion electrodes have been investigated and enzyme coverage optimised. The porous structure of the gas diffusion electrode can increase the mass loading of the biocatalyst and overcome the mass transport limitation of gases. In 2016, Kano group used an O_2-tolerant MBH from *Hydrogenovibrio marinus* and an O_2-sensitive [NiFe]-hydrogenase from *Desulfovibrio vulgaris* "Miyazaki F" to create DET-type gas diffusion electrodes [167]. The authors did not specifically comment on whether their enzyme purification methods might have affected the presence of the cytochrome *b* anchor subunits of MBH. The MBHs were isolated through two different procedures. The large and small subunits of the O_2-tolerant MBH were isolated through detergent solubilisation and maintained in 0.025% Triton-X. The O_2-sensitive MBH went through a trypsinisation process which could lead to the separation of the transmembrane cytochrome *b* anchor. These H_2 oxidation electrodes generated a current density of 10 mA·cm^{-2} in the half-cell configuration. Contrary to the O_2-tolerant MBH, the O_2-sensitive MBH did not show overpotential for H_2 oxidation and, on this basis, was selected by the authors as bioanode of a H_2/O_2 EBC for a further study. Coupling this O_2 sensitive MBH anode with bilirubin oxidase (BOD) from *Myrothecium verrucaria* immobilised on Ketjen black-modified waterproof carbon papers (KB/WPCC) electrode, a dual gas-diffusion membrane- and mediator-less H_2/air-breathing biofuel cell was constructed (Figure 3b) which showed maximum power density in the range of 6.1 mW·cm^{-2} at 0.72 V [168].

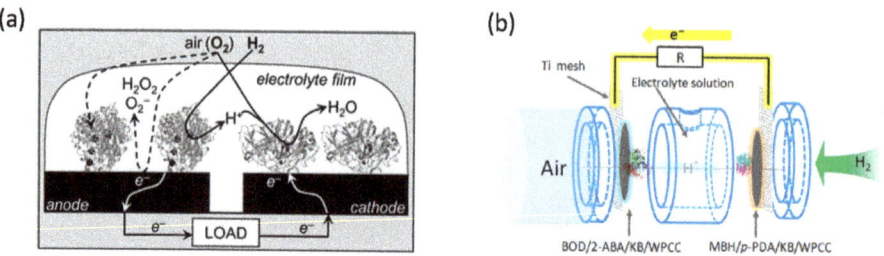

Figure 3. Schematic representations of enzymatic biofuel cells (EBC). (**a**) EBC comprises of graphite electrodes modified with O_2-tolerant MBH of *R. metallidurans* CH34 (anode) and fungal laccase (cathode) in aqueous electrolyte exposed to 3% H_2 in air. Reprinted with the permission from ref [166]. Copyright (2006), The Royal Society of Chemistry. (**b**) A dual gas diffusion membrane-free H_2/air powered EBC comprises of a [NiFe]-MBH (anode) and a bilirubin oxidase (cathode). Reprinted with the permission from ref [168]. Copyright (2016), Elsevier B.V.

The O_2-reduction biocathodes for EBC are usually based on multi-copper oxidases like bilirubin oxidase or laccase which can reduce O_2 almost without overpotential [169]. Among the membrane proteins, cytochrome *c* oxidase has been studied as O_2 reduction catalyst. Katz and Willner assembled a membrane-less glucose/O_2 biofuel cell with cytochrome *c*/cytochrome *c* oxidase as O_2 reducing cathode [170]. The cytochrome *c* was assembled on a maleimide modified gold electrode through a cysteine residue to link cytochrome *c* oxidase. In a follow-up work, the authors developed an electroswitchable and tunable biofuel cell. In this case, the cathode was modified with poly(acrylic acid) loaded with Cu^{2+} to covalently attach cytochrome *c* and link the latter to cytochrome *c* oxidase [171]. Although cytochrome *c* was able mediated electron transfer to cytochrome *c* oxidase for O_2 reduction, only an OCV of 0.12 V was obtained, likely limited by the redox potential of cytochrome *c* (~0.25 V vs. SHE), which is much lower than that of H_2O/O_2 (0.82 V vs. SHE at pH 7). Although cytochrome *c* oxidase is not commonly used for O_2 reducing cathode in biofuel cell, the recent work with gold NPs by the group of Hellwig has shown renewed possibilities for cytochrome *c* oxidase as catalyst (see Section 3.3). Recently, Wang et al. [172] reported that cytochrome *c* oxidase from acidophilic bacterium *Acidithiobacillus ferrooxidans* can reduce O_2 at exceptionally high electrode potentials (+700 to +540 mV vs. NHE). The low overpotential for O_2 reduction of this cytochrome *c* oxidase makes it an attractive biocatalyst as cathode of biofuel cells in the future.

Besides the use of H_2 as fuel for EBC, formate is also a valuable feedstock for biofuel cells because its redox potential is similar to H_2/H^+. The Kano group reported a mediated electron transfer type formate/O_2 biofuel cell by coupling formate dehydrogenase modified bioanode with BOD modified biocathode [173]. In nature, some formate dehydrogenases can catalyse the inverse reaction to reduce CO_2 to formate [174]. However, to the best of our knowledge, the research in this field has been conducted only on soluble formate dehydrogenases. The ability of CO_2 reduction and the applications of membrane-bound formate dehydrogenase need to be further explored.

Respiratory nitrate reductase (Nar) catalyses NO_3^- reduction to NO_2^-, this is an essential step to produce NH_3. Today there is no enzyme known to reduce NO_3^- to NH_3 directly [175]. DET with nitrate reductase has been shown to support the electrochemical reduction of NO_3^- to NO_2^- [34,35]. A full reduction of NO_3^- to NH_3 was demonstrated by a cascade electrocatalysis process which combined nitrate reductase and noble metal catalyst to reduce NO_2^- to NH_3 [58].

5.2. Biophotoelectrocatalysis (PEC)

Biophotoelectrodes fabricated on planar carbon or SAM-modified metal electrode usually show low photocurrent density which limits their applications in bioelectrocatalysis. However, the development of redox polymer electrodes, layer-by-layer assembly and 3D architectures has enhanced the performance of biophotoelectrodes [176,177]. PSII is the only natural protein able to catalyse the photooxidation of water. The electron flow can be blocked by herbicide compounds since they bind to the terminal plastoquinone Q_B of PSII. This inhibition effect can be exploited for designing PSII light-driven biosensor to detect herbicides [178]. A similar approach to detect herbicides was taken with purple bacteria RCs [179].

As water oxidation biophotocatalyst, PSII attracts more attention for solar energy conversion to generate electricity or fuels. As mentioned above, PSII photoanodes have been connected to PSI photocathodes (both in Os redox polymers) to mimic natural photosynthesis Z-scheme for solar-to-electricity generation [62,63]. In a similar approach using a benzoquinone redox polymer, a PSII photoanode (O_2 evolution) has been connected to a bilirubin oxidase cathode for O_2 reduction [180]. Recently, PSII was integrated together with PbS quantum dots within a TiO_2 inverse opal electrode to perform H_2O photooxidation. This was combined with an inverse opal antimony-doped tin oxide (ATO) cathode modified with bilirubin oxidase to catalyse oxygen reduction. This system achieved a high open-circuit voltage of about 1V under illumination (Figure 4a,b) [181].

A full water splitting process can be realised by combining a PSII photoanode with a cathode modified with hydrogenases. In contrast to EBCs (Section 5.1), water-soluble hydrogenases (typically [FeFe] hydrogenases, but also [NiFe] hydrogenases; H2ases) are most commonly used for water splitting systems. When combining PSII and H2ases, there is an energy gap between the terminal electron acceptor Q_B within PSII and one of the FeS cluster within H2ase. A biophotoelectrochemical cell with a PSII photoanode and a H2ase cathode thus requires an applied bias voltage of 0.8 V to drive H_2O splitting [93]. This was improved by wiring H2ase and p-Si on an inverse opal TiO_2 photocathode, lowering the required applied bias voltage to 0.4 V for H_2O splitting [182]. Finally, by integrating PSII on a diketopyrrolopyrrole dye-sensitised TiO_2 photoanode and connecting it with a H2ase cathode, a bias-free photoelectrochemical cell for H_2O splitting was developed by the group of Reisner [183]. Coupling a dye-sensitised PSII photoanode with a W-dependent formate dehydrogenase (FDH) cathode, a biophotoelectrochemical cell was constructed for CO_2 reduction at a small bias voltage of 0.3 V (Figure 4c,d) [184]. The latter study showed the possibility for rational design of biophotoelectrochemical cells for value-added chemicals generation beyond H_2.

Unlike PSII, PSI does not directly catalyse technologically valuable reactions such as water oxidation. However, photoexcitation of PSI provides the reductive potential to drive reactions with other catalysts, such as Pt and H2ase [185]. Recent studies show that it is possible to drive H_2 production from light with PSI and H2ase by electrode design. Photoelectrodes were manufactured in a 'layered' fashion using an Os redox polymer, PSI and, finally, a polymer/H2ase mix [186]. Photoelectrochemical H_2 production is achieved at an onset potential of +0.38 V vs. SHE. In this study, the PSI was randomly orientated and did not form a compact layer, likely limiting efficiency by charge recombination between the carrier and mediator or electrode. In a more recent study, an anisotropically oriented PSI monolayer was formed using Langmuir-Blodgett deposition [187]. A compact and oriented PSI layer minimises charge recombination and enables unidirectional electron transfer to H2ase for H_2 evolution. Combining this PSI/H2ase photocathode with a PSII photoanode created a system able of bias-free light-driven water splitting [187]. Langmuir-Blodgett deposition transfers only a monolayer of PSI and this might limit the performance of the biophotoelectrode as this limits the loading or coverage of PSI.

One of the limitations with biophotoelectrodes is the limited lifetime of the isolated proteins, especially PSII [154]. Light-induced formation of reactive oxygen species can further limit the lifespan of proteins [115]. Another limitation of biophotoelectrodes is that photosynthetic proteins only use a limited range of the solar spectrum which reduces solar conversion efficiency. The absorption spectral range can be enhanced by attaching complementary chromophores to light-harvesting complex proteins [188]. It also can be improved by integrating biological light-harvesting antenna complexes or organic dyes/synthetic compounds to the RCs [189,190].

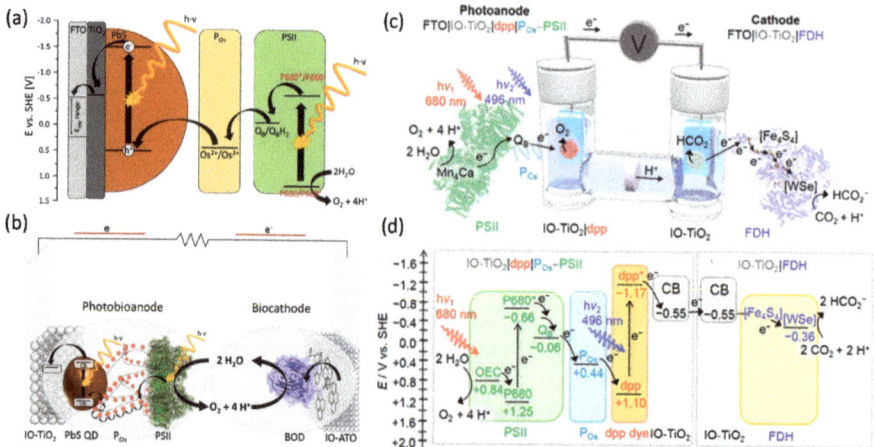

Figure 4. Photoelectrochemical (PEC) cell for electricity or biofuels generation. (**a**) Schematic of the electron transfer steps and energetic level of the components of the light-driven signal chain composed of TiO$_2$, PbS QDs, redox polymer (POs), and PSII. (**b**) A scheme of a PEC cell consisting of an IO-TiO$_2$|PbS|POs|PSII anode and an IO-ATO|PC|BOD cathode. Reprinted with the permission from ref [181]. Copyright (2019), Wiley. (**c**) A schematic representation of a semi-artificial photosynthetic tandem PEC cell coupling CO$_2$ reduction to water oxidation. A blend of POs and PSII adsorbed on a dpp-sensitized photoanode (IO-TiO$_2$|dpp|POs-PSII) is wired to an IO-TiO$_2$|FDH cathode. (**d**) Energy level diagram showing the electron-transfer pathway between PSII, the redox polymer (POs), the dye (dpp), the conduction band (CB) of IO-TiO$_2$ electrodes, four [Fe4S4] clusters, and the [WSe]-active site in FDH. All potentials are reported vs. SHE at pH 6.5. Reprinted with the permission from ref [184]. Copyright (2018), American Chemical Society.

6. Emerging Methods and Future Considerations

6.1. Extending the Lifetime of Membrane Enzymes

A major drawback of enzymes in EBC is their limited active lifetime, which usually ranges from few hours to several days [191]. This applies particularly when membrane enzymes in detergent solutions are used, and functional reconstitution of membrane enzymes into an amphiphilic bilayer, such as liposome or polymersome vesicles, could represent a strategy to extend the enzyme lifetime. Liposomes offer great biocompatibility because they mimic the natural environment of membrane proteins, but they lack chemical and physical long-term stability [192]. Polymersomes offer a more robust amphiphilic polymer environment with increased chemical and physical stability [193,194]. However, this non-native polymeric environment might be limiting the functional incorporation of a wider range of membrane proteins [195]. Recently, hybrid vesicle systems composed of a mixture of lipids and block copolymers, have been developed with the rationale to provide a compromise between the biocompatibility of liposomes and the stability and robustness of polymersomes. In 2016 we showed that hybrid vesicles, composed of biocompatible lipids and stable PBd–PEO copolymer, supported higher activity of reconstituted cytochrome *bo*$_3$ than the proteopolymersomes and significantly extended the functional lifetime of the membrane enzyme when compared to standard proteoliposomes [196] (Figure 5). In 2018, we achieved increased stability of cytochrome *bo*$_3$ reconstituted in hybrid vesicles up to 500 days [197]. Similarly, recent work from Dimova group showed that functional integration of cytochrome *bo*$_3$ oxidase in synthetic membranes made of PDMS-*g*-PEO was capable of lumen acidification and such reconstituted system showed to increase the active lifetime and resistance to free radicals [198]. In 2017, Otrin et al. [199] demonstrated a similar ability to store gradients by reconstituting cytochrome bo_3 together with an ATP synthase in hybrid vesicles constituted of the same copolymer, PDMS-g-PEO. In 2018, Smirnova et al. presented a method that allowed transfer of a functional membrane protein, cytochrome c oxidase (cytochrome aa_3 or yeast Complex IV), with a disc of its native lipids into pre-formed liposomes of well-defined lipid composition and size using amphipathic styrene maleic acid (SMA) copolymer [200]. This recent advance in using SMA copolymer for membrane enzymes isolation and reconstitution offers the advantage to maintain the native phospholipids environment surrounding the proteins and, moreover, could reduce time and cost for enzyme isolation and reconstitution processes by avoiding detergent mediated extraction [201]. Further research into affordable purification strategies and extending the stability of commercially-relevant membrane enzymes is required for membrane enzymes to find applications in bioelectrocatalysis. An alternative approach would be to omit purification altogether and exploit the regenerative capacity of micro-organisms in microbial electrosynthesis.

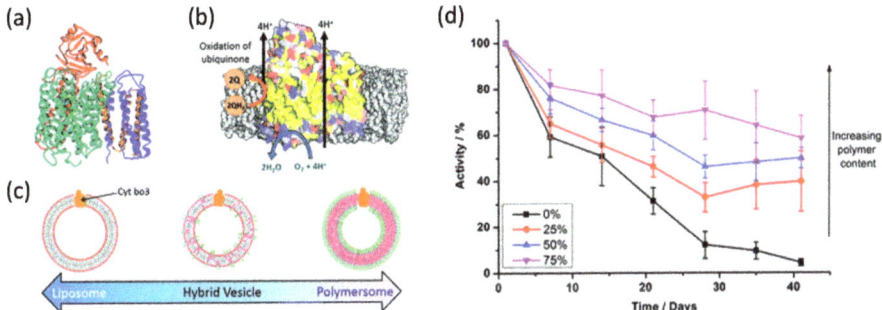

Figure 5. Stability of reconstituted cytochrome bo_3 in hybrid vesicles. (**a**) Ribbon diagram of cytochrome bo_3. (**b**) Schematic representation of cytochrome bo_3 reactions (**c**) schematic representation of proteo-phospholipid/block copolymer hybrid vesicles. (**d**) Comparison of cytochrome bo_3 activity in reconstituted hybrid vesicles with increasing polymer content over a period of 41 days. Reprinted with the permission from ref [196]. Copyright (2016), The Royal Society of Chemistry.

6.2. Microbial Electrosynthesis and Whole-Cell Semi-Artificial Photosynthesis

Microorganisms are of great interest for electrochemical applications because of their self-reproduce nature and diverse metabolic processes. The possibility to use directly the microorganisms overcomes the need to purify and manipulate the proteins. Many microorganisms are exoelectrogens and can communicate with electrodes through extracellular electron transfer (EET), either (1) directly through membrane-bound cytochromes and conductive filaments (nanowires) or (2) via a soluble redox species, such as flavins, either natively secreted by the microorganism or added as a mediator [202].

6.2.1. Microbial Electrosynthesis

Microbial electrosynthesis is an electricity-driven process which generates chemicals using microorganisms as catalysts [203]. To date, the most well-studied exoelectrogens are *Geobacter* spp. and *S. oneidensis* [204]. They are dissimilatory metal-reducing bacteria that can reduce extracellular metal ions (and bulk electrodes) as part of their anaerobic respiration. These electron transfer steps can be reversed to gain electron from electrode for reductive synthesis, some catalysed by membrane enzymes in the microorganisms (Figure 6) [205,206]. It has been shown that *Geobacter* spp. can reduce nitrate to nitrite [207], fumarate to succinate [207] and proton to hydrogen [208] using electrode as the electron donor. *S. oneidensis* was also shown to electrocatalytically reduce fumarate to succinate [209]. It also has been reported that *S. oneidensis* can catalyse CO_2 reduction into formic acid with electron input from the cathode [210]. The cathodic CO_2 reduction in *S. oneidensis* is associated with the electron uptake through outer-membrane c-type cytochromes [211]. The efficiency of the microbial electrosynthesis is mainly limited by the EET. With the genetic engineering of the microorganisms, the EET efficiency can be improved [212–214]. Advances in synthetic biology allow the rational design of non-natural functions in order to increase the diversity of products obtainable from microbial electrosynthesis [215]. For instance, by engineering genes for an ATP-dependent citrate lyase into *Geobacter sulfurreducens*, the microorganism is able to fix CO_2 through a reverse TCA cycle using an electrode as electron donor [216]. A genetically engineered *S. oneidensis* with heterologous Ehrlich pathway genes was shown to produce isobutanol by supplying electricity [217]. It has been demonstrated that *S. oneidensis* can use electrons supplied by an electrode to reduce O_2. This study also showed that the cathodic reaction can reduce NAD^+ via proton-pumping NADH oxidase complex I [218]. With addition of a light-driven proton pump (proteorhodopsin) in the *S. oneidensis* to generate proton-motive force, the electron transferred from cathode to quinone pool can be used to reduce NAD^+ to NADH by native NADH dehydrogenases. This was demonstrated by the reduction of acetoin to 2,3-butanediol via a heterologous butanediol dehydrogenase (Bdh) which is a NADH-dependent enzyme [219].

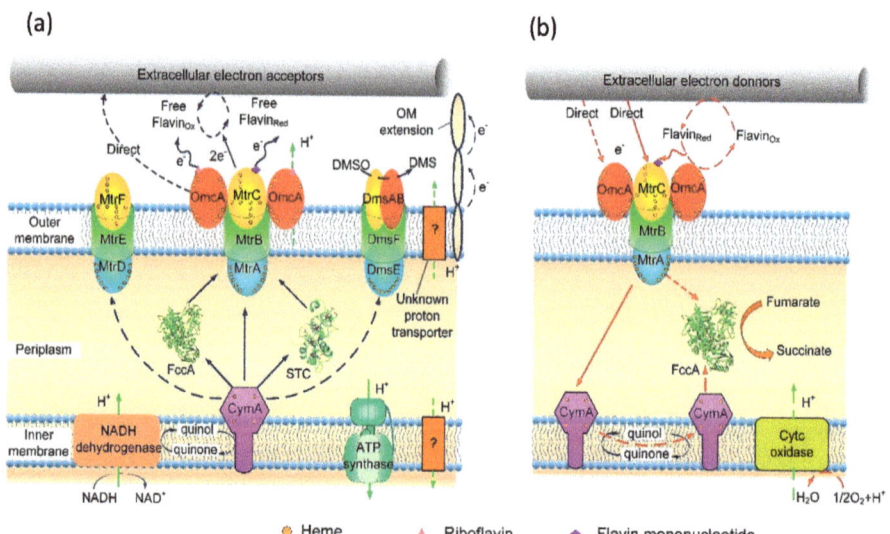

Figure 6. Conceptual model of the bidirectional EET pathways in *S. oneidensis* MR-1. The solid arrows indicate the verified electron flow path, and the dashed arrows indicate the possible electron flow. The blue arrows indicate the outward EET pathway, the red arrows indicate the inward EET pathway, and the green arrows indicate the H^+ flow pathway. (**a**) The outward EET toward extracellular acceptors utilizes the metal-reducing (Mtr) pathway. (**b**) The inward EET toward intracellular terminal reductase involves MtrCBA complex, periplasmic FccA, and inner membrane CymA. The cycle of oxidation (Ox) and reduction (Red) states of self-secreted flavins mediated both outward and inward EET. DMSO, dimethyl sulphoxide; Omc, outer-membrane cytochrome. Reprinted with the permission from ref [220]. Copyright (2020), Taylor & Francis.

6.2.2. Whole-Cell Based Semi-Artificial Photosynthesis

Semi-artificial photosynthesis has attracted great attention for harvesting solar energy for electricity or chemical generation [5]. Instead of isolated photosynthetic proteins (e.g., PSII), photosynthetic microorganisms such as algae [221], cyanobacteria [222–225] and purple bacteria [226], or extracted organelles such as thylakoids [227] have been studied for biophotoelectrochemical systems. The development of 3D structured electrodes increased the loading capacity which enhances photocurrents [228]. However, low EET rates still limit the efficiency of these photoelectrodes. It has been shown that redox polymers can improve electron transfer kinetics between microorganisms and electrode [221,226]. Furthermore improvement in EET kinetics and new electrode designs for accommodating microorganisms are required to boost the performance of photoelectrochemical devices that are based on photosynthetic microorganisms [111].

An emerging direction in the semi-artificial photosynthesis is to interface synthetic light-harvesting materials with non-photosynthetic microorganisms for value-added chemicals [229,230]. As a model microorganism, *S. oneidensis* has been widely studied for integration with artificial light-harvesting materials for solar-driven microbial synthesis. The transmembrane cytochrome MtrCAB, helped by OmcA located on the outside of the membrane, are known to transfer electrons between the interior of the cell and extracellular materials (Figure 6) [231,232]. In the presence of a sacrificial electron donor, MtrC and OmcA can be photoreduced by water-soluble photosensitisers including eosin Y, fluorescein, proflavine, flavin, and adenine dinucleotide, riboflavin and flavin mononucleotide [233]. In an in vitro approach, it was showed that dye-sensitised TiO_2 nanoparticles can photoreduce MtrC or OmcA, either in solution or one an electrode surface [233–235] and it might be possible to extend this process to reductive photosynthesis in *S. oneidensis* [236]. To provide a proof-of-concept, we recently reconstituted MtrCAB into proteoliposomes encapsulating a redox dye, Reactive Red 120 (RR120), which can be reductively bleached. Using light-harvesting nanoparticles including dye-sensitised TiO_2 nanoparticles, amorphous carbon dots and nitrogen doped graphitic carbon dots, we observed RR120 reductive decomposition inside the lumen of the MtrCAB proteoliposomes, confirming that electrons were transferred from the nanoparticles via transmembrane MtrCAB complex into the liposome lumen [237]. These results show that the rational design of light-harvesting nanoparticles and protein hybrids can lead to the development of semi-artificial photosynthetic systems for solar fuels synthesis.

7. Conclusions

Membrane proteins play an important role for biofuel conversion and photosynthesis, and the catalytic and electron transfer ability of the membrane proteins makes them attractive for applications in bioelectrocatalysis. Recent developments in bioelectrochemistry have provided the means to control and improve electron transfer rate between immobilised proteins and electrodes, stabilising the activity of immobilised enzymes. High surface areas of so-called 3D electrodes are able to increase the protein loading, which is desired for bioelectrocatalysis. Despite these enhanced performances, there is still the need to further optimise this technology for practical use in bioelectrocatalysis devices, especially for membrane enzymes. Combining several immobilisation strategies would provide possible solutions. Advances in characterisation techniques enable detailed characterisation of the structure of the electrode, the electron transfer process, catalytic activity and protein structure, which in turn provides an in-depth understanding of the catalytic reaction mechanism, aiding the rational design of the electrode interface for bioelectrocatalysis. Membrane proteins are known to be difficult to study and manipulate due to their amphiphilic nature. Preserving their stability and functionality is crucial for successful application of this technology. Recently developed redox polymers and hybrid vesicles, composed of lipids and block copolymers, are promising platforms to stabilise membrane proteins on the electrode surface. Using exoeletrogenic microorganisms for electrosynthesis and semi-artificial photosynthesis is an encouraging research direction, which will simplify time and cost related with protein purification processes. The ability of microorganisms to regenerate and self-replicate (at an energy cost), also removes the need to stabilise the biocatalyst. With synthetic biology, the microorganisms can be further engineered for diverse biofuel and chemical synthesis. However, the extracellular electron transfer kinetics might still be limiting the efficiency of the whole-cell based systems, especially when organisms are used that have not been optimised by evolution for extracellular electron transfer.

Funding: This research was funded by the Biotechnology and Biological Sciences Research Council (BBSRC), grant numbers BB/T000546/1 and BB/S000704/1.

Conflicts of Interest: The authors declare no conflict of interest.

References

1. von Heijne, G. Membrane-protein topology. *Nat. Rev. Mol. Cell Biol.* **2006**, *7*, 909–918. [CrossRef] [PubMed]
2. Hedin, L.E.; Illergård, K.; Elofsson, A. An introduction to membrane proteins. *J. Proteome Res.* **2011**, *10*, 3324–3331. [CrossRef] [PubMed]
3. Tan, S.; Tan, H.T.; Chung, M.C.M. Membrane proteins and membrane proteomics. *Proteomics* **2008**, *8*, 3924–3932. [CrossRef] [PubMed]
4. Kumar, G.; Kim, S.-H.; Lay, C.-H.; Ponnusamy, V.K. Recent developments on alternative fuels, energy and environment for sustainability. *Bioresour. Technol.* **2020**, *317*, 124010. [CrossRef] [PubMed]
5. Kornienko, N.; Zhang, J.Z.; Sakimoto, K.K.; Yang, P.; Reisner, E. Interfacing nature's catalytic machinery with synthetic materials for semi-artificial photosynthesis. *Nat. Nanotechnol.* **2018**, *13*, 890–899. [CrossRef]
6. Vignais, P.M.; Billoud, B.; Meyer, J. Classification and phylogeny of hydrogenases. *FEMS Microbiol. Rev.* **2001**, *25*, 455–501. [CrossRef]
7. Shima, S.; Pilak, O.; Vogt, S.; Schick, M.; Stagni, M.S.; Meyer-Klaucke, W.; Warkentin, E.; Thauer, R.K.; Ermler, U. The crystal structure of Fe-hydrogenase reveals the geometry of the active site. *Science* **2008**, *321*, 572–575. [CrossRef]
8. Fritsch, J.; Lenz, O.; Friedrich, B. Structure, function and biosynthesis of O_2-tolerant hydrogenases. *Nat. Rev. Microbiol.* **2013**, *11*, 106–114. [CrossRef]
9. Sousa, F.L.; Alves, R.J.; Ribeiro, M.A.; Pereira-Leal, J.B.; Teixeira, M.; Pereira, M.M. The superfamily of heme-copper oxygen reductases: Types and evolutionary considerations. *Biochim. Biophys. Acta Bioenerg.* **2012**, *1817*, 629–637. [CrossRef]
10. Borisov, V.B.; Gennis, R.B.; Hemp, J.; Verkhovsky, M.I. The cytochrome *bd* respiratory oxygen reductases. *Biochim. Biophys. Acta Bioenerg.* **2011**, *1807*, 1398–1413. [CrossRef]
11. Vanlerberghe, G.C. Alternative oxidase: A mitochondrial respiratory pathway to maintain metabolic and signaling homeostasis during abiotic and biotic stress in plants. *Int. J. Mol. Sci.* **2013**, *14*, 6805–6847. [CrossRef] [PubMed]
12. García-Horsman, J.A.; Barquera, B.; Rumbley, J.; Ma, J.; Gennis, R.B. The superfamily of heme-copper respiratory oxidases. *J. Bacteriol.* **1994**, *176*, 5587–5600. [CrossRef] [PubMed]
13. Safarian, S.; Hahn, A.; Mills, D.J.; Radloff, M.; Eisinger, M.L.; Nikolaev, A.; Meier-Credo, J.; Melin, F.; Miyoshi, H.; Gennis, R.B.; et al. Active site rearrangement and structural divergence in prokaryotic respiratory oxidases. *Science* **2019**, *366*, 100–104. [CrossRef] [PubMed]
14. Moreno-Vivián, C.; Cabello, P.; Martínez-Luque, M.; Blasco, R.; Castillo, F. Prokaryotic nitrate reduction: Molecular properties and functional distinction among bacterial nitrate reductases. *J. Bacteriol.* **1999**, *181*, 6573–6584. [CrossRef] [PubMed]
15. Berks, B.C.; Ferguson, S.J.; Moir, J.W.B.; Richardson, D.J. Enzymes and associated electron transport systems that catalyse the respiratory reduction of nitrogen oxides and oxyanions. *Biochim. Biophys. Acta Bioenerg.* **1995**, *1232*, 97–173. [CrossRef]
16. Bertero, M.G.; Rothery, R.A.; Palak, M.; Hou, C.; Lim, D.; Blasco, F.; Weiner, J.H.; Strynadka, N.C.J. Insights into the respiratory electron transfer pathway from the structure of nitrate reductase A. *Nat. Struct. Biol.* **2003**, *10*, 681–687. [CrossRef]
17. Nelson, N.; Ben-Shem, A. The complex architecture of oxygenic photosynthesis. *Nat. Rev. Mol. Cell Biol.* **2004**, *5*, 971–982. [CrossRef]
18. Gao, J.; Wang, H.; Yuan, Q.; Feng, Y. Structure and Function of the Photosystem Supercomplexes. *Front. Plant. Sci.* **2018**, *9*, 357. [CrossRef]
19. Hu, X.; Damjanović, A.; Ritz, T.; Schulten, K. Architecture and mechanism of the light-harvesting apparatus of purple bacteria. *Proc. Natl. Acad. Sci. USA* **1998**, *95*, 5935–5941. [CrossRef]
20. Bahatyrova, S.; Frese, R.N.; Siebert, C.A.; Olsen, J.D.; van der Werf, K.O.; van Grondelle, R.; Niederman, R.A.; Bullough, P.A.; Otto, C.; Hunter, C.N. The native architecture of a photosynthetic membrane. *Nature* **2004**, *430*, 1058–1062. [CrossRef]

21. Mirkovic, T.; Ostroumov, E.E.; Anna, J.M.; van Grondelle, R.; Govindjee; Scholes, G.D. Light Absorption and Energy Transfer in the Antenna Complexes of Photosynthetic Organisms. *Chem. Rev.* **2017**, *117*, 249–293. [CrossRef] [PubMed]
22. Plumeré, N.; Nowaczyk, M.M. Biophotoelectrochemistry of Photosynthetic Proteins. In *Biophotoelectrochemistry: From Bioelectrochemistry to Biophotovoltaics*; Jeuken, L., Ed.; Springer: Cham, Switzerland, 2016; Volume 158, pp. 111–136.
23. Yehezkeli, O.; Tel-Vered, R.; Michaeli, D.; Willner, I.; Nechushtai, R. Photosynthetic reaction center-functionalized electrodes for photo-bioelectrochemical cells. *Photosynth. Res.* **2014**, *120*, 71–85. [CrossRef] [PubMed]
24. Chen, H.; Simoska, O.; Lim, K.; Grattieri, M.; Yuan, M.; Dong, F.; Lee, Y.S.; Beaver, K.; Weliwatte, S.; Gaffney, E.M.; et al. Fundamentals, Applications, and Future Directions of Bioelectrocatalysis. *Chem. Rev.* **2020**. [CrossRef] [PubMed]
25. Carpenter, E.P.; Beis, K.; Cameron, A.D.; Iwata, S. Overcoming the challenges of membrane protein crystallography. *Curr. Opin. Struct. Biol.* **2008**, *18*, 581–586. [CrossRef]
26. Seddon, A.M.; Curnow, P.; Booth, P.J. Membrane proteins, lipids and detergents: Not just a soap opera. *Biochim. Biophys. Acta* **2004**, *1666*, 105–117. [CrossRef]
27. Palazzo, G. Colloidal aspects of photosynthetic membrane proteins. *Curr. Opin. Colloid Interface Sci.* **2006**, *11*, 65–73. [CrossRef]
28. Rawlings, A.E. Membrane proteins: Always an insoluble problem? *Biochem. Soc. Trans.* **2016**, *44*, 790–795. [CrossRef]
29. Gillam, E.M.J. Engineering cytochrome P450 enzymes. *Chem. Res. Toxicol.* **2008**, *21*, 220–231. [CrossRef]
30. Hunte, C.; Richers, S. Lipids and membrane protein structures. *Curr. Opin. Struct. Biol.* **2008**, *18*, 406–411. [CrossRef]
31. Mazurenko, I.; Hitaishi, V.P.; Lojou, E. Recent advances in surface chemistry of electrodes to promote direct enzymatic bioelectrocatalysis. *Curr. Opin. Electrochem.* **2020**, *19*, 113–121. [CrossRef]
32. Yates, N.D.J.; Fascione, M.A.; Parkin, A. Methodologies for "Wiring" Redox Proteins/Enzymes to Electrode Surfaces. *Chem. Eur. J.* **2018**, *24*, 12164–12182. [CrossRef] [PubMed]
33. Sucheta, A.; Ackrell, B.A.; Cochran, B.; Armstrong, F.A. Diode-like behaviour of a mitochondrial electron-transport enzyme. *Nature* **1992**, *356*, 361–362. [CrossRef] [PubMed]
34. Anderson, L.J.; Richardson, D.J.; Butt, J.N. Catalytic protein film voltammetry from a respiratory nitrate reductase provides evidence for complex electrochemical modulation of enzyme activity. *Biochemistry* **2001**, *40*, 11294–11307. [CrossRef] [PubMed]
35. Elliott, S.J.; Hoke, K.R.; Heffron, K.; Palak, M.; Rothery, R.A.; Weiner, J.H.; Armstrong, F.A. Voltammetric studies of the catalytic mechanism of the respiratory nitrate reductase from *Escherichia coli*: How nitrate reduction and inhibition depend on the oxidation state of the active site. *Biochemistry* **2004**, *43*, 799–807. [CrossRef]
36. den Hollander, M.-J.; Magis, J.G.; Fuchsenberger, P.; Aartsma, T.J.; Jones, M.R.; Frese, R.N. Enhanced photocurrent generation by photosynthetic bacterial reaction centers through molecular relays, light-harvesting complexes, and direct protein-gold interactions. *Langmuir* **2011**, *27*, 10282–10294. [CrossRef]
37. Kamran, M.; Delgado, J.D.; Friebe, V.; Aartsma, T.J.; Frese, R.N. Photosynthetic protein complexes as bio-photovoltaic building blocks retaining a high internal quantum efficiency. *Biomacromolecules* **2014**, *15*, 2833–2838. [CrossRef]
38. Yasuda, Y.; Sugino, H.; Toyotama, H.; Hirata, Y.; Hara, M.; Miyake, J. Control of protein orientation in molecular photoelectric devices using Langmuir—Blodgett films of photosynthetic reaction centers from *Rhodopseudomonas viridis*. *Bioelectrochem. Bioenerg.* **1994**, *34*, 135–139. [CrossRef]
39. Uphaus, R.A.; Fang, J.Y.; Picorel, R.; Chumanov, G.; Wang, J.Y.; Cotton, T.M.; Seibert, M. Langmuir-Blodgett and X-ray diffraction studies of isolated photosystem II reaction centers in monolayers and multilayers: Physical dimensions of the complex. *Photochem. Photobiol.* **1997**, *65*, 673–679. [CrossRef]
40. Kernen, P.; Gruszecki, W.I.; Matuła, M.; Wagner, P.; Ziegler, U.; Krupa, Z. Light-harvesting complex II in monocomponent and mixed lipid-protein monolayers. *Biochim. Biophys. Acta Biomembr.* **1998**, *1373*, 289–298. [CrossRef]
41. Pepe, I.M.; Nicolini, C. Langmuir-Blodgett films of photosensitive proteins. *J. Photochem. Photobiol. B* **1996**, *33*, 191–200. [CrossRef]

42. Ko, B.S.; Babcock, B.; Jennings, G.K.; Tilden, S.G.; Peterson, R.R.; Cliffel, D.; Greenbaum, E. Effect of surface composition on the adsorption of photosystem I onto alkanethiolate self-assembled monolayers on gold. *Langmuir* **2004**, *20*, 4033–4038. [CrossRef] [PubMed]
43. Lee, I.; Lee, J.W.; Greenbaum, E. Biomolecular Electronics: Vectorial Arrays of Photosynthetic Reaction Centers. *Phys. Rev. Lett.* **1997**, *79*, 3294–3297. [CrossRef]
44. Trammell, S.A.; Wang, L.; Zullo, J.M.; Shashidhar, R.; Lebedev, N. Orientated binding of photosynthetic reaction centers on gold using Ni-NTA self-assembled monolayers. *Biosens. Bioelectron.* **2004**, *19*, 1649–1655. [CrossRef] [PubMed]
45. Nakamura, C.; Hasegawa, M.; Yasuda, Y.; Miyake, J. Self-Assembling Photosynthetic Reaction Centers on Electrodes for Current Generation. In *Twenty-First Symposium on Biotechnology for Fuels and Chemicals*; Finkelstein, M., Davison, B.H., Eds.; Humana Press: Totowa, NJ, USA, 2000; Volume 84–86, pp. 401–408.
46. Maly, J.; Krejci, J.; Ilie, M.; Jakubka, L.; Masojídek, J.; Pilloton, R.; Sameh, K.; Steffan, P.; Stryhal, Z.; Sugiura, M. Monolayers of photosystem II on gold electrodes with enhanced sensor response-effect of porosity and protein layer arrangement. *Anal. Bioanal. Chem.* **2005**, *381*, 1558–1567. [CrossRef]
47. Faulkner, C.J.; Lees, S.; Ciesielski, P.N.; Cliffel, D.E.; Jennings, G.K. Rapid assembly of photosystem I monolayers on gold electrodes. *Langmuir* **2008**, *24*, 8409–8412. [CrossRef]
48. Friebe, V.M.; Millo, D.; Swainsbury, D.J.K.; Jones, M.R.; Frese, R.N. Cytochrome *c* Provides an Electron-Funneling Antenna for Efficient Photocurrent Generation in a Reaction Center Biophotocathode. *ACS Appl. Mater. Interfaces* **2017**, *9*, 23379–23388. [CrossRef]
49. Badura, A.; Kothe, T.; Schuhmann, W.; Rögner, M. Wiring photosynthetic enzymes to electrodes. *Energy Environ. Sci.* **2011**, *4*, 3263. [CrossRef]
50. Xiao, Y.; Patolsky, F.; Katz, E.; Hainfeld, J.F.; Willner, I. "Plugging into Enzymes": Nanowiring of redox enzymes by a gold nanoparticle. *Science* **2003**, *299*, 1877–1881. [CrossRef]
51. Fournier, E.; Nikolaev, A.; Nasiri, H.R.; Hoeser, J.; Friedrich, T.; Hellwig, P.; Melin, F. Creation of a gold nanoparticle based electrochemical assay for the detection of inhibitors of bacterial cytochrome *bd* oxidases. *Bioelectrochemistry* **2016**, *111*, 109–114. [CrossRef]
52. Meyer, T.; Melin, F.; Richter, O.-M.H.; Ludwig, B.; Kannt, A.; Müller, H.; Michel, H.; Hellwig, P. Electrochemistry suggests proton access from the exit site to the binuclear center in *Paracoccus denitrificans* cytochrome *c* oxidase pathway variants. *FEBS Lett.* **2015**, *589*, 565–568. [CrossRef]
53. Meyer, T.; Melin, F.; Xie, H.; von der Hocht, I.; Choi, S.K.; Noor, M.R.; Michel, H.; Gennis, R.B.; Soulimane, T.; Hellwig, P. Evidence for distinct electron transfer processes in terminal oxidases from different origin by means of protein film voltammetry. *J. Am. Chem. Soc.* **2014**, *136*, 10854–10857. [CrossRef] [PubMed]
54. Melin, F.; Meyer, T.; Lankiang, S.; Choi, S.K.; Gennis, R.B.; Blanck, C.; Schmutz, M.; Hellwig, P. Direct electrochemistry of cytochrome *bo₃* oxidase at a series of gold nanoparticles-modified electrodes. *Electrochem. Commun.* **2013**, *26*, 105–108. [CrossRef] [PubMed]
55. Nikolaev, A.; Makarchuk, I.; Thesseling, A.; Hoeser, J.; Friedrich, T.; Melin, F.; Hellwig, P. Stabilization of the Highly Hydrophobic Membrane Protein, Cytochrome *bd* Oxidase, on Metallic Surfaces for Direct Electrochemical Studies. *Molecules* **2020**, *25*, 3240. [CrossRef]
56. Asadian, E.; Ghalkhani, M.; Shahrokhian, S. Electrochemical sensing based on carbon nanoparticles: A review. *Sens. Actuators B* **2019**, *293*, 183–209. [CrossRef]
57. Vincent, K.A.; Li, X.; Blanford, C.F.; Belsey, N.A.; Weiner, J.H.; Armstrong, F.A. Enzymatic catalysis on conducting graphite particles. *Nat. Chem. Biol.* **2007**, *3*, 761–762. [CrossRef]
58. Duca, M.; Weeks, J.R.; Fedor, J.G.; Weiner, J.H.; Vincent, K.A. Combining Noble Metals and Enzymes for Relay Cascade Electrocatalysis of Nitrate Reduction to Ammonia at Neutral pH. *ChemElectroChem* **2015**, *2*, 1086–1089. [CrossRef]
59. Ruff, A. Redox polymers in bioelectrochemistry: Common playgrounds and novel concepts. *Curr. Opin. Electrochem.* **2017**, *5*, 66–73. [CrossRef]
60. Badura, A.; Guschin, D.; Kothe, T.; Kopczak, M.J.; Schuhmann, W.; Rögner, M. Photocurrent generation by photosystem 1 integrated in crosslinked redox hydrogels. *Energy Environ. Sci.* **2011**, *4*, 2435. [CrossRef]
61. Badura, A.; Guschin, D.; Esper, B.; Kothe, T.; Neugebauer, S.; Schuhmann, W.; Rögner, M. Photo-Induced Electron Transfer Between Photosystem 2 via Cross-linked Redox Hydrogels. *Electroanalysis* **2008**, *20*, 1043–1047. [CrossRef]

62. Kothe, T.; Plumeré, N.; Badura, A.; Nowaczyk, M.M.; Guschin, D.A.; Rögner, M.; Schuhmann, W. Combination of a photosystem 1-based photocathode and a photosystem 2-based photoanode to a Z-scheme mimic for biophotovoltaic applications. *Angew. Chem. Int. Ed.* **2013**, *52*, 14233–14236. [CrossRef]
63. Hartmann, V.; Kothe, T.; Pöller, S.; El-Mohsnawy, E.; Nowaczyk, M.M.; Plumeré, N.; Schuhmann, W.; Rögner, M. Redox hydrogels with adjusted redox potential for improved efficiency in Z-scheme inspired biophotovoltaic cells. *Phys. Chem. Chem. Phys.* **2014**, *16*, 11936–11941. [CrossRef] [PubMed]
64. Kothe, T.; Pöller, S.; Zhao, F.; Fortgang, P.; Rögner, M.; Schuhmann, W.; Plumeré, N. Engineered electron-transfer chain in photosystem 1 based photocathodes outperforms electron-transfer rates in natural photosynthesis. *Chem. Eur. J.* **2014**, *20*, 11029–11034. [CrossRef] [PubMed]
65. Laftsoglou, T.; Jeuken, L.J.C. Supramolecular electrode assemblies for bioelectrochemistry. *Chem. Commun.* **2017**, *53*, 3801–3809. [CrossRef] [PubMed]
66. Jeuken, L.J.C. Electrodes for integral membrane enzymes. *Nat. Prod. Rep.* **2009**, *26*, 1234–1240. [CrossRef] [PubMed]
67. Alvarez-Malmagro, J.; García-Molina, G.; López De Lacey, A. Electrochemical Biosensors Based on Membrane-Bound Enzymes in Biomimetic Configurations. *Sensors* **2020**, *20*, 3393. [CrossRef] [PubMed]
68. Burgess, J.D.; Rhoten, M.C.; Hawkridge, F.M. Cytochrome *c* Oxidase Immobilized in Stable Supported Lipid Bilayer Membranes. *Langmuir* **1998**, *14*, 2467–2475. [CrossRef]
69. Rhoten, M.C.; Hawkridge, F.M.; Wilczek, J. The reaction of cytochrome *c* with bovine and *Bacillus stearothermophilus* cytochrome *c* oxidase immobilized in electrode-supported lipid bilayer membranes. *J. Electroanal. Chem.* **2002**, *535*, 97–106. [CrossRef]
70. Kong, J.; Lu, Z.; Lvov, Y.M.; Desamero, R.Z.B.; Frank, H.A.; Rusling, J.F. Direct Electrochemistry of Cofactor Redox Sites in a Bacterial Photosynthetic Reaction Center Protein. *J. Am. Chem. Soc.* **1998**, *120*, 7371–7372. [CrossRef]
71. Munge, B.; Pendon, Z.; Frank, H.A.; Rusling, J.F. Electrochemical reactions of redox cofactors in *Rhodobacter sphaeroides* reaction center proteins in lipid films. *Bioelectrochemistry* **2001**, *54*, 145–150. [CrossRef]
72. Munge, B.; Das, S.K.; Ilagan, R.; Pendon, Z.; Yang, J.; Frank, H.A.; Rusling, J.F. Electron transfer reactions of redox cofactors in spinach photosystem I reaction center protein in lipid films on electrodes. *J. Am. Chem. Soc.* **2003**, *125*, 12457–12463. [CrossRef]
73. Alcantara, K.; Munge, B.; Pendon, Z.; Frank, H.A.; Rusling, J.F. Thin film voltammetry of spinach photosystem II. Proton-gated electron transfer involving the Mn_4 cluster. *J. Am. Chem. Soc.* **2006**, *128*, 14930–14937. [CrossRef] [PubMed]
74. Noji, T.; Matsuo, M.; Takeda, N.; Sumino, A.; Kondo, M.; Nango, M.; Itoh, S.; Dewa, T. Lipid-Controlled Stabilization of Charge-Separated States ($P^+Q_B^-$) and Photocurrent Generation Activity of a Light-Harvesting-Reaction Center Core Complex (LH1-RC) from *Rhodopseudomonas palustris*. *J. Phys. Chem. B* **2018**, *122*, 1066–1080. [CrossRef] [PubMed]
75. Jeuken, L.J.C.; Connell, S.D.; Henderson, P.J.F.; Gennis, R.B.; Evans, S.D.; Bushby, R.J. Redox enzymes in tethered membranes. *J. Am. Chem. Soc.* **2006**, *128*, 1711–1716. [CrossRef] [PubMed]
76. Radu, V.; Frielingsdorf, S.; Evans, S.D.; Lenz, O.; Jeuken, L.J.C. Enhanced oxygen-tolerance of the full heterotrimeric membrane-bound NiFe-hydrogenase of *Ralstonia eutropha*. *J. Am. Chem. Soc.* **2014**, *136*, 8512–8515. [CrossRef]
77. Pelster, L.N.; Minteer, S.D. Mitochondrial Inner Membrane Biomimic for the Investigation of Electron Transport Chain Supercomplex Bioelectrocatalysis. *ACS Catal.* **2016**, *6*, 4995–4999. [CrossRef]
78. Ataka, K.; Giess, F.; Knoll, W.; Naumann, R.; Haber-Pohlmeier, S.; Richter, B.; Heberle, J. Oriented attachment and membrane reconstitution of His-tagged cytochrome *c* oxidase to a gold electrode: In situ monitoring by surface-enhanced infrared absorption spectroscopy. *J. Am. Chem. Soc.* **2004**, *126*, 16199–16206. [CrossRef]
79. Ataka, K.; Richter, B.; Heberle, J. Orientational control of the physiological reaction of cytochrome *c* oxidase tethered to a gold electrode. *J. Phys. Chem. B* **2006**, *110*, 9339–9347. [CrossRef]
80. Friedrich, M.G.; Giebeta, F.; Naumann, R.; Knoll, W.; Ataka, K.; Heberle, J.; Hrabakova, J.; Murgida, D.H.; Hildebrandt, P. Active site structure and redox processes of cytochrome *c* oxidase immobilised in a novel biomimetic lipid membrane on an electrode. *Chem. Commun.* **2004**, 2376–2377. [CrossRef]
81. Lisdat, F. Trends in the layer-by-layer assembly of redox proteins and enzymes in bioelectrochemistry. *Curr. Opin. Electrochem.* **2017**, *5*, 165–172. [CrossRef]

82. Stieger, K.R.; Ciornii, D.; Kölsch, A.; Hejazi, M.; Lokstein, H.; Feifel, S.C.; Zouni, A.; Lisdat, F. Engineering of supramolecular photoactive protein architectures: The defined co-assembly of photosystem I and cytochrome *c* using a nanoscaled DNA-matrix. *Nanoscale* **2016**, *8*, 10695–10705. [CrossRef]
83. Zhang, Y.; LaFountain, A.M.; Magdaong, N.; Fuciman, M.; Allen, J.P.; Frank, H.A.; Rusling, J.F. Thin film voltammetry of wild type and mutant reaction center proteins from photosynthetic bacteria. *J. Phys. Chem. B* **2011**, *115*, 3226–3232. [CrossRef] [PubMed]
84. Mallardi, A.; Giustini, M.; Lopez, F.; Dezi, M.; Venturoli, G.; Palazzo, G. Functionality of photosynthetic reaction centers in polyelectrolyte multilayers: Toward an herbicide biosensor. *J. Phys. Chem. B* **2007**, *111*, 3304–3314. [CrossRef] [PubMed]
85. Heath, G.R.; Li, M.; Polignano, I.L.; Richens, J.L.; Catucci, G.; O'Shea, P.; Sadeghi, S.J.; Gilardi, G.; Butt, J.N.; Jeuken, L.J.C. Layer-by-Layer Assembly of Supported Lipid Bilayer Poly-L-Lysine Multilayers. *Biomacromolecules* **2016**, *17*, 324–335. [CrossRef] [PubMed]
86. Heath, G.R.; Li, M.; Rong, H.; Radu, V.; Frielingsdorf, S.; Lenz, O.; Butt, J.N.; Jeuken, L.J.C. Multilayered Lipid Membrane Stacks for Biocatalysis Using Membrane Enzymes. *Adv. Funct. Mater.* **2017**, *27*, 1606265. [CrossRef]
87. Sun, H.; Zhu, J.; Baumann, D.; Peng, L.; Xu, Y.; Shakir, I.; Huang, Y.; Duan, X. Hierarchical 3D electrodes for electrochemical energy storage. *Nat. Rev. Mater.* **2019**, *4*, 45–60. [CrossRef]
88. Friebe, V.M.; Delgado, J.D.; Swainsbury, D.J.K.; Gruber, J.M.; Chanaewa, A.; van Grondelle, R.; von Hauff, E.; Millo, D.; Jones, M.R.; Frese, R.N. Plasmon-Enhanced Photocurrent of Photosynthetic Pigment Proteins on Nanoporous Silver. *Adv. Funct. Mater.* **2016**, *26*, 285–292. [CrossRef]
89. Ciesielski, P.N.; Scott, A.M.; Faulkner, C.J.; Berron, B.J.; Cliffel, D.E.; Jennings, G.K. Functionalized nanoporous gold leaf electrode films for the immobilization of photosystem I. *ACS Nano* **2008**, *2*, 2465–2472. [CrossRef]
90. Lu, Y.; Yuan, M.; Liu, Y.; Tu, B.; Xu, C.; Liu, B.; Zhao, D.; Kong, J. Photoelectric performance of bacteria photosynthetic proteins entrapped on tailored mesoporous WO_3-TiO_2 films. *Langmuir* **2005**, *21*, 4071–4076. [CrossRef]
91. Kato, M.; Cardona, T.; Rutherford, A.W.; Reisner, E. Photoelectrochemical water oxidation with photosystem II integrated in a mesoporous indium-tin oxide electrode. *J. Am. Chem. Soc.* **2012**, *134*, 8332–8335. [CrossRef]
92. Kato, M.; Cardona, T.; Rutherford, A.W.; Reisner, E. Covalent immobilization of oriented photosystem II on a nanostructured electrode for solar water oxidation. *J. Am. Chem. Soc.* **2013**, *135*, 10610–10613. [CrossRef]
93. Mersch, D.; Lee, C.-Y.; Zhang, J.Z.; Brinkert, K.; Fontecilla-Camps, J.C.; Rutherford, A.W.; Reisner, E. Wiring of Photosystem II to Hydrogenase for Photoelectrochemical Water Splitting. *J. Am. Chem. Soc.* **2015**, *137*, 8541–8549. [CrossRef] [PubMed]
94. Sokol, K.P.; Mersch, D.; Hartmann, V.; Zhang, J.Z.; Nowaczyk, M.M.; Rögner, M.; Ruff, A.; Schuhmann, W.; Plumeré, N.; Reisner, E. Rational wiring of photosystem II to hierarchical indium tin oxide electrodes using redox polymers. *Energy Environ. Sci.* **2016**, *9*, 3698–3709. [CrossRef]
95. Stieger, K.R.; Feifel, S.C.; Lokstein, H.; Hejazi, M.; Zouni, A.; Lisdat, F. Biohybrid architectures for efficient light-to-current conversion based on photosystem I within scalable 3D mesoporous electrodes. *J. Mater. Chem. A* **2016**, *4*, 17009–17017. [CrossRef]
96. Wolfe, K.D.; Dervishogullari, D.; Stachurski, C.D.; Passantino, J.M.; Kane Jennings, G.; Cliffel, D.E. Photosystem I Multilayers within Porous Indium Tin Oxide Cathodes Enhance Mediated Electron Transfer. *ChemElectroChem* **2020**, *7*, 596–603. [CrossRef]
97. Armstrong, F.A.; Butt, J.N.; Sucheta, A. [18] Voltammetric studies of redox-active centers in metalloproteins adsorbed on electrodes. *Methods Enzymol.* **1993**, *227*, 479–500. [PubMed]
98. Léger, C.; Elliott, S.J.; Hoke, K.R.; Jeuken, L.J.C.; Jones, A.K.; Armstrong, F.A. Enzyme electrokinetics: Using protein film voltammetry to investigate redox enzymes and their mechanisms. *Biochemistry* **2003**, *42*, 8653–8662. [CrossRef]
99. Armstrong, F.A. Recent developments in dynamic electrochemical studies of adsorbed enzymes and their active sites. *Curr. Opin. Chem. Biol.* **2005**, *9*, 110–117. [CrossRef]
100. Melin, F.; Hellwig, P. Recent advances in the electrochemistry and spectroelectrochemistry of membrane proteins. *Biol. Chem.* **2013**, *394*, 593–609. [CrossRef]
101. Vacek, J.; Zatloukalova, M.; Novak, D. Electrochemistry of membrane proteins and protein–lipid assemblies. *Curr. Opin. Electrochem.* **2018**, *12*, 73–80. [CrossRef]

102. Armstrong, F.A.; Heering, H.A.; Hirst, J. Reaction of complex metalloproteins studied by protein-film voltammetry. *Chem. Soc. Rev.* **1997**, *26*, 169–179. [CrossRef]
103. Léger, C.; Bertrand, P. Direct electrochemistry of redox enzymes as a tool for mechanistic studies. *Chem. Rev.* **2008**, *108*, 2379–2438. [CrossRef] [PubMed]
104. Butt, J.N.; Armstrong, F.A. Voltammetry of Adsorbed Redox Enzymes: Mechanisms in the Potential Dimension. In *Bioinorganic Electrochemistry*; Hammerich, O., Ulstrup, J., Eds.; Springer: Dordrecht, The Netherlands, 2008; Volume 101, pp. 91–128.
105. Fourmond, V.; Léger, C. Protein Electrochemistry: Questions and Answers. In *Biophotoelectrochemistry: From Bioelectrochemistry to Biophotovoltaics*; Jeuken, L., Ed.; Springer: Cham, Switzerland, 2016; Volume 158, pp. 1–41.
106. Fourmond, V.; Léger, C. Modelling the voltammetry of adsorbed enzymes and molecular catalysts. *Curr. Opin. Electrochem.* **2017**, *1*, 110–120. [CrossRef]
107. Lang, H.; Duschl, C.; Vogel, H. A new class of thiolipids for the attachment of lipid bilayers on gold surfaces. *Langmuir* **1994**, *10*, 197–210. [CrossRef]
108. Millo, D.; Bonifacio, A.; Moncelli, M.R.; Sergo, V.; Gooijer, C.; van der Zwan, G. Characterization of hybrid bilayer membranes on silver electrodes as biocompatible SERS substrates to study membrane-protein interactions. *Colloids Surf. B* **2010**, *81*, 212–216. [CrossRef] [PubMed]
109. Kato, M.; Zhang, J.Z.; Paul, N.; Reisner, E. Protein film photoelectrochemistry of the water oxidation enzyme photosystem II. *Chem. Soc. Rev.* **2014**, *43*, 6485–6497. [CrossRef] [PubMed]
110. Kornienko, N.; Ly, K.H.; Robinson, W.E.; Heidary, N.; Zhang, J.Z.; Reisner, E. Advancing Techniques for Investigating the Enzyme-Electrode Interface. *Acc. Chem. Res.* **2019**, *52*, 1439–1448. [CrossRef] [PubMed]
111. Zhang, J.Z.; Reisner, E. Advancing photosystem II photoelectrochemistry for semi-artificial photosynthesis. *Nat. Rev. Chem.* **2020**, *4*, 6–21. [CrossRef]
112. Kornienko, N.; Zhang, J.Z.; Sokol, K.P.; Lamaison, S.; Fantuzzi, A.; van Grondelle, R.; Rutherford, A.W.; Reisner, E. Oxygenic Photoreactivity in Photosystem II Studied by Rotating Ring Disk Electrochemistry. *J. Am. Chem. Soc.* **2018**, *140*, 17923–17931. [CrossRef]
113. Zhao, F.; Conzuelo, F.; Hartmann, V.; Li, H.; Nowaczyk, M.M.; Plumeré, N.; Rögner, M.; Schuhmann, W. Light Induced H_2 Evolution from a Biophotocathode Based on Photosystem 1–Pt Nanoparticles Complexes Integrated in Solvated Redox Polymers Films. *J. Phys. Chem. B* **2015**, *119*, 13726–13731. [CrossRef]
114. Zhao, F.; Plumeré, N.; Nowaczyk, M.M.; Ruff, A.; Schuhmann, W.; Conzuelo, F. Interrogation of a PS1-Based Photocathode by Means of Scanning Photoelectrochemical Microscopy. *Small* **2017**, *13*, 1604093. [CrossRef]
115. Zhao, F.; Hardt, S.; Hartmann, V.; Zhang, H.; Nowaczyk, M.M.; Rögner, M.; Plumeré, N.; Schuhmann, W.; Conzuelo, F. Light-induced formation of partially reduced oxygen species limits the lifetime of photosystem 1-based biocathodes. *Nat. Commun.* **2018**, *9*, 1973. [CrossRef] [PubMed]
116. Demirel, G.; Usta, H.; Yilmaz, M.; Celik, M.; Alidagi, H.A.; Buyukserin, F. Surface-enhanced Raman spectroscopy (SERS): An adventure from plasmonic metals to organic semiconductors as SERS platforms. *J. Mater. Chem. C* **2018**, *6*, 5314–5335. [CrossRef]
117. Grytsyk, N.; Boubegtiten-Fezoua, Z.; Javahiraly, N.; Omeis, F.; Devaux, E.; Hellwig, P. Surface-enhanced resonance Raman spectroscopy of heme proteins on a gold grid electrode. *Spectrochim. Acta Part A* **2020**, *230*, 118081. [CrossRef] [PubMed]
118. Hrabakova, J.; Ataka, K.; Heberle, J.; Hildebrandt, P.; Murgida, D.H. Long distance electron transfer in cytochrome *c* oxidase immobilised on electrodes. A surface enhanced resonance Raman spectroscopic study. *Phys. Chem. Chem. Phys.* **2006**, *8*, 759–766. [CrossRef]
119. Sezer, M.; Frielingsdorf, S.; Millo, D.; Heidary, N.; Utesch, T.; Mroginski, M.-A.; Friedrich, B.; Hildebrandt, P.; Zebger, I.; Weidinger, I.M. Role of the HoxZ subunit in the electron transfer pathway of the membrane-bound NiFe-hydrogenase from *Ralstonia eutropha* immobilized on electrodes. *J. Phys. Chem. B* **2011**, *115*, 10368–10374. [CrossRef]
120. Ataka, K.; Stripp, S.T.; Heberle, J. Surface-enhanced infrared absorption spectroscopy (SEIRAS) to probe monolayers of membrane proteins. *Biochim. Biophys. Acta* **2013**, *1828*, 2283–2293. [CrossRef]
121. Wiebalck, S.; Kozuch, J.; Forbrig, E.; Tzschucke, C.C.; Jeuken, L.J.C.; Hildebrandt, P. Monitoring the Transmembrane Proton Gradient Generated by Cytochrome bo_3 in Tethered Bilayer Lipid Membranes Using SEIRA Spectroscopy. *J. Phys. Chem. B* **2016**, *120*, 2249–2256. [CrossRef]

122. López-Lorente, Á.I.; Kranz, C. Recent advances in biomolecular vibrational spectroelectrochemistry. *Curr. Opin. Electrochem.* **2017**, *5*, 106–113. [CrossRef]
123. Zhai, Y.; Zhu, Z.; Zhou, S.; Zhu, C.; Dong, S. Recent advances in spectroelectrochemistry. *Nanoscale* **2018**, *10*, 3089–3111. [CrossRef]
124. Melin, F.; Hellwig, P. Redox Properties of the Membrane Proteins from the Respiratory Chain. *Chem. Rev.* **2020**, *120*, 10244–10297. [CrossRef]
125. Białek, R.; Friebe, V.; Ruff, A.; Jones, M.R.; Frese, R.; Gibasiewicz, K. In situ spectroelectrochemical investigation of a biophotoelectrode based on photoreaction centers embedded in a redox hydrogel. *Electrochim. Acta* **2020**, *330*, 135190. [CrossRef]
126. Ash, P.A.; Vincent, K.A. Spectroscopic analysis of immobilised redox enzymes under direct electrochemical control. *Chem. Commun.* **2012**, *48*, 1400–1409. [CrossRef] [PubMed]
127. Ritter, M.; Anderka, O.; Ludwig, B.; Mäntele, W.; Hellwig, P. Electrochemical and FTIR spectroscopic characterization of the cytochrome bc_1 complex from *Paracoccus denitrificans*: Evidence for protonation reactions coupled to quinone binding. *Biochemistry* **2003**, *42*, 12391–12399. [CrossRef] [PubMed]
128. Baymann, F.; Robertson, D.E.; Dutton, P.L.; Mäntele, W. Electrochemical and spectroscopic investigations of the cytochrome bc_1 complex from *Rhodobacter capsulatus*. *Biochemistry* **1999**, *38*, 13188–13199. [CrossRef]
129. Kato, Y.; Sugiura, M.; Oda, A.; Watanabe, T. Spectroelectrochemical determination of the redox potential of pheophytin a, the primary electron acceptor in photosystem II. *Proc. Natl. Acad. Sci. USA* **2009**, *106*, 17365–17370. [CrossRef]
130. Nakamura, A.; Suzawa, T.; Kato, Y.; Watanabe, T. Species dependence of the redox potential of the primary electron donor P700 in photosystem I of oxygenic photosynthetic organisms revealed by spectroelectrochemistry. *Plant. Cell Physiol.* **2011**, *52*, 815–823. [CrossRef]
131. Haas, A.S.; Pilloud, D.L.; Reddy, K.S.; Babcock, G.T.; Moser, C.C.; Blasie, J.K.; Dutton, P.L. Cytochrome *c* and Cytochrome *c* Oxidase: Monolayer Assemblies and Catalysis. *J. Phys. Chem. B* **2001**, *105*, 11351–11362. [CrossRef]
132. Lee, C.-Y.; Reuillard, B.; Sokol, K.P.; Laftsoglou, T.; Lockwood, C.W.J.; Rowe, S.F.; Hwang, E.T.; Fontecilla-Camps, J.C.; Jeuken, L.J.C.; Butt, J.N.; et al. A decahaem cytochrome as an electron conduit in protein-enzyme redox processes. *Chem. Commun.* **2016**, *52*, 7390–7393. [CrossRef]
133. Nowak, C.; Laredo, T.; Gebert, J.; Lipkowski, J.; Gennis, R.B.; Ferguson-Miller, S.; Knoll, W.; Naumann, R.L.C. 2D-SEIRA spectroscopy to highlight conformational changes of the cytochrome *c* oxidase induced by direct electron transfer. *Metallomics* **2011**, *3*, 619–627. [CrossRef]
134. Steininger, C.; Reiner-Rozman, C.; Schwaighofer, A.; Knoll, W.; Naumann, R.L.C. Kinetics of cytochrome *c* oxidase from *R. sphaeroides* initiated by direct electron transfer followed by tr-SEIRAS. *Bioelectrochemistry* **2016**, *112*, 1–8. [CrossRef]
135. Bogner, A.; Jouneau, P.-H.; Thollet, G.; Basset, D.; Gauthier, C. A history of scanning electron microscopy developments: Towards "wet-STEM" imaging. *Micron* **2007**, *38*, 390–401. [CrossRef] [PubMed]
136. Monsalve, K.; Roger, M.; Gutierrez-Sanchez, C.; Ilbert, M.; Nitsche, S.; Byrne-Kodjabachian, D.; Marchi, V.; Lojou, E. Hydrogen bioelectrooxidation on gold nanoparticle-based electrodes modified by *Aquifex aeolicus* hydrogenase: Application to hydrogen/oxygen enzymatic biofuel cells. *Bioelectrochemistry* **2015**, *106*, 47–55. [CrossRef] [PubMed]
137. Alsteens, D.; Gaub, H.E.; Newton, R.; Pfreundschuh, M.; Gerber, C.; Müller, D.J. Atomic force microscopy-based characterization and design of biointerfaces. *Nat. Rev. Mater.* **2017**, *2*, 17008. [CrossRef]
138. Connell, S.D.; Smith, D.A. The atomic force microscope as a tool for studying phase separation in lipid membranes. *Mol. Membr. Biol.* **2006**, *23*, 17–28. [CrossRef] [PubMed]
139. Goodchild, J.A.; Walsh, D.L.; Connell, S.D. Nanoscale Substrate Roughness Hinders Domain Formation in Supported Lipid Bilayers. *Langmuir* **2019**, *35*, 15352–15363. [CrossRef] [PubMed]
140. Aufderhorst-Roberts, A.; Chandra, U.; Connell, S.D. Three-Phase Coexistence in Lipid Membranes. *Biophys. J.* **2017**, *112*, 313–324. [CrossRef] [PubMed]
141. Frederix, P.L.T.M.; Bosshart, P.D.; Engel, A. Atomic force microscopy of biological membranes. *Biophys. J.* **2009**, *96*, 329–338. [CrossRef]
142. Carvalho, F.A.; Connell, S.; Miltenberger-Miltenyi, G.; Pereira, S.V.; Tavares, A.; Ariëns, R.A.S.; Santos, N.C. Atomic force microscopy-based molecular recognition of a fibrinogen receptor on human erythrocytes. *ACS Nano* **2010**, *4*, 4609–4620. [CrossRef]

143. Heath, G.R.; Roth, J.; Connell, S.D.; Evans, S.D. Diffusion in low-dimensional lipid membranes. *Nano Lett.* **2014**, *14*, 5984–5988. [CrossRef]
144. Heath, G.R.; Scheuring, S. Advances in high-speed atomic force microscopy (HS-AFM) reveal dynamics of transmembrane channels and transporters. *Curr. Opin. Struct. Biol.* **2019**, *57*, 93–102. [CrossRef]
145. Hao, X.; Zhang, J.; Christensen, H.E.M.; Wang, H.; Ulstrup, J. Electrochemical single-molecule AFM of the redox metalloenzyme copper nitrite reductase in action. *ChemPhysChem* **2012**, *13*, 2919–2924. [CrossRef] [PubMed]
146. Tang, J.; Yan, X.; Huang, W.; Engelbrekt, C.; Duus, J.Ø.; Ulstrup, J.; Xiao, X.; Zhang, J. Bilirubin oxidase oriented on novel type three-dimensional biocathodes with reduced graphene aggregation for biocathode. *Biosens. Bioelectron.* **2020**, *167*, 112500. [CrossRef] [PubMed]
147. González Arzola, K.; Gimeno, Y.; Arévalo, M.C.; Falcón, M.A.; Hernández Creus, A. Electrochemical and AFM characterization on gold and carbon electrodes of a high redox potential laccase from *Fusarium proliferatum*. *Bioelectrochemistry* **2010**, *79*, 17–24. [CrossRef] [PubMed]
148. Ciaccafava, A.; Infossi, P.; Ilbert, M.; Guiral, M.; Lecomte, S.; Giudici-Orticoni, M.T.; Lojou, E. Electrochemistry, AFM, and PM-IRRA spectroscopy of immobilized hydrogenase: Role of a hydrophobic helix in enzyme orientation for efficient H_2 oxidation. *Angew. Chem. Int. Ed.* **2012**, *51*, 953–956. [CrossRef]
149. Gutiérrez-Sanz, O.; Olea, D.; Pita, M.; Batista, A.P.; Alonso, A.; Pereira, M.M.; Vélez, M.; de Lacey, A.L. Reconstitution of respiratory complex I on a biomimetic membrane supported on gold electrodes. *Langmuir* **2014**, *30*, 9007–9015. [CrossRef]
150. Neupane, S.; de Smet, Y.; Renner, F.U.; Losada-Pérez, P. Quartz Crystal Microbalance with Dissipation Monitoring: A Versatile Tool to Monitor Phase Transitions in Biomimetic Membranes. *Front. Mater.* **2018**, *5*, 46. [CrossRef]
151. Giess, F.; Friedrich, M.G.; Heberle, J.; Naumann, R.L.; Knoll, W. The protein-tethered lipid bilayer: A novel mimic of the biological membrane. *Biophys. J.* **2004**, *87*, 3213–3220. [CrossRef]
152. Lam, K.B.; Irwin, E.F.; Healy, K.E.; Lin, L. Bioelectrocatalytic self-assembled thylakoids for micro-power and sensing applications. *Sens. Actuators B* **2006**, *117*, 480–487. [CrossRef]
153. Chen, H.; Dong, F.; Minteer, S.D. The progress and outlook of bioelectrocatalysis for the production of chemicals, fuels and materials. *Nat. Catal.* **2020**, *3*, 225–244. [CrossRef]
154. Ruff, A.; Conzuelo, F.; Schuhmann, W. Bioelectrocatalysis as the basis for the design of enzyme-based biofuel cells and semi-artificial biophotoelectrodes. *Nat. Catal.* **2020**, *3*, 214–224. [CrossRef]
155. Grattieri, M.; Beaver, K.; Gaffney, E.M.; Dong, F.; Minteer, S.D. Advancing the fundamental understanding and practical applications of photo-bioelectrocatalysis. *Chem. Commun.* **2020**, *56*, 8553–8568. [CrossRef] [PubMed]
156. Milton, R.D.; Minteer, S.D. Direct enzymatic bioelectrocatalysis: Differentiating between myth and reality. *J. R. Soc. Interface* **2017**, *14*, 20170253. [CrossRef] [PubMed]
157. Xiao, X.; Xia, H.-Q.; Wu, R.; Bai, L.; Yan, L.; Magner, E.; Cosnier, S.; Lojou, E.; Zhu, Z.; Liu, A. Tackling the Challenges of Enzymatic (Bio)Fuel Cells. *Chem. Rev.* **2019**, *119*, 9509–9558. [CrossRef] [PubMed]
158. Cooney, M.J.; Svoboda, V.; Lau, C.; Martin, G.; Minteer, S.D. Enzyme catalysed biofuel cells. *Energy Environ. Sci.* **2008**, *1*, 320–337. [CrossRef]
159. Minteer, S.D.; Liaw, B.Y.; Cooney, M.J. Enzyme-based biofuel cells. *Curr. Opin. Biotechnol.* **2007**, *18*, 228–234. [CrossRef]
160. Sheldon, R.A.; Woodley, J.M. Role of Biocatalysis in Sustainable Chemistry. *Chem. Rev.* **2018**, *118*, 801–838. [CrossRef]
161. Mazurenko, I.; Wang, X.; de Poulpiquet, A.; Lojou, E. H_2/O_2 enzymatic fuel cells: From proof-of-concept to powerful devices. *Sustain. Energy Fuels* **2017**, *1*, 1475–1501. [CrossRef]
162. Vincent, K.A.; Parkin, A.; Armstrong, F.A. Investigating and exploiting the electrocatalytic properties of hydrogenases. *Chem. Rev.* **2007**, *107*, 4366–4413. [CrossRef]
163. Plumeré, N.; Rüdiger, O.; Oughli, A.A.; Williams, R.; Vivekananthan, J.; Pöller, S.; Schuhmann, W.; Lubitz, W. A redox hydrogel protects hydrogenase from high-potential deactivation and oxygen damage. *Nat. Chem.* **2014**, *6*, 822–827. [CrossRef]
164. Ruff, A.; Szczesny, J.; Marković, N.; Conzuelo, F.; Zacarias, S.; Pereira, I.A.C.; Lubitz, W.; Schuhmann, W. A fully protected hydrogenase/polymer-based bioanode for high-performance hydrogen/glucose biofuel cells. *Nat. Commun.* **2018**, *9*, 3675. [CrossRef]

165. Vincent, K.A.; Cracknell, J.A.; Lenz, O.; Zebger, I.; Friedrich, B.; Armstrong, F.A. Electrocatalytic hydrogen oxidation by an enzyme at high carbon monoxide or oxygen levels. *Proc. Natl. Acad. Sci. USA* **2005**, *102*, 16951–16954. [CrossRef] [PubMed]
166. Vincent, K.A.; Cracknell, J.A.; Clark, J.R.; Ludwig, M.; Lenz, O.; Friedrich, B.; Armstrong, F.A. Electricity from low-level H_2 in still air-an ultimate test for an oxygen tolerant hydrogenase. *Chem. Commun.* **2006**, 5033–5035. [CrossRef] [PubMed]
167. So, K.; Kitazumi, Y.; Shirai, O.; Nishikawa, K.; Higuchi, Y.; Kano, K. Direct electron transfer-type dual gas diffusion H_2/O_2 biofuel cells. *J. Mater. Chem. A* **2016**, *4*, 8742–8749. [CrossRef]
168. Xia, H.-Q.; So, K.; Kitazumi, Y.; Shirai, O.; Nishikawa, K.; Higuchi, Y.; Kano, K. Dual gas-diffusion membrane- and mediatorless dihydrogen/air-breathing biofuel cell operating at room temperature. *J. Power Sources* **2016**, *335*, 105–112. [CrossRef]
169. Mano, N.; de Poulpiquet, A. O_2 Reduction in Enzymatic Biofuel Cells. *Chem. Rev.* **2018**, *118*, 2392–2468. [CrossRef]
170. Katz, E.; Willner, I.; Kotlyar, A.B. A non-compartmentalized glucose|O_2 biofuel cell by bioengineered electrode surfaces. *J. Electroanal. Chem.* **1999**, *479*, 64–68. [CrossRef]
171. Katz, E.; Willner, I. A biofuel cell with electrochemically switchable and tunable power output. *J. Am. Chem. Soc.* **2003**, *125*, 6803–6813. [CrossRef]
172. Wang, X.; Clément, R.; Roger, M.; Bauzan, M.; Mazurenko, I.; de Poulpiquet, A.; Ilbert, M.; Lojou, E. Bacterial Respiratory Chain Diversity Reveals a Cytochrome *c* Oxidase Reducing O_2 at Low Overpotentials. *J. Am. Chem. Soc.* **2019**, *141*, 11093–11102. [CrossRef]
173. Sakai, K.; Kitazumi, Y.; Shirai, O.; Takagi, K.; Kano, K. High-Power Formate/Dioxygen Biofuel Cell Based on Mediated Electron Transfer Type Bioelectrocatalysis. *ACS Catal.* **2017**, *7*, 5668–5673. [CrossRef]
174. Reda, T.; Plugge, C.M.; Abram, N.J.; Hirst, J. Reversible interconversion of carbon dioxide and formate by an electroactive enzyme. *Proc. Natl. Acad. Sci. USA* **2008**, *105*, 10654–10658. [CrossRef]
175. Milton, R.D.; Minteer, S.D. Enzymatic Bioelectrosynthetic Ammonia Production: Recent Electrochemistry of Nitrogenase, Nitrate Reductase, and Nitrite Reductase. *ChemPlusChem* **2017**, *82*, 513–521. [CrossRef] [PubMed]
176. Teodor, A.H.; Bruce, B.D. Putting Photosystem I to Work: Truly Green Energy. *Trends Biotechnol.* **2020**, *38*, 1329–1342. [CrossRef] [PubMed]
177. Friebe, V.M.; Frese, R.N. Photosynthetic reaction center-based biophotovoltaics. *Curr. Opin. Electrochem.* **2017**, *5*, 126–134. [CrossRef]
178. Koblizek, M.; Masojidek, J.; Komenda, J.; Kucera, T.; Pilloton, R.; Mattoo, A.K.; Giardi, M.T. A sensitive photosystem II-based biosensor for detection of a class of herbicides. *Biotechnol. Bioeng.* **1998**, *60*, 664–669. [CrossRef]
179. Swainsbury, D.J.K.; Friebe, V.M.; Frese, R.N.; Jones, M.R. Evaluation of a biohybrid photoelectrochemical cell employing the purple bacterial reaction centre as a biosensor for herbicides. *Biosens. Bioelectron.* **2014**, *58*, 172–178. [CrossRef]
180. Yehezkeli, O.; Tel-Vered, R.; Wasserman, J.; Trifonov, A.; Michaeli, D.; Nechushtai, R.; Willner, I. Integrated photosystem II-based photo-bioelectrochemical cells. *Nat. Commun.* **2012**, *3*, 781. [CrossRef]
181. Riedel, M.; Wersig, J.; Ruff, A.; Schuhmann, W.; Zouni, A.; Lisdat, F. A Z-Scheme-Inspired Photobioelectrochemical H_2O/O_2 Cell with a 1 V Open-Circuit Voltage Combining Photosystem II and PbS Quantum Dots. *Angew. Chem. Int. Ed.* **2019**, *58*, 801–805. [CrossRef]
182. Nam, D.H.; Zhang, J.Z.; Andrei, V.; Kornienko, N.; Heidary, N.; Wagner, A.; Nakanishi, K.; Sokol, K.P.; Slater, B.; Zebger, I.; et al. Solar Water Splitting with a Hydrogenase Integrated in Photoelectrochemical Tandem Cells. *Angew. Chem. Int. Ed.* **2018**, *57*, 10595–10599. [CrossRef]
183. Sokol, K.P.; Robinson, W.E.; Warnan, J.; Kornienko, N.; Nowaczyk, M.M.; Ruff, A.; Zhang, J.Z.; Reisner, E. Bias-free photoelectrochemical water splitting with photosystem II on a dye-sensitized photoanode wired to hydrogenase. *Nat. Energy* **2018**, *3*, 944–951. [CrossRef]
184. Sokol, K.P.; Robinson, W.E.; Oliveira, A.R.; Warnan, J.; Nowaczyk, M.M.; Ruff, A.; Pereira, I.A.C.; Reisner, E. Photoreduction of CO_2 with a Formate Dehydrogenase Driven by Photosystem II Using a Semi-artificial Z-Scheme Architecture. *J. Am. Chem. Soc.* **2018**, *140*, 16418–16422. [CrossRef]
185. Lubner, C.E.; Grimme, R.; Bryant, D.A.; Golbeck, J.H. Wiring photosystem I for direct solar hydrogen production. *Biochemistry* **2010**, *49*, 404–414. [CrossRef] [PubMed]

186. Tapia, C.; Milton, R.D.; Pankratova, G.; Minteer, S.D.; Åkerlund, H.-E.; Leech, D.; De Lacey, A.L.; Pita, M.; Gorton, L. Wiring of Photosystem I and Hydrogenase on an Electrode for Photoelectrochemical H$_2$ Production by using Redox Polymers for Relatively Positive Onset Potential. *ChemElectroChem* **2017**, *4*, 90–95. [CrossRef]
187. Zhao, F.; Wang, P.; Ruff, A.; Hartmann, V.; Zacarias, S.; Pereira, I.A.C.; Nowaczyk, M.M.; Rögner, M.; Conzuelo, F.; Schuhmann, W. A photosystem I monolayer with anisotropic electron flow enables Z-scheme like photosynthetic water splitting. *Energy Environ. Sci.* **2019**, *12*, 3133–3143. [CrossRef]
188. Hancock, A.M.; Meredith, S.A.; Connell, S.D.; Jeuken, L.J.C.; Adams, P.G. Proteoliposomes as energy transferring nanomaterials: Enhancing the spectral range of light-harvesting proteins using lipid-linked chromophores. *Nanoscale* **2019**, *11*, 16284–16292. [CrossRef]
189. Liu, J.; Friebe, V.M.; Frese, R.N.; Jones, M.R. Polychromatic solar energy conversion in pigment-protein chimeras that unite the two kingdoms of (bacterio)chlorophyll-based photosynthesis. *Nat. Commun.* **2020**, *11*, 1542. [CrossRef]
190. Liu, J.; Mantell, J.; Jones, M.R. Minding the Gap between Plant and Bacterial Photosynthesis within a Self-Assembling Biohybrid Photosystem. *ACS Nano* **2020**, *14*, 4536–4549. [CrossRef]
191. Moehlenbrock, M.J.; Minteer, S.D. Extended lifetime biofuel cells. *Chem. Soc. Rev.* **2008**, *37*, 1188–1196. [CrossRef]
192. Rideau, E.; Dimova, R.; Schwille, P.; Wurm, F.R.; Landfester, K. Liposomes and polymersomes: A comparative review towards cell mimicking. *Chem. Soc. Rev.* **2018**, *47*, 8572–8610. [CrossRef]
193. Discher, D.E.; Eisenberg, A. Polymer vesicles. *Science* **2002**, *297*, 967–973. [CrossRef]
194. Poschenrieder, S.T.; Schiebel, S.K.; Castiglione, K. Stability of polymersomes with focus on their use as nanoreactors. *Eng. Life Sci.* **2018**, *18*, 101–113. [CrossRef]
195. Beales, P.A.; Khan, S.; Muench, S.P.; Jeuken, L.J.C. Durable vesicles for reconstitution of membrane proteins in biotechnology. *Biochem. Soc. Trans.* **2017**, *45*, 15–26. [CrossRef] [PubMed]
196. Khan, S.; Li, M.; Muench, S.P.; Jeuken, L.J.C.; Beales, P.A. Durable proteo-hybrid vesicles for the extended functional lifetime of membrane proteins in bionanotechnology. *Chem. Commun.* **2016**, *52*, 11020–11023. [CrossRef] [PubMed]
197. Seneviratne, R.; Khan, S.; Moscrop, E.; Rappolt, M.; Muench, S.P.; Jeuken, L.J.C.; Beales, P.A. A reconstitution method for integral membrane proteins in hybrid lipid-polymer vesicles for enhanced functional durability. *Methods* **2018**, *147*, 142–149. [CrossRef]
198. Marušič, N.; Otrin, L.; Zhao, Z.; Lira, R.B.; Kyrilis, F.L.; Hamdi, F.; Kastritis, P.L.; Vidaković-Koch, T.; Ivanov, I.; Sundmacher, K.; et al. Constructing artificial respiratory chain in polymer compartments: Insights into the interplay between bo_3 oxidase and the membrane. *Proc. Natl. Acad. Sci. USA* **2020**, *117*, 15006–15017. [CrossRef] [PubMed]
199. Otrin, L.; Marušič, N.; Bednarz, C.; Vidaković-Koch, T.; Lieberwirth, I.; Landfester, K.; Sundmacher, K. Toward Artificial Mitochondrion: Mimicking Oxidative Phosphorylation in Polymer and Hybrid Membranes. *Nano Lett.* **2017**, *17*, 6816–6821. [CrossRef]
200. Smirnova, I.A.; Ädelroth, P.; Brzezinski, P. Extraction and liposome reconstitution of membrane proteins with their native lipids without the use of detergents. *Sci. Rep.* **2018**, *8*, 14950. [CrossRef]
201. Lee, S.C.; Knowles, T.J.; Postis, V.L.G.; Jamshad, M.; Parslow, R.A.; Lin, Y.-P.; Goldman, A.; Sridhar, P.; Overduin, M.; Muench, S.P.; et al. A method for detergent-free isolation of membrane proteins in their local lipid environment. *Nat. Protoc.* **2016**, *11*, 1149–1162. [CrossRef]
202. Kumar, A.; Hsu, L.H.-H.; Kavanagh, P.; Barrière, F.; Lens, P.N.L.; Lapinsonnière, L.; Lienhard V, J.H.; Schröder, U.; Jiang, X.; Leech, D. The ins and outs of microorganism–electrode electron transfer reactions. *Nat. Rev. Chem.* **2017**, *1*, 24. [CrossRef]
203. Rabaey, K.; Rozendal, R.A. Microbial electrosynthesis—Revisiting the electrical route for microbial production. *Nat. Rev. Microbiol.* **2010**, *8*, 706–716. [CrossRef]
204. Logan, B.E.; Rossi, R.; Ragab, A.; Saikaly, P.E. Electroactive microorganisms in bioelectrochemical systems. *Nat. Rev. Microbiol.* **2019**, *17*, 307–319. [CrossRef]
205. Jiang, Y.; Zeng, R.J. Bidirectional extracellular electron transfers of electrode-biofilm: Mechanism and application. *Bioresour. Technol.* **2019**, *271*, 439–448. [CrossRef] [PubMed]
206. Shi, M.; Jiang, Y.; Shi, L. Electromicrobiology and biotechnological applications of the exoelectrogens *Geobacter* and *Shewanella* spp. *Sci. China Technol. Sci.* **2019**, *62*, 1670–1678. [CrossRef]

207. Gregory, K.B.; Bond, D.R.; Lovley, D.R. Graphite electrodes as electron donors for anaerobic respiration. *Environ. Microbiol.* **2004**, *6*, 596–604. [CrossRef] [PubMed]
208. Geelhoed, J.S.; Stams, A.J.M. Electricity-assisted biological hydrogen production from acetate by *Geobacter sulfurreducens*. *Environ. Sci. Technol.* **2011**, *45*, 815–820. [CrossRef] [PubMed]
209. Ross, D.E.; Flynn, J.M.; Baron, D.B.; Gralnick, J.A.; Bond, D.R. Towards electrosynthesis in *shewanella*: Energetics of reversing the Mtr pathway for reductive metabolism. *PLoS ONE* **2011**, *6*, e16649. [CrossRef]
210. Le, Q.A.T.; Kim, H.G.; Kim, Y.H. Electrochemical synthesis of formic acid from CO_2 catalyzed by *Shewanella oneidensis* MR-1 whole-cell biocatalyst. *Enzyme Microb. Technol.* **2018**, *116*, 1–5. [CrossRef]
211. La Cava, E.; Guionet, A.; Saito, J.; Okamoto, A. Involvement of Proton Transfer for Carbon Dioxide Reduction Coupled with Extracellular Electron Uptake in *Shewanella oneidensis* MR-1. *Electroanalysis* **2020**, *32*, 1659–1663. [CrossRef]
212. Shin, H.J.; Jung, K.A.; Nam, C.W.; Park, J.M. A genetic approach for microbial electrosynthesis system as biocommodities production platform. *Bioresour. Technol.* **2017**, *245*, 1421–1429. [CrossRef]
213. Min, D.; Cheng, L.; Zhang, F.; Huang, X.-N.; Li, D.-B.; Liu, D.-F.; Lau, T.-C.; Mu, Y.; Yu, H.-Q. Enhancing Extracellular Electron Transfer of *Shewanella oneidensis* MR-1 through Coupling Improved Flavin Synthesis and Metal-Reducing Conduit for Pollutant Degradation. *Environ. Sci. Technol.* **2017**, *51*, 5082–5089. [CrossRef]
214. Yang, Y.; Ding, Y.; Hu, Y.; Cao, B.; Rice, S.A.; Kjelleberg, S.; Song, H. Enhancing Bidirectional Electron Transfer of *Shewanella oneidensis* by a Synthetic Flavin Pathway. *ACS Synth. Biol.* **2015**, *4*, 815–823. [CrossRef]
215. Glaven, S.M. Bioelectrochemical systems and synthetic biology: More power, more products. *Microb. Biotechnol.* **2019**, *12*, 819–823. [CrossRef] [PubMed]
216. Ueki, T.; Nevin, K.P.; Woodard, T.L.; Aklujkar, M.A.; Holmes, D.E.; Lovley, D.R. Construction of a *Geobacter* Strain With Exceptional Growth on Cathodes. *Front. Microbiol.* **2018**, *9*, 1512. [CrossRef] [PubMed]
217. La, J.A.; Jeon, J.-M.; Sang, B.-I.; Yang, Y.-H.; Cho, E.C. A Hierarchically Modified Graphite Cathode with Au Nanoislands, Cysteamine, and Au Nanocolloids for Increased Electricity-Assisted Production of Isobutanol by Engineered *Shewanella oneidensis* MR-1. *ACS Appl. Mater. Interfaces* **2017**, *9*, 43563–43574. [CrossRef] [PubMed]
218. Rowe, A.R.; Rajeev, P.; Jain, A.; Pirbadian, S.; Okamoto, A.; Gralnick, J.A.; El-Naggar, M.Y.; Nealson, K.H. Tracking Electron Uptake from a Cathode into *Shewanella* Cells: Implications for Energy Acquisition from Solid-Substrate Electron Donors. *mBio* **2018**, *9*, e02203-17. [CrossRef]
219. Tefft, N.M.; TerAvest, M.A. Reversing an Extracellular Electron Transfer Pathway for Electrode-Driven Acetoin Reduction. *ACS Synth. Biol.* **2019**, *8*, 1590–1600. [CrossRef]
220. Xie, Q.; Lu, Y.; Tang, L.; Zeng, G.; Yang, Z.; Fan, C.; Wang, J.; Atashgahi, S. The mechanism and application of bidirectional extracellular electron transport in the field of energy and environment. *Crit. Rev. Environ. Sci. Technol.* **2020**, *6*, 1–46.
221. Hasan, K.; Çevik, E.; Sperling, E.; Packer, M.A.; Leech, D.; Gorton, L. Photoelectrochemical Wiring of *Paulschulzia pseudovolvox* (Algae) to Osmium Polymer Modified Electrodes for Harnessing Solar Energy. *Adv. Energy Mater.* **2015**, *5*, 1501100. [CrossRef]
222. Bombelli, P.; Müller, T.; Herling, T.W.; Howe, C.J.; Knowles, T.P.J. A High Power-Density, Mediator-Free, Microfluidic Biophotovoltaic Device for Cyanobacterial Cells. *Adv. Energy Mater.* **2015**, *5*, 1–6. [CrossRef]
223. McCormick, A.J.; Bombelli, P.; Scott, A.M.; Philips, A.J.; Smith, A.G.; Fisher, A.C.; Howe, C.J. Photosynthetic biofilms in pure culture harness solar energy in a mediatorless bio-photovoltaic cell (BPV) system. *Energy Environ. Sci.* **2011**, *4*, 4699. [CrossRef]
224. Sekar, N.; Umasankar, Y.; Ramasamy, R.P. Photocurrent generation by immobilized cyanobacteria via direct electron transport in photo-bioelectrochemical cells. *Phys. Chem. Chem. Phys.* **2014**, *16*, 7862–7871. [CrossRef]
225. Zhang, J.Z.; Bombelli, P.; Sokol, K.P.; Fantuzzi, A.; Rutherford, A.W.; Howe, C.J.; Reisner, E. Photoelectrochemistry of Photosystem II in Vitro vs in Vivo. *J. Am. Chem. Soc.* **2018**, *140*, 6–9. [CrossRef] [PubMed]
226. Grattieri, M.; Patterson, S.; Copeland, J.; Klunder, K.; Minteer, S.D. Purple Bacteria and 3D Redox Hydrogels for Bioinspired Photo-bioelectrocatalysis. *ChemSusChem* **2020**, *13*, 230–237. [CrossRef] [PubMed]
227. Pankratov, D.; Pankratova, G.; Gorton, L. Thylakoid membrane–based photobioelectrochemical systems: Achievements, limitations, and perspectives. *Curr. Opin. Electrochem.* **2020**, *19*, 49–54. [CrossRef]
228. Wenzel, T.; Härtter, D.; Bombelli, P.; Howe, C.J.; Steiner, U. Porous translucent electrodes enhance current generation from photosynthetic biofilms. *Nat. Commun.* **2018**, *9*, 1299. [CrossRef]

229. Sakimoto, K.K.; Wong, A.B.; Yang, P. Self-photosensitization of nonphotosynthetic bacteria for solar-to-chemical production. *Science* **2016**, *351*, 74–77. [CrossRef]
230. Cestellos-Blanco, S.; Zhang, H.; Kim, J.M.; Shen, Y.-X.; Yang, P. Photosynthetic semiconductor biohybrids for solar-driven biocatalysis. *Nat. Catal.* **2020**, *3*, 245–255. [CrossRef]
231. Hartshorne, R.S.; Reardon, C.L.; Ross, D.; Nuester, J.; Clarke, T.A.; Gates, A.J.; Mills, P.C.; Fredrickson, J.K.; Zachara, J.M.; Shi, L.; et al. Characterization of an electron conduit between bacteria and the extracellular environment. *Proc. Natl. Acad. Sci. USA* **2009**, *106*, 22169–22174. [CrossRef]
232. Shi, L.; Deng, S.; Marshall, M.J.; Wang, Z.; Kennedy, D.W.; Dohnalkova, A.C.; Mottaz, H.M.; Hill, E.A.; Gorby, Y.A.; Beliaev, A.S.; et al. Direct involvement of type II secretion system in extracellular translocation of *Shewanella oneidensis* outer membrane cytochromes MtrC and OmcA. *J. Bacteriol.* **2008**, *190*, 5512–5516. [CrossRef]
233. Ainsworth, E.V.; Lockwood, C.W.J.; White, G.F.; Hwang, E.T.; Sakai, T.; Gross, M.A.; Richardson, D.J.; Clarke, T.A.; Jeuken, L.J.C.; Reisner, E.; et al. Photoreduction of *Shewanella oneidensis* Extracellular Cytochromes by Organic Chromophores and Dye-Sensitized TiO_2. *Chembiochem* **2016**, *17*, 2324–2333. [CrossRef]
234. Hwang, E.T.; Sheikh, K.; Orchard, K.L.; Hojo, D.; Radu, V.; Lee, C.-Y.; Ainsworth, E.; Lockwood, C.; Gross, M.A.; Adschiri, T.; et al. A Decaheme Cytochrome as a Molecular Electron Conduit in Dye-Sensitized Photoanodes. *Adv. Funct. Mater.* **2015**, *25*, 2308–2315. [CrossRef]
235. Hwang, E.T.; Orchard, K.L.; Hojo, D.; Beton, J.; Lockwood, C.W.J.; Adschiri, T.; Butt, J.N.; Reisner, E.; Jeuken, L.J.C. Exploring Step-by-Step Assembly of Nanoparticle:Cytochrome Biohybrid Photoanodes. *ChemElectroChem* **2017**, *4*, 1959–1968. [CrossRef] [PubMed]
236. Rowe, S.F.; Le Gall, G.; Ainsworth, E.V.; Davies, J.A.; Lockwood, C.W.J.; Shi, L.; Elliston, A.; Roberts, I.N.; Waldron, K.W.; Richardson, D.J.; et al. Light-Driven H_2 Evolution and C=C or C=O Bond Hydrogenation by *Shewanella oneidensis*: A Versatile Strategy for Photocatalysis by Nonphotosynthetic Microorganisms. *ACS Catal.* **2017**, *7*, 7558–7566. [CrossRef]
237. Stikane, A.; Hwang, E.T.; Ainsworth, E.V.; Piper, S.E.H.; Critchley, K.; Butt, J.N.; Reisner, E.; Jeuken, L.J.C. Towards compartmentalized photocatalysis: Multihaem proteins as transmembrane molecular electron conduits. *Faraday Discuss.* **2019**, *215*, 26–38. [CrossRef] [PubMed]

Publisher's Note: MDPI stays neutral with regard to jurisdictional claims in published maps and institutional affiliations.

© 2020 by the authors. Licensee MDPI, Basel, Switzerland. This article is an open access article distributed under the terms and conditions of the Creative Commons Attribution (CC BY) license (http://creativecommons.org/licenses/by/4.0/).

Review

Recent Progress in Applications of Enzymatic Bioelectrocatalysis

Taiki Adachi, Yuki Kitazumi, Osamu Shirai and Kenji Kano *,†

Division of Applied Life Sciences, Graduate School of Agriculture, Kyoto University, Sakyo, Kyoto 606-8502, Japan; adachi.taiki.62s@st.kyoto-u.ac.jp (T.A.); kitazumi.yuki.7u@kyoto-u.ac.jp (Y.K.); shirai.osamu.3x@kyoto-u.ac.jp (O.S.)
* Correspondence: kano.kenji.5z@kyoto-u.ac.jp
† Present address: Center for Advanced Science and Innovation, Kyoto University, Gokasho, Uji, Kyoto 611-0011, Japan.

Received: 15 November 2020; Accepted: 1 December 2020; Published: 3 December 2020

Abstract: Bioelectrocatalysis has become one of the most important research fields in electrochemistry and provided a firm base for the application of important technology in various bioelectrochemical devices, such as biosensors, biofuel cells, and biosupercapacitors. The understanding and technology of bioelectrocatalysis have greatly improved with the introduction of nanostructured electrode materials and protein-engineering methods over the last few decades. Recently, the electroenzymatic production of renewable energy resources and useful organic compounds (bioelectrosynthesis) has attracted worldwide attention. In this review, we summarize recent progress in the applications of enzymatic bioelectrocatalysis.

Keywords: bioelectrocatalysis; nanostructured electrodes; protein engineering; bioelectrosynthesis; photo-bioelectrocatalysis

---

1. Introduction

Oxidoreductases catalyze redox reactions between two sets of redox substrate couples and are considered industrially useful catalysts due to their high activities and substrate specificities under mild conditions (room temperature, normal pressure, and neutral pH). However, most oxidoreductases, in addition to nicotinamide cofactor (NAD(P))-dependent enzymes, show low substrate specificities for one of the substrates. Such redox enzymes can accept or donate electrons from or to electrodes directly or via artificial redox mediators. The coupled reaction is called bioelectrocatalysis, and the catalytic function of the redox enzymes provides a variety of specific and strong catalytic activities to nonspecific electrode reactions. [1–5]. Bioelectrocatalysis provides a firm base for characterizing redox enzyme reactions and applying the concept and related technologies to useful bioelectrochemical devices such as biosensors [6–14], biofuel cells [11,15–21], biosupercapacitors [22], and other bioreactors [23].

Bioelectrocatalytic reactions are classified into two types according to the mode of the electron transfer described above: direct electron transfer (DET) and mediated electron transfer (MET), as shown in Figure 1. These reactions proceed in the following schemes in the simple case of the substrate's oxidation. On the other hand, we note here that some redox enzymes that have a catalytic site alone, without any other redox site(s), are also able to show DET-type bioelectrocatalysis, e.g., cytochrome c peroxidase [24], horseradish peroxidase (HRP) [25–27], ferredoxin-NADP$^+$ reductase [28], flavin adenine dinucleotide (FAD)-dependent glucose oxidase (FAD-GOD) [29,30], and FAD-dependent glucose dehydrogenase (FAD-GDH) [31].

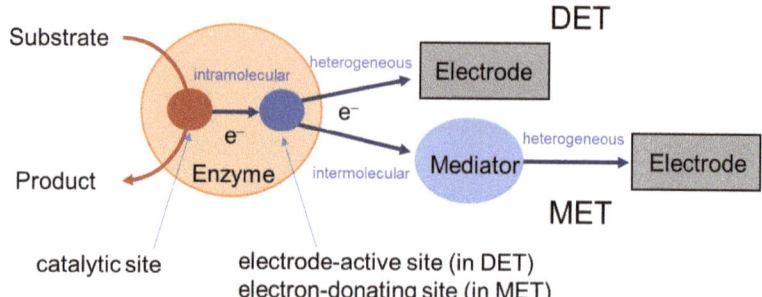

Figure 1. Schematic of electron transfer processes in direct electron transfer (DET)- and mediated electron transfer (MET)-type bioelectrocatalysis for substrate oxidation. In this scheme, the enzyme is assumed to have both catalytic and electron-donating sites in the molecules.

The DET-type reaction is given by Equations (1) and (2):

$$S + \frac{n_S}{n_E}E_O \xrightarrow{\text{Enzyme}} \frac{n_S}{n_P}P + \frac{n_S}{n_E}E_R, \tag{1}$$

and

$$E_R \underset{}{\overset{\text{Electrode}}{\rightleftarrows}} E_O + n_E e^-, \tag{2}$$

On the other hand, the MET-type reaction is given by Equations (3) and (4):

$$S + \frac{n_S}{n_M}M_O \xrightarrow{\text{Enzyme}} \frac{n_S}{n_P}P + \frac{n_S}{n_M}M_R, \tag{3}$$

and

$$M_R \underset{}{\overset{\text{Electrode}}{\rightleftarrows}} M_O + n_M e^-, \tag{4}$$

where S, P, E, and M indicate a substrate, a product, an enzyme, and a mediator, respectively. n_X is the number of electrons of the chemical species X. X_O and X_R are the oxidized and reduced forms of X, respectively. In DET-type bioelectrocatalysis, it is easy to construct relatively simple systems with minimum overpotentials in the electron transfer between the electrode-active redox center of an enzyme and an electrode from the thermodynamic perspectives. In this review, the electrode-active redox center means the site that can directly communicate with electrodes and is assigned to the catalytic active site (especially for redox enzymes that have the catalytic site alone) or an electron-donating/accepting site (other than the catalytic center) that constitutes the intramolecular electron transfer. However, the reported redox enzymes enable DET-type reactions are increasing but still limited in number because the electrochemical communication between an enzyme and an electrode occurs only when the electrode-active cofactor of the enzyme is close in distance to the electrode surface [32–35]. In MET-type bioelectrocatalysis, on the other hand, a variety of enzymes can be used in principle, and a suitable selection of mediators makes it possible to construct realistic systems. In addition, once both enzymes and mediators are stably immobilized on electrodes, the measurement systems work as pseudo-DET-type systems [10,11,36]. Particularly, redox polymers anchoring osmium complexes [3,37–45], ferricyanide [46,47], metallocenes [48–51], and viologen units [52–55] are constructed as polymeric mediators immobilized on electrodes. In summary, a DET-type system is often more ideal than a MET-type system, whereas it seems to be practical to utilize an MET-type system for several objectives.

The overpotential in bioelectrocatalysis has two components: (1) thermodynamics; the difference in the formal potential between the substrate and the electron-donating site, and (2) kinetics; slow kinetics in the heterogeneous electron transfer (Figure 1). There is no way to avoid the problem concerning the first issue as long as one utilizes natural enzymes. Slow kinetics in heterogeneous electron transfer is compensated by the so-called overpotential of the electrode in the DET-type reaction and by the large driving force (that is, increased difference in the formal potential) between the enzyme and mediator. In this review, we will describe several techniques for improving the performance of enzymatic bioelectrocatalytic systems and summarize recent studies on their applications.

2. Tuning of Nanostructured Electrodes and Protein-Engineering of Redox Enzymes

2.1. Electrode Nanomaterials

In DET-type and enzyme/mediator-immobilized MET-type bioelectrocatalysis, immobilization of as many amounts of enzymes and mediators as possible on the electrode surface is, in principle, able to increase the current density of bioelectrocatalytic systems. Therefore, nanostructured electrodes with a large ratio of the effective surface area against the projective surface area are useful and frequently utilized for bioelectrocatalytic systems. In addition, it is suggested that the heterogeneous electron transfer kinetics at the top edge of the microstructures of these electrodes is accelerated by the electric field strengthened by the expansion of the electric double layer [56] and the charge accumulation as expected by the Poisson equation [4]. This effect is very useful to decrease the overpotential due to the heterogeneous electron transfer in the DET-type reaction.

Electrode nanomaterials utilized in bioelectrocatalysis are roughly divided into two classes: carbon and metal. Carbon nanomaterials, such as carbon nanotubes [57–59], carbon blacks [60], carbon cryogels [61,62], and MgO-templated carbons [43,63,64], have properties to physically adsorb a large amount of enzymes and mediators at hydrophobic sites and are generally used as platforms favorable for bioelectrocatalysis. On the other hand, nanoporous gold constructed by anodization [27,65–67] or dealloying [68–70] and metallic nanoparticles of gold [57,71–78], silver [79–81], platinum [29–31], titanium oxide (TiO_2) [80,81], iron oxide (Fe_2O_3) [82,83], and indium tin oxide (ITO) [84] are also widely used. Compared to carbon nanomaterials, the pore and particle sizes of metallic nanomaterials can be easily controlled according to several manufacturing methods. In addition, conductive supports, such as polymer hydrogels, act as nanostructured electrodes [85].

Furthermore, nanostructured electrodes exert another positive effect on DET-type bioelectrocatalysis from the viewpoint of the enzyme's orientation. The limited catalytic current density (j_{cat}) in DET-type reactions is expressed by Equation (5):

$$j_{cat} = \pm n_E F k_{c,DET} \Gamma_{E,eff}, \tag{5}$$

where F is the Faraday constant, $k_{c,DET}$ is the catalytic constant in DET-type bioelectrocatalysis (with $k_{c,DET} = (n_S/n_E)k_{c,DET(1)}$, $k_{c,DET(1)}$ being the catalytic constant for single turnover of the enzyme), and $\Gamma_{E,eff}$ is the surface concentration of the effective enzyme. Here, "effective" means being able to electrochemically communicate with electrodes. In other words, enzymes of which the electrode-active sites face away from the electrode are not included because the long-range electron transfer kinetic constant ($k°$) exponentially decreases with an increase in the distance between the electrode surface and the electrode-active site in an enzyme (d) [86–88], as described by Equation (6):

$$k° = k°_{max} \exp(-\beta d), \tag{6}$$

where $k°_{max}$ is the rate constant at the closest approach (d = d_{min}) and β is the decay coefficient. Based on the random orientation model in which enzymes are assumed to be randomly oriented on electrodes [32,89,90], the enzymes can penetrate into nanostructured electrodes with mesopores, and the mesoporous electrodes can adsorb a large amount of enzymes suitable for DET-type reactions compared with planar electrodes, as illustrated in Figure 2. This is called the curvature effect of mesoporous electrodes for DET-type bioelectrocatalysis and is also very effective in decreasing the overpotential of DET-type bioelectrocatalysis. In practical applications, it is necessary to select and optimize electrode nanomaterials based on the estimated size, shape, and hydrophobicity of enzymes.

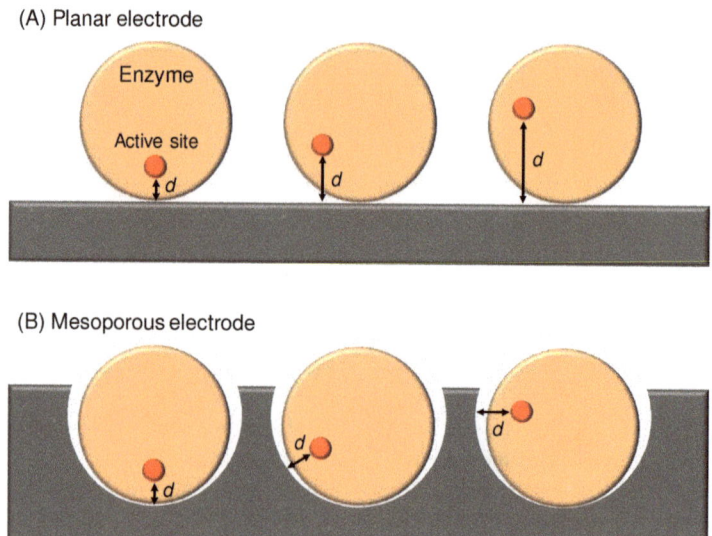

Figure 2. Schematic of the enzyme orientations at the (A) planar and (B) mesoporous electrodes.

On the other hand, chemical modifications of electrode surfaces are very effective in controlling the enzyme orientation by electrostatic or other specific interactions between the electrode-active site and the electrode surface; therefore, they are useful to increase the population of the enzyme orientations with minimum distances between the electrode-active site and the electrode surface [91]. Such favorite orientations also contribute to decreasing the overpotential in the DET-type reaction. For example, [NiFe]-hydrogenase (H_2ase) from *Desulfovibrio vulgaris* Miyazaki F and copper efflux oxidase (CueO) expressed in *Escherichia coli* showed strong DET-type bioelectrocatalysis activity at (positively charged) *p*-phenylenediamine-functionalized Ketjen-Black-modified glassy carbon electrodes compared with Ketjen-Black-modified electrode without any chemical functionalization and artificially introduced charged groups. The surfaces near the electrode-active sites of the enzymes are estimated to be negatively charged and the attractive electrostatic interaction with positively charged electrode surfaces increases the probability of enzyme orientations favorable for the interfacial electron transfer [92,93].

In contrast, bilirubin oxidase (BOD) from *Myrothecium verrucaria* showed strong DET-type bioelectrocatalytic activity at negatively charged electrodes. The surface near the electrode-active site of the enzyme is positively charged and the electrostatic interaction with negatively charged electrodes increases the probability of favorable orientations of the enzyme [94]. On the other hand, the DET-type bioelectrocatalysis of BOD was also improved by modifying an electrode with bilirubin (as the natural electron donor), which seemed to attractively interact with the electrode-active type I copper site as the electron-accepting site. The interaction increases the probability of favorable orientations of the enzyme [95]. In addition, membrane-bound D-fructose dehydrogenase (FDH) from *Gluconobacter japonicus* NBRC3260 showed strong DET-type bioelectrocatalytic activity at a 4-mercaptophenol-modified porous gold electrode, probably due to the stabilization of the enzyme layer by the hydroxy groups of 4-mercaptophenol [96]. FDH also showed the attractive and specific interaction with methoxy substituents on the electrode surface, which resulted in the favorable orientation [97].

In addition, gas-diffusion bioelectrodes are effective for enzymatic bioelectrocatalysis in which gaseous substrates such as dihydrogen, dioxygen, and carbon dioxide were used [98]. Since these gasses have low water solubility, the bioelectrocatalytic currents are often controlled and limited by diffusion processes of the substrates at usual electrodes. On the other hand, gas-diffusion bioelectrodes realize direct supplies of gaseous substrates from the gas phase to the reaction layer due to their suitable conductivity, hydrophobicity/hydrophilicity balance, and gas permeability, as illustrated in Figure 3.

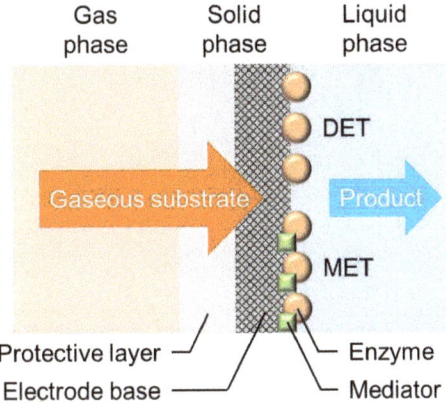

✓ Suitable conductivity
✓ Hydrophobicity/hydrophilicity balance
✓ Gas permeability

Figure 3. Schematic of a bio-three-phase interface of a gas-diffusion bioelectrode.

2.2. Protein-Engineering Approaches

Bioelectrocatalysis can be improved not only by functionalizing electrodes but also by engineering enzymes. Based on the crystal structure of enzymes, various mutations can be introduced into redox enzymes by protein-engineering methods [21,99].

2.2.1. Formal Potential Shift of Electrode-Active Sites

The negative shift of the formal potential of the electron-donating site for the substrate oxidation process (vice versa for the substrate reduction) by mutation is very effective in reducing the thermodynamic overpotential in the intramolecular electron transfer in redox enzymes with multiredox sites (Figure 1). The formal potentials of metallic cofactors such as hemes and copper clusters are predominantly controlled by coordinated amino acid residues. Particularly, the point mutation of the axial ligands of hemes and blue copper clusters can easily tune their formal potentials. The direction of the potential shift depends in part on the electron-donating/accepting characteristics of the mutated amino acid side chains [100,101]. Briefly in general, an axial ligand with a relatively strong electron-donating character shifts the formal potential of the cofactor in the negative potential direction due to the stabilization of the oxidized state of the cofactor and vice versa. The formal potentials of redox enzymes are also greatly interfered with by conformational changes caused by mutated amino acid residues.

FDH is among the accepted model enzymes to investigate the effects of protein engineering on DET-type bioelectrocatalysis. The enzyme is a heterotrimeric enzyme composed of a FAD-containing large catalytic subunit, a three-heme c (called hemes 1c, 2c, and 3c) from the N-terminus-containing cytochrome subunit, and a small subunit and proceeds with a DET-type reaction by transferring electrons from FAD, heme 3c, and heme 2c to an electrode in this order [102–104]. The sixth axial ligand (methionine 450) of heme 2c was then replaced by glutamine with electron-donating characteristics to shift the formal potential of heme 2c in the direction of the negative potential. The FDH variant (M450Q_FDH) provided DET-type bioelectrocatalytic waves for the oxidation of fructose at a half-wave potential of approximately 50 mV, more negative than that of the recombinant native enzyme, as shown in Figure 4A [103,104]. A drastic change in the limiting catalytic current was not observed. This suggests that the rate constant of the intramolecular electron transfer from FAD to heme 2c in the enzyme is sufficiently large compared with the catalytic reaction rate constant at the catalytic center (FAD); the latter predominantly determines $k_{c,DET}$ in Equation (5).

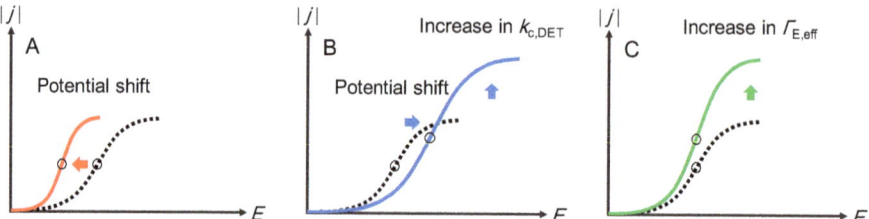

Figure 4. Illustrative drawing of the effect of mutations on steady-state bioelectrocatalytic waves in the DET-type bioelectrocatalysis of "substrate oxidation". (**A**) Negative potential shift in the formal potential of the electron-donating site of the DET-type redox enzyme reactions, in which the intramolecular electron transfer is not the rate-determining process in the enzymatic catalytic reaction. (**B**) Positive potential shift in the formal potential of the electron-donating site of the DET-type redox enzyme reactions, in which the intramolecular electron transfer is the rate-determining process in the enzymatic catalytic reaction, and (**C**) downsizing without any change in the intrinsic enzymatic activity.

On the other hand, such replacement is also effective for the axial ligand of the electron-accepting site (type I blue copper site) to negatively shift the formal potential. Replacement of the axial ligand methionine 467 in the type I copper site of BOD with glutamine (M467Q_BOD) caused a large negative shift (approximately 0.23 V) in the half-wave potential of the DET-type bioelectrocatalytic waves. This means an increase in the overpotential in the intramolecular electron transfer in the enzyme. Fortunately, in this case, the catalytic limiting current density increased compared with that of the recombinant native BOD, as shown in Figure 4B (note here that the catalytic wave in Figure 4B is illustrated for substrate oxidation) [105]. Most probably, the catalytic rate constant of the dioxygen reduction at the type II/III catalytic site is sufficiently large compared with the rate constant of the intramolecular electron transfer from the electron-accepting type I site to the dioxygen-reducing type II/III site; the intramolecular electron transfer is the rate-determining step to determine $k_{c,DET}$ in Equation (5). In addition, the electron transfer kinetics seems to obey the linear free energy relationship (LFER); the increase in the formal potential difference between type I and II/III sites (that is, the driving force of the reaction) results in an increase in the intramolecular electron transfer rate constant.

Even in MET-type bioelectrocatalysis, the redox potential shift of the electron-donating site (for substrate oxidation) can tune the overpotential in the intramolecular electron transfer process. In addition, the intermolecular electron transfer kinetics between the electron-donating site in the enzyme and the mediator seem to be improved in theory by the potential shift mutation of the electron-donating site (for substrate oxidation) on the basis of the concept of LFER, though there is no report on this matter. Therefore, the mutational tuning of the redox potential of the electron-donating site (for substrate oxidation) may also expand the strategy of the mediator selection for MET-type bioelectrocatalysis.

2.2.2. Downsizing

As described by Equation (5), an increase in $\Gamma_{E,eff}$ is essential to getting a large bioelectrocatalytic current density. $\Gamma_{E,eff}$ can be then increased by downsizing enzymes from which the regions not deeply involved in bioelectrocatalysis are deleted, as shown in Figure 4C.

Downsizing effects were also investigated in FDH. Heme 1c of FDH is suggested to be uninvolved in the DET-type reaction, and the downsized FDH without the heme 1c region (Δ1c_FDH) showed a larger DET-type bioelectrocatalytic current density than the recombinant native FDH (r_FDH) [104,106]. In addition, a further downsized FDH without the heme 1c and 2c regions (Δ1c2c_FDH) also showed DET-type bioelectrocatalytic activity with reduced overpotential due to direct electrical communication between an electrode and heme 3c with the most negative formal potential [104,107]. However, whereas $\Gamma_{E,eff}$ was suggested to be increased, j_{cat} of Δ1c2c_FDH was as large as that of r_FDH (smaller than that of Δ1c_FDH), which seemed to be ascribed to a decrease in enzymatic activity ($\approx k_{c,DET}$) of Δ1c2c_FDH by excessive deletion [104,107]. Thus, it is important to avoid conformational changes due to deletion and retain enzymatic activity as high as possible. Furthermore, a double mutant of downsizing and the potential shift (M450QΔ1c_FDH) accomplished both an increase in j_{cat} and an overpotential reduction [104,108,109].

2.2.3. Surface Amino Acid Mutation

It is desirable to tightly immobilize enzymes on electrodes in both DET- and MET-type bioelectrocatalytic reactions. Cross-linkers such as glutaraldehyde [110], carbodiimide [29], and maleimide [71] are often used for the covalent immobilization of enzymes [111]. The mutation of amino acid residue(s) located on the enzyme surface also enhances the cross-coupling reactions and can control the orientation of the enzyme on a suitable electrode surface. For example, cysteine introduction onto the enzyme surface close to the electrode-active site can increase the enzyme orientations suitable for DET-type reactions at thiol- and maleimide-functionalized electrodes by forming (di)sulfide bonds. Holland et al. reported DET-type bioelectrocatalysis of cysteine-introduced GOD conjugated with maleimide-modified gold nanoparticles on a gold electrode [71]. In addition, Ferapontova et al. revealed that cysteine mutation of HRP was effective for its orientation on gold electrodes to improve DET-type bioelectrocatalytic properties [112].

2.2.4. Fusion Protein

DET-type bioelectrocatalysis is sometimes achieved by introducing an electrode-active domain into a native enzyme. Cellobiose dehydrogenase (CDH) is often used as a model of DET-type fusion enzymes. CDH has two domains: a larger catalytic dehydrogenase domain and a smaller electrode-active cytochrome domain. The domains are linked by a flexible polypeptide [113]. The fused cytochrome domain mediates the electron transfer from the catalytic domain to electrodes, such as a "built-in mediator." Utilizing the fusion protein-engineering methods, DET-type reactions by FAD-GDH [114,115], pyrroloquinoline quinone (PQQ)-dependent GDH [116], and flavodoxin [117] were reported.

3. Novel Bioelectrochemical System

Major applications in enzymatic bioelectrocatalysis are biosensors and biofuel cells. Biosensors are utilized as analytical devices for food analyses and clinical diagnoses [6–14], and biofuel cells are ecofriendly energy conversion devices in which the chemical energy of fuels is electrically extracted [11,15–21]. The characteristics, properties, and progress of these devices were previously summarized in the corresponding reviews cited above. Recently, in addition, novel methods for bioelectrocatalytic applications are reported as described below.

3.1. Biosupercapacitor

Biosupercapacitors are self-powered energy storage devices using bioelectrocatalysis to charge capacitors [22]. In the biosupercapacitor, electric power is generated by a biocathode and/or a bioanode. The fundamental difference between biosupercapacitors and biofuel cells is the separation of the output of the current and bioelectrocatalytic reaction. Expressed as an equivalent circuit, a biosupercapacitor is a biofuel cell and a capacitor connected in parallel. Regardless of the connection to the external circuit, the bioelectrocatalytic reaction injects the electrons into the supercapacitor. When the external circuit requires the current, the biosupercapacitor provides the current. An advantage of biosupercapacitors is that a large current can flow even if the kinetics of the bioelectrocatalysis is inadequate. Additionally, biosupercapacitors are compatible with the self-powered biosensors [118,119].

The supercapacitor is classified into two types: the electrical double layer capacitor (EDLC) and the electrochemical pseudocapacitor (EPC). EDLC is the electrode with a large surface area, the electric charge is accumulated in the electrical double layer at the interface between the electrode and the electrolyte solution. EPC is constructed with the electrode and the reversible electrode-active redox species. The externally applied voltage shifts the electrode potential at the anode and a cathode in the negative and positive directions, respectively, as expressed in the Nernst equation.

Generally, redox-active species are immobilized at the electrode surface. Since an electrochemical capacitor contains two electrodes and an electrolyte solution, electrochemical capacitors are also classified into three types: EDLC-EDLC, EPC-EPC, and EDLC-EPC [120]. Therefore, biosupercapacitors are basically classified into the three types. The EDLC is compatible with DET-type bioelectrocatalysis and MET-type bioelectrocatalysis is adapted to the EPC. In order to avoid the mixing of mediators, EPCs in the biosupercapacitor are frequently employed redox polymers as mediators to immobilize at the electrode surface.

Electrochemically inactive porous electrodes in an electrolyte solution work as the EDLC. Additionally, DET-type bioelectrocatalytic activity is required for the EDLC-type biosupercapacitor. The reported materials suitable for the EDLC and the bioelectrode are carbon nanotubes [121–123], porous gold [124], gold nanoparticles [125], and indium tin oxide nanoparticles [126]. Porous electrodes often show high activity for DET-type bioelectrocatalysis. Therefore, it is considered that the investigation of bioelectrodes for DET-type biofuel cells will be directly useful for the biosupercapacitor.

The separation of oxidant and reductant is important in EPCs, since the mixing of the two species causes the cross-reaction discharge. Therefore, in the case of biosupercapacitors using the MET-type bioelectrocatalysis, careful attention is required to head off the outflow of mediators. As mentioned above, redox polymers are employed as mediators in biosupercapacitors [42,127–129]. The immobilization of redox-active proteins at the electrode surface has been investigated for the improvement of the capacitance of EPC-type biosupercapacitors [130,131].

3.2. Bioelectrosynthesis

Bioelectrosynthesis is a coined term for the generation of renewable energy resources and useful organic compounds [132]. Particularly, stereoselective enzymatic bioelectrocatalysis is desirable for pharmaceutical use because pharmaceutical precursors are required to have high enantiomeric purity [133,134].

Some useful compounds are reduced forms of redox couples, and reductive conversion is often the opposite reaction in metabolisms in vivo. In bioelectrocatalysis, on the other hand, several enzymes can proceed bidirectional electroenzymatic reactions (both the oxidation and reduction of substrate redox couples), in which the driving forces are reversed depending on pH. For example, H_2ase [135,136], formate dehydrogenase (FoDH) [75,136–139], carbon monoxide dehydrogenase (CODH) [136], diaphorase (DI) [136,140], ferredoxine-$NADP^+$ reductase (FNR) [28], and some $NAD(P)^+$-dependent dehydrogenases [28,140–143] show bidirectional bioelectrocatalytic activities. The characteristic of these enzymes is that the formal potentials of the substrate redox couples and the cofactors in the enzymes are relatively close to each other, and thus, the small uphill intramolecular electron transfer proceeds in acceptable velocity [34]. Bidirectional bioelectrocatalysis is a key reaction for constructing bioelectrochemical energy/compound conversion systems.

There are mainly two types of bioelectrosynthetic systems: a fuel-cell-type [142,144–149] and an electrolysis-type [28,53,141,143,150–154]. The former realizes a spontaneous production of compounds without any external power supplies, whereas the latter proceeds relatively rapid reactions due to an optimally controlled electrode potential to realize diffusion-controlled conditions. We will show examples of bioelectrosyntheses in the following sections.

3.2.1. Dihydrogen Production

Dihydrogen (H_2) is a typical clean energy source that emits no harmful products in combustion. H_2ase, a common anodic bioelectrocatalyst for H_2 oxidation, can also catalyze proton (H^+) reduction [135,136]. The reversible reaction is biased to favor H_2 oxidation, and it is difficult to realize a large cathodic current density of H^+ reduction in DET-type systems. H^+ reduction with a large current density was achieved in a H_2ase MET-type system using a viologen polymer as a redox mediator [53]. On the other hand, a dual DET-type bioelectrocatalytic water–gas shift reaction using CODH and H_2ase ($H_2O + CO \rightarrow H_2 + CO_2$) was reported [146]. The system requires no external power supplies.

3.2.2. Formate Production

Formate/formic acid ($HCOO^-$/$HCOOH$) is a liquid energy resource with a high energy density due to its property of being completely oxidized to CO_2. FoDH interconverts $HCOO^-$ and CO_2 both in DET- and MET-type systems [75,136–139]. Sakai et al. reported efficient bioelectrocatalytic CO_2 reduction using FoDH from *Methylobacterium extorquens* AM1 and a synthesized redox mediator with the low formal potential on a gas-diffusion electrode [150]. In addition, spontaneous interconversion between $HCOO^-$ and H_2 without any external power supplies was achieved by coupling two bioelectrocatalyses by FoDH and H_2ase ($H_2 + CO_2 \rightarrow HCOO^- + H^+$) [147]. The direction of the reaction is controlled by pH- and concentration-dependent equilibrium potentials of the $2H^+/H_2$ and $CO_2/HCOO^-$ couples.

3.2.3. Ammonia Production

Ammonia (NH_3) is an industrially beneficial compound used as a chemical raw material, fuel, hydrogen storage, and so on [155,156]. Nitrogenase (N_2ase) is an enzyme that catalyzes a nitrogen fixation from dinitrogen (N_2) to NH_3 using adenosine triphosphate (ATP) hydrolysis energy in microorganisms under an anaerobic condition [155,156]. Particularly, molybdenum-dependent N_2ase, which comprises a catalytic MoFe protein and a homodimeric MgATP-binding Fe protein, has been mainly investigated in view of bioelectrocatalysis [156,157]. N_2ase reduces not only N_2 but also nitrite (NO_2^-) and azide (N_3^-) to NH_3 on electrodes, and the MoFe protein disassociated with the Fe protein shows the DET-type bioelectrocatalytic activity without ATP [154]. Furthermore, ATP-independent NH_3 bioelectrosynthesis was improved in the MET-type system using N_2ase and a cobaltocene-functionalized polymeric mediator [51]. On the other hand, Milton et al. reported a NH_3-producing H_2/N_2 biofuel cell using MET-type reactions of N_2ase and H_2ase [148]. If the fatal weakness of N_2ase, a lack of oxygen tolerance, is improved, N_2ase will be expected to have further industrial applications.

3.2.4. NAD(P)$^+$-Dependent Bioelectrosynthesis

Nicotinamide coenzymes (NAD(P)$^+$/NAD(P)H) play essential roles in the function of many NAD(P)$^+$-dependent dehydrogenases, but the dehydrogenases need suitable redox mediators in single-step enzymatic bioelectrocatalysis because the NAD(P)$^+$/NAD(P)H interconversion requires high overpotentials at conventional electrodes due to its hydride ion-transfer-type characteristics that are completely different from two-step single-electron transfer-type electrode reactions. On the other hand, the introduction of dipaphorase (DI) and FNR, which are bioelectrocatalysts for the interconversion of the NAD$^+$/NADH and NADP$^+$/NADPH redox couples, respectively, can realize various NAD(P)$^+$-dependent bioelectrocatalyses with relatively low overpotentials [28,140].

The bienzymatic MET-type system of NAD$^+$-dependent dehydrogenase and DI is kinetically and thermodynamically investigated and often applied to biosensors [40,140,158,159]. In addition, the coulometric electrooxidation of organic substances was achieved in the same multienzymatic system mimicking the tricarboxylic acid cycle [151]. On the other hand, reductive production of chiral compounds is useful in many cases. Minteer's group reported various NAD$^+$-dependent multienzymatic cascade reactions for generating chiral compounds, as shown in Figure 5 [149,152,153].

They also demonstrated the bioelectrosynthesis of polyhydroxybutarate with NADH regeneration by DI [143].

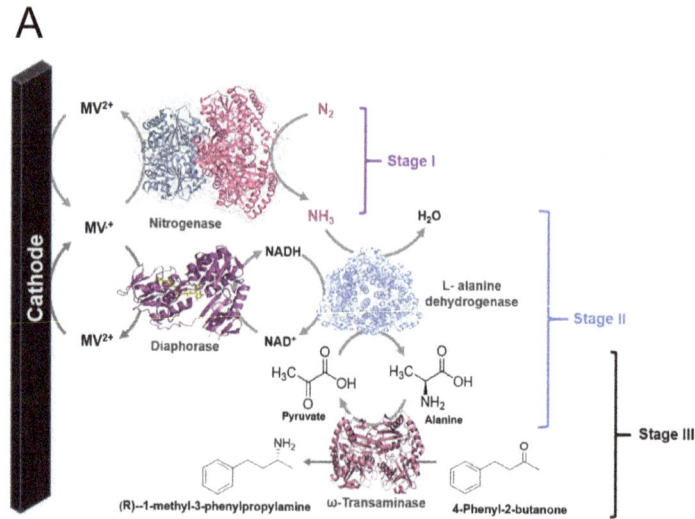

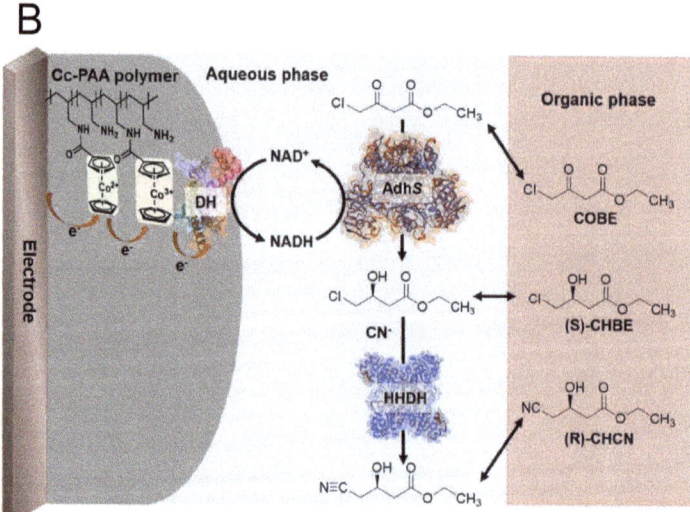

Figure 5. Schematic representation of multienzymatic bioelectrosyntheses of (**A**) chiral amine and (**B**) β-hydroxy nitrile. Abbreviations; DH: diaphorase, AdhS: alcohol dehydrogenase, HHDH: halohydrin dehalogenase, COBE: 4-chloroacetoacetate, CHBE: 4-chloro-3-hydroxybutanoate, CHCN: ethyl-4-cyano-3-hydroxybutyrate. Reprinted from Ref. [152,153], Copyright (2019,2020), with permission from ACS Publications.

In an $NADP^+$-dependent system, on the other hand, FNR showed DET-type bioelectrocatalytic activity [28]. Particularly, Armstrong's group investigated $NADP^+$-dependent bienzymatic bioelectrocatalysis using FNR. They reported the reductive carboxylation of pyruvate to malate by malate dehydrogenase [142], reductive amination of 2-oxoglutamate to L-glutamate by glutamate dehydrogenase [28,144], and enantioselective interconversion of the secondary alcohol/ketone couple by engineered alcohol dehydrogenase [141,145].

3.3. Photo-Bioelectrocatalysis

Photo-bioelectrocatalysis enables reducing a substrate with a lower formal potential by oxidizing a sacrificial reagent with a higher formal potential, which cannot spontaneously occur without solar energy. Electrons that photosensitizers accept from sacrificial reagents are excited by solar energy, and the electric potential is shifted in the negative direction, corresponding to the wavelength of the adsorbed light. The electrons are then donated to substrates via enzymes and mediators, as shown in Figure 6. Photosensitizers such as TiO_2 [80,81,160–162], PbS quantum dots [162], silver nanoclusters [80,81], and organic dyes [160,161] are incorporated in anodes of transparent electrode bases (ITO in general) with or without other catalysts. Particularly, in addition, the photosystem II (PSII) complex in the thylakoid membrane of cyanobacteria and higher plants is often used as a water-splitting anodic photo-bioelectrocatalyst [129,160–181]. Biosolar cells [163–173] and solar biosupercapacitors [129,164,168,174], using PSII/I, thylakoid membranes, or cyanobacteria in anodes, and BOD or laccase in cathodes, realized the conversion from solar to electric energy without any sacrificial reagents in total. On the other hand, photo-bioelectrosyntheses, also called artificial photosyntheses, are reported. H_2 is generated by H_2ase [80,160], and CO_2 is fixed by FoDH and CODH [81,161]. These cathodic reactions proceed at quite low potentials. Thus, in order to improve the Faradaic efficiency of these photo-bioelectrosyntheses, it is essential to reduce electron leakage to dissolved oxygen as much as possible, especially when oxygen-generating photosynthetic proteins or organisms are used as anodic photo-bioelectrocatalysts.

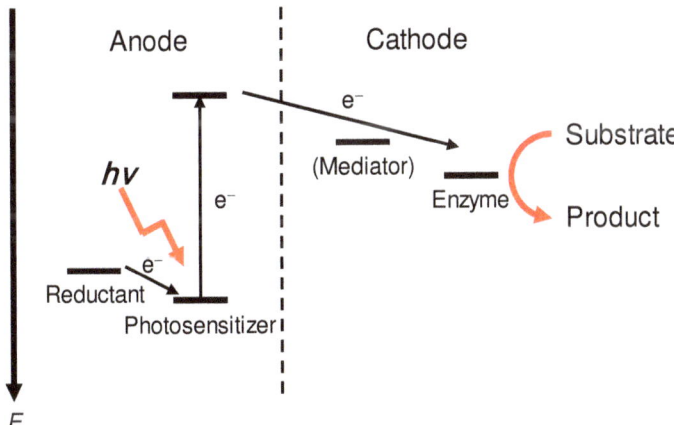

Figure 6. Potential profile in a simple photo-bioelectrocatalytic system.

4. Conclusions

Enzymatic bioelectrocatalysis is advancing day by day due to the finding of novel enzymes and the development of electrode nanomaterials. However, there are still issues to be solved such as thermostability, oxygen tolerance, and physical fragility, compared to inorganic electrocatalysts. Further understanding and consideration of the electrode interface and the interaction between enzymes and electrodes from kinetics and thermodynamics are required. Other perspectives, such as the improvement of redox enzymes by protein-engineering approaches, the selection of electrode materials and mediators (including redox hydrogels), the immobilization of enzymes, and the layout of electrodes will be also required for the improvement of present bioelectrochemical devices.

Author Contributions: Conceptualization, T.A. and K.K.; data curation, T.A.; writing—original draft preparation, T.A.; writing—review and editing, Y.K., O.S. and K.K.; supervision, K.K.; project administration, K.K. All authors have read and agreed to the published version of the manuscript.

Funding: This research received no external funding.

Conflicts of Interest: The authors declare no conflict of interest.

References

1. Bartlett, P.N. *Bioelectrochemistry: Fundamentals, Experimental Techniques and Applications*; John Wiley & Sons: Chichester, UK, 2008.
2. Wilson, G.S.; Johnson, M.A. In-Vivo Electrochemistry: What Can We Learn about Living Systems? *Chem. Rev.* **2008**, *108*, 2462–2481. [CrossRef]
3. Heller, A. Electrical connection of enzyme redox centers to electrodes. *J. Phys. Chem.* **1992**, *96*, 3579–3587. [CrossRef]
4. Kano, K. Fundamentals and Applications of Redox Enzyme-functionalized Electrode Reactions. *Electrochemistry* **2019**, *87*, 301–311. [CrossRef]
5. Scheller, F.; Kirstein, D.; Schubert, F.; Wollenberger, U.; Ollson, B.; Gorton, L.; Johansson, G. Enzyme electrodes and their application. *Philos. Trans. R. Soc. B Biol. Sci.* **1987**, *316*, 85–94. [CrossRef]
6. Thevenot, D.R.; Tóth, K.; Durst, R.A.; Wilson, G.S. Electrochemical Biosensors: Recommended Definitions and Classification. *Pure Appl. Chem.* **1999**, *71*, 2333–2348. [CrossRef]
7. Martinkova, P.; Kostelnik, A.; Valek, T.; Pohanska, M. Main streams in the Construction of Biosensors and Their Applications. *Int. J. Electrochem. Sci.* **2017**, *12*, 7386–7403. [CrossRef]
8. Bollella, P.; Gorton, L. Enzyme based amperometric biosensors. *Curr. Opin. Electrochem.* **2018**, *10*, 157–173. [CrossRef]
9. Kucherenko, I.S.; Soldatkin, A.; Kucherenko, D.Y.; Soldatkina, O.V.; Dzyadevych, S.V. Advances in nanomaterial application in enzyme-based electrochemical biosensors: A review. *Nanoscale Adv.* **2019**, *1*, 4560–4577. [CrossRef]
10. Nguyen, H.H.; Lee, S.H.; Lee, U.J.; Fermin, C.D.; Kim, M. Immobilized Enzymes in Biosensor Applications. *Materials* **2019**, *12*, 121. [CrossRef]
11. Pinyou, P.; Blay, V.; Muresan, L.M.; Noguer, T. Enzyme-modified electrodes for biosensors and biofuel cells. *Mater. Horiz.* **2019**, *6*, 1336–1358. [CrossRef]
12. Bollella, P.; Katz, E. Enzyme-Based Biosensors: Tackling Electron Transfer Issues. *Sensors* **2020**, *20*, 3517. [CrossRef] [PubMed]
13. Willner, I.; Katz, E.; Willner, B. Electrical contact of redox enzyme layers associated with electrodes: Routes to amperometric biosensors. *Electroanalysis* **1997**, *9*, 965–977. [CrossRef]
14. Scheller, F.; Schubert, F.; Pfeiffer, D.; Hintsche, R.; Dransfeld, I.; Renneberg, R.; Wollenberger, U.; Riedel, K.; Pavlova, M.; Kühn, M.; et al. Research and development of biosensors. A review. *Analyst* **1989**, *114*, 653–662. [CrossRef] [PubMed]
15. De Poulpiquet, A.; Ranava, D.; Monsalve, K.; Giudici-Orticoni, M.-T.; Lojou, E. Biohydrogen for a New Generation of H2/O2 Biofuel Cells: A Sustainable Energy Perspective. *ChemElectroChem* **2014**, *1*, 1724–1750. [CrossRef]
16. Barton, S.C.; Gallaway, J.; Atanassov, P. Enzymatic Biofuel Cells for Implantable and Microscale Devices. *Chem. Rev.* **2004**, *104*, 4867–4886. [CrossRef]
17. Cracknell, J.A.; Vincent, K.A.; Armstrong, F.A. Enzymes as Working or Inspirational Electrocatalysts for Fuel Cells and Electrolysis. *Chem. Rev.* **2008**, *108*, 2439–2461. [CrossRef]
18. Meredith, M.T.; Minteer, S.D. Biofuel Cells: Enhanced Enzymatic Bioelectrocatalysis. *Annu. Rev. Anal. Chem.* **2012**, *5*, 157–179. [CrossRef]
19. Mazurenko, I.; De Poulpiquet, A.; Lojou, E. Recent developments in high surface area bioelectrodes for enzymatic fuel cells. *Curr. Opin. Electrochem.* **2017**, *5*, 74–84. [CrossRef]
20. Mano, N.; De Poulpiquet, A. O_2 Reduction in Enzymatic Biofuel Cells. *Chem. Rev.* **2018**, *118*, 2392–2468. [CrossRef]
21. Xiao, X.; Xia, H.-Q.; Wu, R.; Bai, L.; Yan, L.; Magner, E.; Cosnier, S.; Lojou, E.; Zhu, Z.; Liu, A. Tackling the Challenges of Enzymatic (Bio)Fuel Cells. *Chem. Rev.* **2019**, *119*, 9509–9558. [CrossRef]

22. Shleev, S.; González-Arribas, E.; Falk, M. Biosupercapacitors. *Curr. Opin. Electrochem.* **2017**, *5*, 226–233. [CrossRef]
23. Krieg, T.; Sydow, A.; Schröder, U.; Schrader, J.; Holtmann, D. Reactor concepts for bioelectrochemical syntheses and energy conversion. *Trends Biotechnol.* **2014**, *32*, 645–655. [CrossRef] [PubMed]
24. Paddock, R.M.; Bowden, E.F. Electrocatalytic reduction of hydrogen peroxide via direct electron transfer from pyrolytic graphite electrodes to irreversibly adsorbed cytochrome c peroxidase. *J. Electroanal. Chem. Interfacial Electrochem.* **1989**, *260*, 487–494. [CrossRef]
25. Xia, H.; Kitazumi, Y.; Shirai, O.; Kano, K. Direct Electron Transfer-type Bioelectrocatalysis of Peroxidase at Mesoporous Carbon Electrodes and Its Application for Glucose Determination Based on Bienzyme System. *Anal. Sci.* **2017**, *33*, 839–844. [CrossRef] [PubMed]
26. Gu, T.; Wang, J.; Xia, H.; Wang, S.; Yu, X. Direct Electrochemistry and Electrocatalysis of Horseradish Peroxidase Immobilized in a DNA/Chitosan-Fe3O4 Magnetic Nanoparticle Bio-Complex Film. *Materials* **2014**, *7*, 1069–1083. [CrossRef]
27. Sakai, K.; Kitazumi, Y.; Shirai, O.; Kano, K. Nanostructured Porous Electrodes by the Anodization of Gold for an Application as Scaffolds in Direct-electron-transfer-type Bioelectrocatalysis. *Anal. Sci.* **2018**, *34*, 1317–1322. [CrossRef]
28. Siritanaratkul, B.; Megarity, C.F.; Roberts, T.G.; Samuels, T.O.M.; Winkler, M.; Warner, J.H.; Happe, T.; Armstrong, F.A. Transfer of photosynthetic NADP+/NADPH recycling activity to a porous metal oxide for highly specific, electrochemically-driven organic synthesis. *Chem. Sci.* **2017**, *8*, 4579–4586. [CrossRef]
29. Yehezkeli, O.; Raichlin, S.; Tel-Vered, R.; Kesselman, E.; Danino, D.; Willner, I. Biocatalytic Implant of Pt Nanoclusters into Glucose Oxidase: A Method to Electrically Wire the Enzyme and to Transform It from an Oxidase to a Hydrogenase. *J. Phys. Chem. Lett.* **2010**, *1*, 2816–2819. [CrossRef]
30. Trifonov, A.; Stemmer, A.; Tel-Vered, R. Enzymatic self-wiring in nanopores and its application in direct electron transfer biofuel cells. *Nanoscale Adv.* **2019**, *1*, 347–356. [CrossRef]
31. Adachi, T.; Fujii, T.; Honda, M.; Kitazumi, Y.; Shirai, O.; Kano, K. Direct electron transfer-type bioelectrocatalysis of FAD-dependent glucose dehydrogenase using porous gold electrodes and enzymatically implanted platinum nanoclusters. *Bioelectrochemistry* **2020**, *133*, 107457. [CrossRef]
32. Léger, C.; Bertrand, P. Direct Electrochemistry of Redox Enzymes as a Tool for Mechanistic Studies. *Chem. Rev.* **2008**, *108*, 2379–2438. [CrossRef] [PubMed]
33. Milton, R.D.; Minteer, S.D. Direct enzymatic bioelectrocatalysis: Differentiating between myth and reality. *J. R. Soc. Interface* **2017**, *14*, 20170253. [CrossRef] [PubMed]
34. Adachi, T.; Kitazumi, Y.; Shirai, O.; Kano, K. Direct Electron Transfer-Type Bioelectrocatalysis of Redox Enzymes at Nanostructured Electrodes. *Catalysts* **2020**, *10*, 236. [CrossRef]
35. Kitazumi, Y.; Shirai, O.; Kano, K. *Catalyst Materials for Bioelectrochemical Systems: Fundamentals and Applications*; ACS Publications: Washington, DC, USA, 2020; pp. 147–163.
36. Gorton, L.; Jönsson-Pettersson, G.; Csöregi, E.; Johansson, K.; Domínguez, E.; Marko-Varga, G. Amperometric biosensors based on an apparent direct electron transfer between electrodes and immobilized peroxidases. Plenary lecture. *Analyst* **1992**, *117*, 1235–1241. [CrossRef]
37. Degani, Y.; Heller, A. Electrical communication between redox centers of glucose oxidase and electrodes via electrostatically and covalently bound redox polymers. *J. Am. Chem. Soc.* **1989**, *111*, 2357–2358. [CrossRef]
38. Ohara, T.J.; Rajagopalan, R.; Heller, A. Glucose Electrodes Based on Cross-Linked (Os(bpy)2Cl)+/2+ Complexed Poly(1-vinylimidazole) Films. *Anal. Chem.* **1993**, *65*, 3512–3517. [CrossRef]
39. Timur, S.; Yigzaw, Y.; Gorton, L. Electrical wiring of pyranose oxidase with osmium redox polymers. *Sens. Actuators B Chem.* **2006**, *113*, 684–691. [CrossRef]
40. Nikitina, O.; Shleev, S.; Gayda, G.; Demkiv, O.; Gonchar, M.; Gorton, L.; Csöregi, E.; Nistor, M. Bi-enzyme biosensor based on NAD+- and glutathione-dependent recombinant formaldehyde dehydrogenase and diaphorase for formaldehyde assay. *Sens. Actuators B Chem.* **2007**, *125*, 1–9. [CrossRef]
41. Nieh, C.-H.; Kitazumi, Y.; Shirai, O.; Yamamoto, M.; Kano, K. Potentiometric coulometry based on charge accumulation with a peroxidase/osmium polymer-immobilized electrode for sensitive determination of hydrogen peroxide. *Electrochem. Commun.* **2013**, *33*, 135–137. [CrossRef]
42. Alsaoub, S.; Conzuelo, F.; Gounel, S.; Mano, N.; Schuhmann, W.; Ruff, A. Introducing Pseudocapavitive Bioelectrodes into a Biofuel Cell/Biosupercapacitor Hybrid Device for Optimized Open Circuit Voltage. *ChemElectroChem* **2019**, *6*, 2080–2087. [CrossRef]

43. Tsujimura, S.; Takeuchi, S. Toward an Ideal Platform Structure Based on MgO-Templated Carbon for Flavin Adenine Dinucleotide-Dependent Glucose Dehydrogenase-Os Polymer-Hydrogel Electrodes. *Electrochim. Acta* **2020**, *343*, 136110. [CrossRef]
44. Xiao, X.; Conghaile, P.Ó.; Leech, D.; Magner, E. Use of Polymer Coatings to Enhance the Response of Redox-Polymer-Mediated Electrodes. *ChemElectroChem* **2019**, *6*, 1344–1349. [CrossRef]
45. Čėnas, N.K.; Pocius, A.K.; Kulys, J.J. Electron Exchange between Flavin- and Heme-Containing Enzymes and Electrodes Modified by Redox Polymers. *Bioelectrochem. Bioenerg.* **1983**, *11*, 61–73. [CrossRef]
46. Nieh, C.-H.; Kitazumi, Y.; Shirai, O.; Kano, K. Sensitive d-amino acid biosensor based on oxidase/peroxidase system mediated by pentacyanoferrate-bound polymer. *Biosens. Bioelectron.* **2013**, *47*, 350–355. [CrossRef] [PubMed]
47. Nieh, C.-H.; Tsujimura, S.; Shirai, O.; Kano, K. Amperometric biosensor based on reductive H2O2 detection using pentacyanoferrate-bound polymer for creatinine determination. *Anal. Chim. Acta* **2013**, *767*, 128–133. [CrossRef]
48. Merchant, S.A.; Meredith, M.T.; Tran, T.O.; Brunski, D.B.; Johnson, M.B.; Glatzhofer, D.T.; Schmidtke, D.W. Effect of Mediator Spacing on Electrochemical and Enzymatic Response of Ferrocene Redox Polymers. *J. Phys. Chem. C* **2010**, *114*, 11627–11634. [CrossRef]
49. Tran, T.O.; Lammert, E.G.; Chen, J.; Merchant, S.A.; Brunski, D.B.; Keay, J.C.; Johnson, M.B.; Glatzhofer, D.T.; Schmidtke, D.W. Incorporation of Single-Walled Carbon Nanotubes into Ferrocene-Modified Linear Polyethylenimine Redox Polymer Films. *Langmuir* **2011**, *27*, 6201–6210. [CrossRef]
50. Milton, R.D.; Abdellaoui, S.; Khadka, N.; Dean, D.R.; Leech, D.; Seefeldt, L.C.; Minteer, S.D. Nitrogenase bioelectrocatalysis: Heterogeneous ammonia and hydrogen production by MoFe protein. *Energy Environ. Sci.* **2016**, *9*, 2550–2554. [CrossRef]
51. Lee, Y.S.; Yuan, M.; Cai, R.; Lim, K.; Minteer, S.D. Nitrogenase Bioelectrocatalysis: ATP-Independent Ammonia Production Using a Redox Polymer/MoFe Protein System. *ACS Catal.* **2020**, *10*, 6854–6861. [CrossRef]
52. Eng, L.H.; Elmgren, M.; Komlos, P.; Nordling, M.; Lindquist, S.-E.; Neujahr, H.Y. Viologen-Based Redox Polymer for Contacting the Low-Potential Redox Enzyme Hydrogenase at an Electrode Surface. *J. Phys. Chem.* **1994**, *98*, 7068–7072. [CrossRef]
53. Shiraiwa, S.; So, K.; Sugimoto, Y.; Kitazumi, Y.; Shirai, O.; Kano, K. Reactivation of Standard (NiFe)-Hydrogenase and Bioelectrochemical Catalysis of Proton Reduction and Hydrogen Oxidation in a Mediated-Electron-Transfer System. *Bioelectrochemistry* **2018**, *123*, 156–161. [CrossRef]
54. Sakai, K.; Kitazumi, Y.; Shirai, O.; Kano, K. Bioelectrocatalytic formate oxidation and carbon dioxide reduction at high current density and low overpotential with tungsten-containing formate dehydrogenase and mediators. *Electrochem. Commun.* **2016**, *65*, 31–34. [CrossRef]
55. Sakai, K.; Kitazumi, Y.; Shirai, O.; Takagi, K.; Kano, K. High-Power Formate/Dioxygen Biofuel Cell Based on Mediated Electron Transfer Type Bioelectrocatalysis. *ACS Catal.* **2017**, *7*, 5668–5673. [CrossRef]
56. Kitazumi, Y.; Shirai, O.; Yamamoto, M.; Kano, K. Numerical simulation of diffuse double layer around microporous electrodes based on the Poisson–Boltzmann equation. *Electrochim. Acta* **2013**, *112*, 171–175. [CrossRef]
57. Serafín, V.; Hernández, P.; Agüí, L.; Yáñez-Sedeño, P.; Pingarrón, J.M. Electrochemical Biosensor for Creatinine Based on the Immobilization of Creatininase, Creatinase and Sarcosine Oxidase onto a Ferrocene/Horseradish Peroxidase/Gold Nanoparticles/Multi-Walled Carbon Nanotubes/Teflon Composite Electrode. *Electrochim. Acta* **2013**, *97*, 175–183. [CrossRef]
58. Holzinger, M.; Le Goff, A.; Cosnier, S. Carbon nanotube/enzyme biofuel cells. *Electrochim. Acta* **2012**, *82*, 179–190. [CrossRef]
59. Tominaga, M.; Sasaki, A.; Togami, M. Laccase Bioelectrocatalyst at a Steroid-Type Biosurfactant-Modified Carbon Nanotube Interface. *Anal. Chem.* **2015**, *87*, 5417–5421. [CrossRef]
60. Tsujimura, S.; Nishina, A.; Kamitaka, Y.; Kano, K. Coulometric D-Fructose Biosensor Based on Direct Electron Transfer Using D-Fructose Dehydrogenase. *Anal. Chem.* **2009**, *81*, 9383–9387. [CrossRef]
61. Flexer, V.; Durand, F.; Tsujimura, S.; Mano, N. Efficient Direct Electron Transfer of PQQ-glucose Dehydrogenase on Carbon Cryogel Electrodes at Neutral pH. *Anal. Chem.* **2011**, *83*, 5721–5727. [CrossRef]
62. Hamano, Y.; Tsujimura, S.; Shirai, O.; Kano, K. Micro-cubic monolithic carbon cryogel electrode for direct electron transfer reaction of fructose dehydrogenase. *Bioelectrochemistry* **2012**, *88*, 114–117. [CrossRef]

63. Murata, K.; Akatsuka, W.; Tsujimura, S. Bioelectrocatalytic Oxidation of Glucose on MgO-templated Mesoporous Carbon-modified Electrode. *Chem. Lett.* **2014**, *43*, 928–930. [CrossRef]
64. Mazurenko, I.; Clément, R.; Byrne-Kodjabachian, D.; de Poulpiquet, A.; Tsujimura, S.; Lojou, E. Pore Size Effect of MgO-Templated Carbon on Enzymatic H_2 Oxidation by the Hyperthermophilic Hydrogenase from Aquifex aeolicus. *J. Electroanal. Chem.* **2018**, *812*, 221–226. [CrossRef]
65. Takahashi, Y.; Wanibuchi, M.; Kitazumi, Y.; Shirai, O.; Kano, K. Improved direct electron transfer-type bioelectrocatalysis of bilirubin oxidase using porous gold electrodes. *J. Electroanal. Chem.* **2019**, *843*, 47–53. [CrossRef]
66. Mie, Y.; Yasutake, Y.; Ikegami, M.; Tamura, T. Anodized gold surface enables mediator-free and low-overpotential electrochemical oxidation of NADH: A facile method for the development of an NAD^+-dependent enzyme biosensor. *Sens. Actuators B Chem.* **2019**, *288*, 512–518. [CrossRef]
67. Miyata, M.; Kitazumi, Y.; Shirai, O.; Kataoka, K.; Kano, K. Diffusion-limited biosensing of dissolved oxygen by direct electron transfer-type bioelectrocatalysis of multi-copper oxidases immobilized on porous gold microelectrodes. *J. Electroanal. Chem.* **2020**, *860*, 113895. [CrossRef]
68. Siepenkoetter, T.; Salaj-Kosla, U.; Xiao, X.; Belochapkine, S.; Magner, E. Nanoporous Gold Electrodes with Tuneable Pore Sizes for Bioelectrochemical Applications. *Electroanalysis* **2016**, *28*, 2415–2423. [CrossRef]
69. Siepenkoetter, T.; Salaj-Kosla, U.; Magner, E. The Immobilization of Fructose Dehydrogenase on Nanoporous Gold Electrodes for the Detection of Fructose. *ChemElectroChem* **2017**, *4*, 905–912. [CrossRef]
70. Xiao, X.; Siepenkoetter, T.; Conghaile, P.Ó.; Leech, D.; Magner, E. Nanoporous Gold-Based Biofuel Cells on Contact Lenses. *ACS Appl. Mater. Interfaces* **2018**, *10*, 7107–7116. [CrossRef] [PubMed]
71. Holland, J.T.; Lau, C.; Brozik, S.; Atanassov, P.; Banta, S. Engineering of Glucose Oxidase for Direct Electron Transfer via Site-Specific Gold Nanoparticle Conjugation. *J. Am. Chem. Soc.* **2011**, *133*, 19262–19265. [CrossRef] [PubMed]
72. Gutiérrez-Sánchez, C.; Pita, M.; Vaz-Domínguez, C.; Shleev, S.; De Lacey, A.L. Gold Nanoparticles as Electronic Bridges for Laccase-Based Biocathodes. *J. Am. Chem. Soc.* **2012**, *134*, 17212–17220. [CrossRef]
73. Suzuki, M.; Murata, K.; Nakamura, N.; Ohno, H. The Effect of Particle Size on the Direct Electron Transfer Reactions of Metalloproteins Using Au Nanoparticle-Modified Electrodes. *Electrochemistry* **2012**, *80*, 337–339. [CrossRef]
74. Monsalve, K.; Roger, M.; Gutierrez-Sanchez, C.; Ilbert, M.; Nitsche, S.; Byrne-Kodjabachian, D.; Marchi, V.; Lojou, E. Hydrogen bioelectrooxidation on gold nanoparticle-based electrodes modified by Aquifex aeolicus hydrogenase: Application to hydrogen/oxygen enzymatic biofuel cells. *Bioelectrochemistry* **2015**, *106*, 47–55. [CrossRef] [PubMed]
75. Sakai, K.; Kitazumi, Y.; Shirai, O.; Takagi, K.; Kano, K. Direct Electron Transfer-Type Four-Way Bioelectrocatalysis of CO_2/Formate and NAD^+/NADH Redox Couples by Tungsten-Containing Formate Dehydrogenase Adsorbed on Gold Nanoparticle-Embedded Mesoporous Carbon Electrodes Modified with 4-Mercaptopyridine. *Electrochem. Commun.* **2017**, *84*, 75–79. [CrossRef]
76. Takahashi, Y.; Kitazumi, Y.; Shirai, O.; Kano, K. Improved direct electron transfer-type bioelectrocatalysis of bilirubin oxidase using thiol-modified gold nanoparticles on mesoporous carbon electrode. *J. Electroanal. Chem.* **2019**, *832*, 158–164. [CrossRef]
77. Hitaishi, V.P.; Mazurenko, I.; Murali, A.V.; De Poulpiquet, A.; Coustillier, G.; Delaporte, P.; Lojou, E. Nanosecond Laser-Fabricated Monolayer of Gold Nanoparticles on ITO for Bioelectrocatalysis. *Front. Chem.* **2020**, *8*, 431. [CrossRef]
78. Kizling, M.; Dzwonek, M.; Więckowska, A.; Bilewicz, R. Gold nanoparticles in bioelectrocatalysis—The role of nanoparticle size. *Curr. Opin. Electrochem.* **2018**, *12*, 113–120. [CrossRef]
79. Murata, K.; Suzuki, M.; Nakamura, N.; Ohno, H. Direct evidence of electron flow via the heme c group for the direct electron transfer reaction of fructose dehydrogenase using a silver nanoparticle-modified electrode. *Electrochem. Commun.* **2009**, *11*, 1623–1626. [CrossRef]
80. Zhang, L.; Beaton, S.E.; Carr, S.B.; Armstrong, F.A. Direct visible light activation of a surface cysteine-engineered [NiFe]-hydrogenase by silver nanoclusters. *Energy Environ. Sci.* **2018**, *11*, 3342–3348. [CrossRef]
81. Zhang, L.; Can, M.; Ragsdale, S.W.; Armstrong, F. Fast and Selective Photoreduction of CO_2 to CO Catalyzed by a Complex of Carbon Monoxide Dehydrogenase, TiO_2, and Ag Nanoclusters. *ACS Catal.* **2018**, *8*, 2789–2795. [CrossRef]

82. Nakamura, R.; Kamiya, K.; Hashimoto, K. Direct electron-transfer conduits constructed at the interface between multicopper oxidase and nanocrystalline semiconductive Fe oxides. *Chem. Phys. Lett.* **2010**, *498*, 307–311. [CrossRef]
83. Kizling, M.; Rekorajska, A.; Krysinski, P.; Bilewicz, R. Magnetic-field-induced orientation of fructose dehydrogenase on iron oxide nanoparticles for enhanced direct electron transfer. *Electrochem. Commun.* **2018**, *93*, 66–70. [CrossRef]
84. Rozniecka, E.; Jonsson-Niedziolka, M.; Sobczak, J.W.; Opallo, M. Mediatorless bioelectrocatalysis of dioxygen reduction at indium-doped tin oxide (ITO) and ITO nanoparticulate film electrodes. *Electrochim. Acta* **2011**, *56*, 8739–8745. [CrossRef]
85. Willner, I.; Katz, E. Integration of Layered Redox Proteins and Conductive Supports for Bioelectronic Applications. *Angew. Chem. Int. Ed.* **2000**, *39*, 1180–1218. [CrossRef]
86. Moser, C.C.; Keske, J.M.; Warncke, K.; Farid, R.S.; Dutton, P.L. Nature of biological electron transfer. *Nat. Cell Biol.* **1992**, *355*, 796–802. [CrossRef] [PubMed]
87. Marcus, R.; Sutin, N. Electron transfers in chemistry and biology. *Biochim. Biophys. Acta BBA Rev. Bioenerg.* **1985**, *811*, 265–322. [CrossRef]
88. Marcus, R.A. Electron Transfer Reactions in Chemistry: Theory and Experiment (Nobel Lecture). *Angew. Chem. Int. Ed.* **1993**, *32*, 1111–1121. [CrossRef]
89. Sugimoto, Y.; Takeuchi, R.; Kitazumi, Y.; Shirai, O.; Kano, K. Significance of Mesoporous Electrodes for Noncatalytic Faradaic Process of Randomly Oriented Redox Proteins. *J. Phys. Chem. C* **2016**, *120*, 26270–26277. [CrossRef]
90. Sugimoto, Y.; Kitazumi, Y.; Shirai, O.; Kano, K. Effects of Mesoporous Structures on Direct Electron Transfer-Type Bioelectrocatalysis: Facts and Simulation on a Three-Dimensional Model of Random Orientation of Enzymes. *Electrochemistry* **2017**, *85*, 82–87. [CrossRef]
91. Hitaishi, V.P.; Clement, R.; Bourassin, N.; Baaden, M.; De Poulpiquet, A.; Sacquin-Mora, S.; Ciaccafava, A.; Sacquin-Mora, S. Controlling Redox Enzyme Orientation at Planar Electrodes. *Catalysts* **2018**, *8*, 192. [CrossRef]
92. Xia, H.; So, K.; Kitazumi, Y.; Shirai, O.; Nishikawa, K.; Higuchi, Y.; Kano, K. Dual gas-diffusion membrane- and mediatorless dihydrogen/air-breathing biofuel cell operating at room temperature. *J. Power Sources* **2016**, *335*, 105–112. [CrossRef]
93. Adachi, T.; Kitazumi, Y.; Shirai, O.; Kawano, T.; Kataoka, K.; Kano, K. Effects of Elimination of α Helix Regions on Direct Electron Transfer-type Bioelectrocatalytic Properties of Copper Efflux Oxidase. *Electrochemistry* **2020**, *88*, 185–189. [CrossRef]
94. Xia, H.; Kitazumi, Y.; Shirai, O.; Kano, K. Enhanced direct electron transfer-type bioelectrocatalysis of bilirubin oxidase on negatively charged aromatic compound-modified carbon electrode. *J. Electroanal. Chem.* **2016**, *763*, 104–109. [CrossRef]
95. So, K.; Kawai, S.; Hamano, Y.; Kitazumi, Y.; Shirai, O.; Hibi, M.; Ogawa, J.; Kano, K. Improvement of a direct electron transfer-type fructose/dioxygen biofuel cell with a substrate-modified biocathode. *Phys. Chem. Chem. Phys.* **2014**, *16*, 4823–4829. [CrossRef] [PubMed]
96. Bollella, P.; Hibino, Y.; Kano, K.; Gorton, L.; Antiochia, R. Highly Sensitive Membraneless Fructose Biosensor Based on Fructose Dehydrogenase Immobilized onto Aryl Thiol Modified Highly Porous Gold Electrode: Characterization and Application in Food Samples. *Anal. Chem.* **2018**, *90*, 12131–12136. [CrossRef] [PubMed]
97. Xia, H.; Hibino, Y.; Kitazumi, Y.; Shirai, O.; Kano, K. Interaction between d-fructose dehydrogenase and methoxy-substituent-functionalized carbon surface to increase productive orientations. *Electrochim. Acta* **2016**, *218*, 41–46. [CrossRef]
98. So, K.; Sakai, K.; Kano, K. Gas diffusion bioelectrodes. *Curr. Opin. Electrochem.* **2017**, *5*, 173–182. [CrossRef]
99. Wong, T.S.; Schwaneberg, U. Protein engineering in bioelectrocatalysis. *Curr. Opin. Biotechnol.* **2003**, *14*, 590–596. [CrossRef]
100. Battistuzzi, G.; Borsari, M.; Cowan, J.A.; Ranieri, A.; Sola, M. Control of cytochrome C redox potential: Axial ligation and protein environment effects. *J. Am. Chem. Soc.* **2002**, *124*, 5315–5324. [CrossRef]
101. Li, H.; Webb, S.P.; Ivanic, J.; Jensen, J.H. Determinants of the Relative Reduction Potentials of Type-1 Copper Sites in Proteins. *J. Am. Chem. Soc.* **2004**, *126*, 8010–8019. [CrossRef]
102. Kawai, S.; Yakushi, T.; Matsushita, K.; Kitazumi, Y.; Shirai, O.; Kano, K. The electron transfer pathway in direct electrochemical communication of fructose dehydrogenase with electrodes. *Electrochem. Commun.* **2014**, *38*, 28–31. [CrossRef]

103. Hibino, Y.; Kawai, S.; Kitazumi, Y.; Shirai, O.; Kano, K. Mutation of heme c axial ligands in d-fructose dehydrogenase for investigation of electron transfer pathways and reduction of overpotential in direct electron transfer-type bioelectrocatalysis. *Electrochem. Commun.* **2016**, *67*, 43–46. [CrossRef]
104. Adachi, T.; Kaida, Y.; Kitazumi, Y.; Shirai, O.; Kano, K. Bioelectrocatalytic performance of d-fructose dehydrogenase. *Bioelectrochemistry* **2019**, *129*, 1–9. [CrossRef] [PubMed]
105. Kamitaka, Y.; Tsujimura, S.; Kataoka, K.; Sakurai, T.; Ikeda, T.; Kano, K. Effects of axial ligand mutation of the type I copper site in bilirubin oxidase on direct electron transfer-type bioelectrocatalytic reduction of dioxygen. *J. Electroanal. Chem.* **2007**, *601*, 119–124. [CrossRef]
106. Hibino, Y.; Kawai, S.; Kitazumi, Y.; Shirai, O.; Kano, K. Construction of a protein-engineered variant of d-fructose dehydrogenase for direct electron transfer-type bioelectrocatalysis. *Electrochem. Commun.* **2017**, *77*, 112–115. [CrossRef]
107. Kaida, Y.; Hibino, Y.; Kitazumi, Y.; Shirai, O.; Kano, K. Ultimate downsizing of d-fructose dehydrogenase for improving the performance of direct electron transfer-type bioelectrocatalysis. *Electrochem. Commun.* **2019**, *98*, 101–105. [CrossRef]
108. Hibino, Y.; Kawai, S.; Kitazumi, Y.; Shirai, O.; Kano, K. Protein-Engineering Improvement of Direct Electron Transfer-Type Bioelectrocatalytic Properties of d-Fructose Dehydrogenase. *Electrochemistry* **2019**, *87*, 47–51. [CrossRef]
109. Kaida, Y.; Hibino, Y.; Kitazumi, Y.; Shirai, O.; Kano, K. Discussion on Direct Electron Transfer-Type Bioelectrocatalysis of Downsized and Axial-Ligand Exchanged Variants of d-Fructose Dehydrogenase. *Electrochemistry* **2020**, *88*, 195–199. [CrossRef]
110. Matsui, Y.; Hamamoto, K.; Kitazumi, Y.; Shirai, O.; Kano, K. Diffusion-Controlled Mediated Electron Transfer-Type Bioelectrocatalysis Using Ultrathin-Ring and Microband Electrodes as Ultimate Amperometric Glucose Sensors. *Anal. Sci.* **2017**, *33*, 845–851. [CrossRef]
111. Mazurenko, I.; Hitaishi, V.P.; Lojou, E. Recent advances in surface chemistry of electrodes to promote direct enzymatic bioelectrocatalysis. *Curr. Opin. Electrochem.* **2020**, *19*, 113–121. [CrossRef]
112. Ferapontova, E.E.; Schmengler, K.; Börchers, T.; Ruzgas, T.; Gorton, L. Effect of cysteine mutations on direct electron transfer of horseradish peroxidase on gold. *Biosens. Bioelectron.* **2002**, *17*, 953–963. [CrossRef]
113. Bollella, P.; Ludwig, R.; Gorton, L. Cellobiose dehydrogenase: Insights on the nanostructuration of electrodes for improved development of biosensors and biofuel cells. *Appl. Mater. Today* **2017**, *9*, 319–332. [CrossRef]
114. Algov, I.; Grushka, J.; Zarivach, R.; Alfonta, L. Highly Efficient Flavin–Adenine Dinucleotide Glucose Dehydrogenase Fused to a Minimal Cytochrome C Domain. *J. Am. Chem. Soc.* **2017**, *139*, 17217–17220. [CrossRef] [PubMed]
115. Ito, K.; Okuda-Shimazaki, J.; Mori, K.; Kojima, K.; Tsugawa, W.; Ikebukuro, K.; Lin, C.-E.; La Belle, J.T.; Yoshida, H.; Sode, K. Designer fungus FAD glucose dehydrogenase capable of direct electron transfer. *Biosens. Bioelectron.* **2019**, *123*, 114–123. [CrossRef] [PubMed]
116. Okuda, J.; Sode, K. PQQ glucose dehydrogenase with novel electron transfer ability. *Biochem. Biophys. Res. Commun.* **2004**, *314*, 793–797. [CrossRef]
117. Gilardi, G.; Meharenna, Y.T.; Tsotsou, G.E.; Sadeghi, S.J.; Fairhead, M.; Giannini, S. Molecular Lego: Design of molecular assemblies of P450 enzymes for nanobiotechnology. *Biosens. Bioelectron.* **2002**, *17*, 133–145. [CrossRef]
118. Hanashi, T.; Yamazaki, T.; Tsugawa, W.; Ferri, S.; Nakayama, D.; Tomiyama, M.; Ikebukuro, K.; Sode, K. BioCapacitor-A Novel Category of Biosensor. *Biosens. Bioelectron.* **2009**, *24*, 1837–1842. [CrossRef]
119. Conzuelo, F.; Ruff, A.; Schuhmann, W. Self-powered bioelectrochemical devices. *Curr. Opin. Electrochem.* **2018**, *12*, 156–163. [CrossRef]
120. Falk, M.; Shleev, S. Hybrid Dual-functioning Electrodes for Combined Ambient Energy Harvesting and Charge Strage: Towards Self-Powered Systems. *Biosens. Bioelectron.* **2019**, *126*, 275–291. [CrossRef]
121. Agnès, C.; Holzinger, M.; Le Goff, A.; Reuillard, B.; Elouarzaki, K.; Tingry, S.; Cosnier, S. Supercapacitor/biofuel cell hybrids based on wired enzymes on carbon nanotube matrices: Autonomous reloading after high power pulses in neutral buffered glucose solutions. *Energy Environ. Sci.* **2014**, *7*, 1884–1888. [CrossRef]
122. Pankratov, D.; Blum, Z.; Suyatin, D.B.; Popov, V.O.; Shleev, S. Self-Charging Electrochemical Biocapacitor. *ChemElectroChem* **2013**, *1*, 343–346. [CrossRef]
123. Pankratov, D.; Blum, Z.; Shleev, S. Hybrid Electric Power Biodevices. *ChemElectroChem* **2014**, *1*, 1798–1807. [CrossRef]

124. Xiao, X.; Magner, E. A quasi-solid-state and self-powered biosupercapacitor based on flexible nanoporous gold electrodes. *Chem. Commun.* **2018**, *54*, 5823–5826. [CrossRef] [PubMed]
125. Pankratov, D.; Shen, F.; Ortiz, R.; Toscano, M.D.; Thormann, E.; Zhang, J.; Gorton, L.; Chi, Q. Fuel-independent and membrane-less self-charging biosupercapacitor. *Chem. Commun.* **2018**, *54*, 11801–11804. [CrossRef] [PubMed]
126. Bobrowski, T.; González-Arribas, E.; Ludwig, R.; Toscano, M.D.; Shleev, S.; Schuhmann, W. Rechargeable, Flexible and Mediator-Free Biosupercapacitor Based on Transparent ITO Nanoparticle Modified Electrodes Acting in µM Glucose Containing Buffers. *Biosens. Bioelectron.* **2018**, *101*, 84–89. [CrossRef]
127. Pankratov, D.; Conzuelo, F.; Pinyou, P.; Alsaoub, S.; Schuhmann, W.; Shleev, S. A Nernstian Biosupercapacitor. *Angew. Chem. Int. Ed.* **2016**, *55*, 15434–15438. [CrossRef]
128. Alsaoub, S.; Ruff, A.; Conzuelo, F.; Ventosa, E.; Ludwig, R.; Shleev, S.; Schuhmann, W. An Intrinsic Self-Charging Biosupercapacitor Comprised of a High-Potential Bioanode and a Low-Potential Biocathode. *ChemPlusChem* **2017**, *82*, 576–583. [CrossRef]
129. Zhao, F.; Bobrowski, T.; Ruff, A.; Hartmann, V.; Nowaczyk, M.M.; Rögner, M.; Conzuelo, F.; Schuhmann, W. A light-driven Nernstian biosupercapacitor. *Electrochim. Acta* **2019**, *306*, 660–666. [CrossRef]
130. González-Arribas, E.; Falk, M.; Aleksejeva, O.; Bushnev, S.; Sebastián, P.; Feliu, J.M.; Shleev, S. A Conventional Symmetric Biosupercapacitor Based on Rusticyanin Modified Gold Electrodes. *J. Electroanal. Chem.* **2018**, *816*, 253–258. [CrossRef]
131. Shen, F.; Pankratov, D.; Pankratova, G.; Toscano, M.D.; Zhang, J.; Ulstrup, J.; Chi, Q.; Gorton, L. Supercapacitor/Biofuel Cell Hybrid Device Employing Biomolecules for Energy Conversion and Charge Stroge. *Bioelectrochemistry* **2019**, *128*, 94–99. [CrossRef]
132. Wu, R.; Ma, C.; Zhu, Z. Enzymatic electrosynthesis as an emerging electrochemical synthesis platform. *Curr. Opin. Electrochem.* **2020**, *19*, 1–7. [CrossRef]
133. Schmid, A.I.; Dordick, J.S.; Hauer, B.; Kiener, A.; Wubbolts, M.; Witholt, B. Industrial biocatalysis today and tomorrow. *Nat. Cell Biol.* **2001**, *409*, 258–268. [CrossRef] [PubMed]
134. Zaks, A. Industrial biocatalysis. *Curr. Opin. Chem. Biol.* **2001**, *5*, 130–136. [CrossRef]
135. Vincent, K.A.; Parkin, A.; Armstrong, F.A. Investigating and Exploiting the Electrocatalytic Properties of Hydrogenases. *Chem. Rev.* **2007**, *107*, 4366–4413. [CrossRef] [PubMed]
136. Armstrong, F.; Hirst, J. Reversibility and efficiency in electrocatalytic energy conversion and lessons from enzymes. *Proc. Natl. Acad. Sci. USA* **2011**, *108*, 14049–14054. [CrossRef] [PubMed]
137. Sakai, K.; Hsieh, B.-C.; Maruyama, A.; Kitazumi, Y.; Shirai, O.; Kano, K. Interconversion between formate and hydrogen carbonate by tungsten-containing formate dehydrogenase-catalyzed mediated bioelectrocatalysis. *Sens. Bio Sens. Res.* **2015**, *5*, 90–96. [CrossRef]
138. Reda, T.; Plugge, C.M.; Abram, N.J.; Hirst, J. Reversible interconversion of carbon dioxide and formate by an electroactive enzyme. *Proc. Natl. Acad. Sci. USA* **2008**, *105*, 10654–10658. [CrossRef]
139. Bassegoda, A.; Madden, C.; Wakerley, D.W.; Reisner, E.; Hirst, J. Reversible Interconversion of CO_2 and Formate by a Molybdenum-Containing Formate Dehydrogenase. *J. Am. Chem. Soc.* **2014**, *136*, 15473–15476. [CrossRef]
140. Takagi, K.; Kano, K.; Ikeda, T. Mediated bioelectrocatalysis based on NAD-related enzymes with reversible characteristics. *J. Electroanal. Chem.* **1998**, *445*, 211–219. [CrossRef]
141. Megarity, C.F.; Siritanaratkul, B.; Heath, R.S.; Wan, L.; Morello, G.; Fitzpatrick, S.R.; Booth, R.L.; Sills, A.J.; Robertson, A.W.; Warner, J.H.; et al. Electrocatalytic Volleyball: Rapid Nanoconfined Nicotinamide Cycling for Organic Synthesis in Electrode Pores. *Angew. Chem. Int. Ed.* **2019**, *58*, 4948–4952. [CrossRef]
142. Morello, G.; Siritanaratkul, B.; Megarity, C.F.; Armstrong, F.A. Efficient Electrocatalytic CO_2 Fixation by Nanoconfined Enzymes via a C3-to-C4 Reaction That Is Favored over H_2 Production. *ACS Catal.* **2019**, *9*, 11255–11262. [CrossRef]
143. Alkotaini, B.; Abdellaoui, S.; Hasan, K.; Grattieri, M.; Quah, T.; Cai, R.; Yuan, M.; Minteer, S.D. Sustainable Bioelectrosynthesis of the Bioplastic Polyhydroxybutyrate: Overcoming Substrate Requirement for NADH Regeneration. *ACS Sustain. Chem. Eng.* **2018**, *6*, 4909–4915. [CrossRef]
144. Wan, L.; Megarity, C.F.; Siritanaratkul, B.; Armstrong, F. A hydrogen fuel cell for rapid, enzyme-catalysed organic synthesis with continuous monitoring. *Chem. Commun.* **2018**, *54*, 972–975. [CrossRef] [PubMed]

145. Wan, L.; Heath, R.S.; Siritanaratkul, B.; Megarity, C.F.; Sills, A.J.; Thompson, M.P.; Turner, N.J.; Armstrong, F. Enzyme-catalysed enantioselective oxidation of alcohols by air exploiting fast electrochemical nicotinamide cycling in electrode nanopores. *Green Chem.* **2019**, *21*, 4958–4963. [CrossRef]

146. Lazarus, O.; Woolerton, T.W.; Parkin, A.; Lukey, M.J.; Reisner, E.; Seravalli, J.; Pierce, E.; Ragsdale, S.W.; Sargent, F.; Armstrong, F.A. Water–Gas Shift Reaction Catalyzed by Redox Enzymes on Conducting Graphite Platelets. *J. Am. Chem. Soc.* **2009**, *131*, 14154–14155. [CrossRef] [PubMed]

147. Adachi, T.; Kitazumi, Y.; Shirai, O.; Kano, K. Construction of a bioelectrochemical formate generating system from carbon dioxide and dihydrogen. *Electrochem. Commun.* **2018**, *97*, 73–76. [CrossRef]

148. Milton, R.D.; Cai, R.; Abdellaoui, S.; Leech, D.; De Lacey, A.L.; Pita, M.; Minteer, S.D. Bioelectrochemical Haber-Bosch Process: An Ammonia-Producing H_2/N_2 Fuel Cell. *Angew. Chem. Int. Ed.* **2017**, *56*, 2680–2683. [CrossRef] [PubMed]

149. Chen, H.; Prater, M.B.; Cai, R.; Dong, F.; Chen, H.; Minteer, S.D. Bioelectrocatalytic Conversion from N_2 to Chiral Amino Acids in a H_2/α-Keto Acid Enzymatic Fuel Cell. *J. Am. Chem. Soc.* **2020**, *142*, 4028–4036. [CrossRef]

150. Sakai, K.; Kitazumi, Y.; Shirai, O.; Takagi, K.; Kano, K. Efficient bioelectrocatalytic CO_2 reduction on gas-diffusion-type biocathode with tungsten-containing formate dehydrogenase. *Electrochem. Commun.* **2016**, *73*, 85–88. [CrossRef]

151. Fukuda, J.; Tsujimura, S.; Kano, K. Coulometric bioelectrocatalytic reactions based on NAD-dependent dehydrogenases in tricarboxylic acid cycle. *Electrochim. Acta* **2008**, *54*, 328–333. [CrossRef]

152. Chen, H.; Cai, R.; Patel, J.; Dong, F.; Chen, H.; Minteer, S.D. Upgraded Bioelectrocatalytic N_2 Fixation: From N2 to Chiral Amine Intermediates. *J. Am. Chem. Soc.* **2019**, *141*, 4963–4971. [CrossRef]

153. Dong, F.; Chen, H.; Malapit, C.A.; Prater, M.B.; Li, M.; Yuan, M.; Lim, K.; Minteer, S.D. Biphasic Bioelectrocatalytic Synthesis of Chiral β-Hydroxy Nitriles. *J. Am. Chem. Soc.* **2020**, *142*, 8374–8382. [CrossRef] [PubMed]

154. Hickey, D.P.; Cai, R.; Yang, Z.-Y.; Grunau, K.; Einsle, O.; Seefeldt, L.C.; Minteer, S.D. Establishing a Thermodynamic Landscape for the Active Site of Mo-Dependent Nitrogenase. *J. Am. Chem. Soc.* **2019**, *141*, 17150–17157. [CrossRef] [PubMed]

155. Smith, B.E. STRUCTURE: Nitrogenase Reveals Its Inner Secrets. *Science* **2002**, *297*, 1654–1655. [CrossRef] [PubMed]

156. Milton, R.D.; Minteer, S.D. Enzymatic Bioelectrosynthetic Ammonia Production: Recent Electrochemistry of Nitrogenase, Nitrate Reductase, and Nitrite Reductase. *ChemPlusChem* **2016**, *82*, 513–521. [CrossRef]

157. Cai, R.; Minteer, S.D. Nitrogenase Bioelectrocatalysis: From Understanding Electron-Transfer Mechanisms to Energy Applications. *ACS Energy Lett.* **2018**, *3*, 2736–2742. [CrossRef]

158. Antiochia, R.; Gallina, A.; Lavagnini, I.; Magno, F. Kinetic and Thermodynamic Aspects of NAD-Related Enzyme-Linked Mediated Bioelectrocatalysis. *Electroanalysis* **2002**, *14*, 1256–1261. [CrossRef]

159. Lobo, M.J.; Miranda, A.J.; Tuñón, P. Amperometric Biosensors Based on NAD(P)-Dependent Dehydrogenase Enzymes. *Electroanalysis* **1997**, *9*, 191–201. [CrossRef]

160. Sokol, K.P.; Robinson, W.E.; Warnan, J.; Kornienko, N.; Nowaczyk, M.M.; Ruff, A.; Zhang, J.Z.; Reisner, E. Bias-free photoelectrochemical water splitting with photosystem II on a dye-sensitized photoanode wired to hydrogenase. *Nat. Energy* **2018**, *3*, 944–951. [CrossRef]

161. Sokol, K.P.; Robinson, W.E.; Oliveira, A.R.; Warnan, J.; Nowaczyk, M.M.; Ruff, A.; Pereira, I.A.C.; Reisner, E. Photoreduction of CO_2 with a Formate Dehydrogenase Driven by Photosystem II Using a Semi-artificial Z-Scheme Architecture. *J. Am. Chem. Soc.* **2018**, *140*, 16418–16422. [CrossRef]

162. Riedel, M.; Wersig, J.; Ruff, A.; Schuhmann, W.; Zouni, A.; Lisdat, F. A Z-Scheme-Inspired Photobioelectrochemical H_2O/O_2 Cell with a 1V Open-Circuit Voltage Combining Photosystem II and PbS Quantum Dots. *Angew. Chem. Int. Ed.* **2019**, *58*, 801–805. [CrossRef]

163. Adachi, T.; Kataoka, K.; Kitazumi, Y.; Shirai, O.; Kano, K. A Bio-solar Cell with Thylakoid Membranes and Bilirubin Oxidase. *Chem. Lett.* **2019**, *48*, 686–689. [CrossRef]

164. Pankratov, D.; Pankratova, G.; Dyachkova, T.P.; Falkman, P.; Åkerlund, H.-E.; Toscano, M.D.; Chi, Q.; Gorton, L. Supercapacitive Biosolar Cell Driven by Direct Electron Transfer between Photosynthetic Membranes and CNT Networks with Enhanced Performance. *ACS Energy Lett.* **2017**, *2*, 2635–2639. [CrossRef]

165. Rasmussen, M.; Wingersky, A.; Minteer, S.D. Improved Performance of a Thylakoid Bio-Solar Cell by Incorporation of Carbon Quantum Dots. *ECS Electrochem. Lett.* **2013**, *3*, H1–H3. [CrossRef]

166. Efrati, A.; Tel-Vered, R.; Michaeli, R.; Nechushtai, R.; Willner, I. Cytochrome c-coupled photosystem I and photosystem II (PSI/PSII) photo-bioelectrochemical cells. *Energy Environ. Sci.* **2013**, *6*, 2950. [CrossRef]
167. Kirchhofer, N.D.; Rasmussen, M.; Dahlquist, F.W.; Minteer, S.D.; Bazan, G.C. The photobioelectrochemical activity of thylakoid bioanodes is increased via photocurrent generation and improved contacts by membrane-intercalating conjugated oligoelectrolytes. *Energy Environ. Sci.* **2015**, *8*, 2698–2706. [CrossRef]
168. Pankratova, G.; Pankratov, D.; Hasan, K.; Åkerlund, H.-E.; Albertsson, P.-Å.; Leech, D.; Shleev, S.; Gorton, L. Supercapacitive Photo-Bioanodes and Biosolar Cells: A Novel Approach for Solar Energy Harnessing. *Adv. Energy Mater.* **2017**, *7*, 1602285. [CrossRef]
169. Yehezkeli, O.; Tel-Vered, R.; Wasserman, J.; Trifonov, A.; Michaeli, D.; Nechushtai, R.; Willner, I. Integrated photosystem II-based photo-bioelectrochemical cells. *Nat. Commun.* **2012**, *3*, 742. [CrossRef]
170. Calkins, J.O.; Umasankar, Y.; O'Neill, H.; Ramasamy, R.P. High photo-electrochemical activity of thylakoid–carbon nanotube composites for photosynthetic energy conversion. *Energy Environ. Sci.* **2013**, *6*, 1891–1900. [CrossRef]
171. Tsujimura, S.; Wadano, A.; Kano, K.; Ikeda, T. Photosynthetic bioelectrochemical cell utilizing cyanobacteria and water-generating oxidase. *Enzym. Microb. Technol.* **2001**, *29*, 225–231. [CrossRef]
172. Mimcault, M.; Carpentier, R. Kinetics of Photocurrent Induction by a Thylakoid Containing Elctrochemical Cell. *J. Electroanal. Chem.* **1989**, *276*, 145–158. [CrossRef]
173. Carpentier, R.; Lemieux, S.; Mimeault, M.; Purcell, M.; Goetze, D.C. A Photoelectrochemical Cell Using Immobilized Photosynthetic Membranes. *J. Electroanal. Chem. Interfacial Electrochem.* **1989**, *276*, 391–401. [CrossRef]
174. González-Arribas, E.; Aleksejeva, O.; Bobrowski, T.; Toscano, M.D.; Gorton, L.; Schuhmann, W.; Shleev, S. Solar biosupercapacitor. *Electrochem. Commun.* **2017**, *74*, 9–13. [CrossRef]
175. Rasmussen, M.; Minteer, S.D. Investigating the Mechanism of Thylakoid Direct Electron Transfer for Photocurrent Generation. *Electrochim. Acta* **2014**, *126*, 68–73. [CrossRef]
176. Saboe, P.O.; Conte, E.; Chan, S.; Feroz, H.; Ferlez, B.; Farell, M.; Poyton, M.F.; Sines, I.T.; Yan, H.; Bazan, G.C.; et al. Biomimetic wiring and stabilization of photosynthetic membrane proteins with block copolymer interfaces. *J. Mater. Chem. A* **2016**, *4*, 15457–15463. [CrossRef]
177. Kanso, H.; Pankratova, G.; Bollella, P.; Leech, D.; Hernandez, D.; Gorton, L. Sunlight photocurrent generation from thylakoid membranes on gold nanoparticle modified screen-printed electrodes. *J. Electroanal. Chem.* **2018**, *816*, 259–264. [CrossRef]
178. Hamidi, H.; Hasan, K.; Emek, S.C.; Dilgin, Y.; Åkerlund, H.-E.; Albertsson, P.-. Åke; Leech, D.; Gorton, L. Photocurrent Generation from Thylakoid Membranes on Osmium-Redox-Polymer-Modified Electrodes. *ChemSusChem* **2015**, *8*, 990–993. [CrossRef] [PubMed]
179. Hasan, K.; Milton, R.D.; Grattieri, M.; Wang, T.; Stephanz, M.; Minteer, S.D. Photobioelectrocatalysis of Intact Chloroplasts for Solar Energy Conversion. *ACS Catal.* **2017**, *7*, 2257–2265. [CrossRef]
180. Takeuchi, R.; Suzuki, A.; Sakai, K.; Kitazumi, Y.; Shirai, O.; Kano, K. Construction of photo-driven bioanodes using thylakoid membranes and multi-walled carbon nanotubes. *Bioelectrochemistry* **2018**, *122*, 158–163. [CrossRef]
181. Hasan, K.; Dilgin, Y.; Emek, S.C.; Tavahodi, M.; Åkerlund, H.-E.; Albertsson, P.-Å.; Gorton, L. Photoelectrochemical Communication between Thylakoid Membranes and Gold Electrodes through Different Quinone Derivatives. *ChemElectroChem* **2014**, *1*, 131–139. [CrossRef]

Publisher's Note: MDPI stays neutral with regard to jurisdictional claims in published maps and institutional affiliations.

© 2020 by the authors. Licensee MDPI, Basel, Switzerland. This article is an open access article distributed under the terms and conditions of the Creative Commons Attribution (CC BY) license (http://creativecommons.org/licenses/by/4.0/).

Review

Enzymatic Bioreactors: An Electrochemical Perspective

Simin Arshi [1], Mehran Nozari-Asbemarz [2] and Edmond Magner [1,*]

1. Department of Chemical Sciences and Bernal Institute, University of Limerick, V94 T9PX Limerick, Ireland; simin.arshi@ul.ie
2. Department of Chemistry, University of Mohaghegh Ardabili, Ardabil 5619911367, Iran; m.nozari@uma.ac.ir
* Correspondence: edmond.magner@ul.ie; Tel.: +353-61-202629

Received: 30 September 2020; Accepted: 18 October 2020; Published: 24 October 2020

Abstract: Biocatalysts provide a number of advantages such as high selectivity, the ability to operate under mild reaction conditions and availability from renewable resources that are of interest in the development of bioreactors for applications in the pharmaceutical and other sectors. The use of oxidoreductases in biocatalytic reactors is primarily focused on the use of NAD(P)-dependent enzymes, with the recycling of the cofactor occurring via an additional enzymatic system. The use of electrochemically based systems has been limited. This review focuses on the development of electrochemically based biocatalytic reactors. The mechanisms of mediated and direct electron transfer together with methods of immobilising enzymes are briefly reviewed. The use of electrochemically based batch and flow reactors is reviewed in detail with a focus on recent developments in the use of high surface area electrodes, enzyme engineering and enzyme cascades. A future perspective on electrochemically based bioreactors is presented.

Keywords: bioelectrocatalysts; oxidoreductases; biocatalytic reactors; electrochemical reactors

---

1. Introduction

Biocatalysts represent an alternative to conventional catalysts, providing a number of advantages that include availability from renewable resources, biodegradability and high selectivity [1]. Enzymes are proteins and catalyse a wide variety of reactions that have applications in a range of industrial processes [2,3]. They are extremely effective biological catalysts, highly selective and can operate under mild conditions (ambient temperatures, physiological pH and atmospheric pressure). When used in organic synthesis, enzymes can obviate the need to protect and activate functional groups. Enzymes are generally soluble in water and can avoid the use of organic solvents, resulting in the generation of less waste [4–6]. Oxidoreductases are a class of enzymes that catalyse redox reactions [7] such as oxygenation, dehydrogenation, oxidative bond formation and electron transfer reactions [8]. Over the last four decades, oxidoreductases have been coupled successfully with electrochemistry in biofuel cells [9–12] and biosensors [13–16]. The use of oxidoreductases in biocatalysis has gained significant attention in the synthesis of a variety of chemicals, such as chiral compounds [17–19], biofuels [20] and ammonia [21]. While biocatalysts are most commonly used in batch reactors, the development of biocatalytic flow reactors is of increasing interest as it can bring significant advantages that include improved mass and heat transfer, lower costs and higher yields [22,23]. This review describes oxidoreductases and bioreactors from an electrochemical perspective, with a focus on the electrocatalytic activity of oxidoreductases and methods to improve their use.

2. Bioelectrochemistry and Bioelectrocatalysts

Oxidoreductases undergo electron transfer with the surface of an electrode via two mechanisms [24]; mediated [25–28] and direct electron transfer [29–31] (Figure 1). As the redox active centres of oxidoreductases can be placed deep within the enzymes, the rate of electron transfer between the redox centre and the electrode can be restricted and only a limited number of redox enzymes can undergo direct electron transfer (DET) [32]. Direct electron transfer relies on the appropriate orientation of an enzyme on the electrode surface, the location of the active site within the enzyme and the distance between the redox centre and the electrode. Efficient rates of electron transfer require that the redox active site in the enzyme be close to the surface of the electrode, with the appropriate orientation of the enzyme on the electrode. When the redox site is inaccessible, mediators can be used as electron shuttles between the electrode and the active site of enzymes (Figure 1a) [33,34].

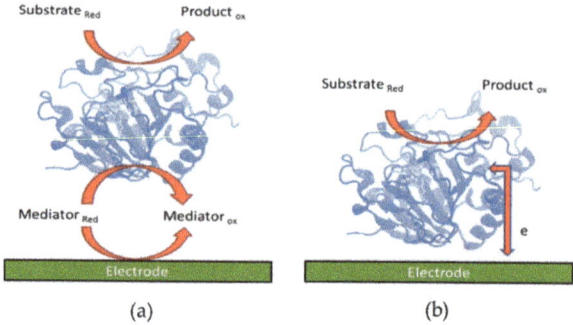

Figure 1. Schematic diagram of (**a**) mediated and (**b**) direct electron transfer between an enzyme and an electrode.

2.1. Mediated Electron Transfer

The redox properties of the mediator and of the enzymes need to match in order to have an efficient rate of electron transfer. The reduction potential of the mediator ($E°_{mediator}$) should be more positive than the redox potential of the prosthetic group of the enzyme ($E°_{enzyme}$) for oxidation to occur, while for reduction, $E°_{mediator}$ should be more negative than $E°_{enzyme}$. The mediator should undergo fast electron transfer both at the surface of the electrode and with the active site of the enzyme [34,35]. Properties such as mediator stability, selectivity and the electrochemical reversibility of the redox couple also need to be considered [33]. Approximately 90% of oxidoreductases utilised nicotinamide coenzymes to catalyse reactions. For example, alcohol dehydrogenase (ADH) uses the cofactor nicotinamide adenine dinucleotide (NAD^+) to catalyse the oxidation of substrates such as cyclohexanol [36], primary and secondary alcohols [37] and methionine [38]. Due to the high cost of $NAD(P)^+$, it is important to use an efficient regeneration system. The direct regeneration of nicotinamide coenzymes at unmodified electrode surfaces requires the use of high overpotentials that can result in the formation of dimers of $NAD(P)^+$ that are enzymatically inactive. Mediators such as methylene green, methylene blue [39–42], 2,2-azino-bis-(3-ethyl-benzo-thiazoline-6-sulfonic acid (ABTS) [43], naphthoquinone [44], ferrocenes [45,46], and viologens [47] are widely used to regenerate NAD^+, etc. For example; an indirect $NAD(P)^+$ regeneration system utilised ABTS to oxidise alcohols with a turnover of 1200 h^{-1} [48]. A number of enzymes such as dihydrolipoamide dehydrogenase (DLD), ferredoxin-$NADP^+$ reductase (FNR) have been used to regenerate NAD [49–51]. Chen et al. used reduced methyl viologen ($MV^{•+}$) and diaphorase for effective NADH regeneration in the production of NH_3 and for asymmetric amination [52]. The mediated electrochemical regeneration of NADH through methyl viologen was also used in the enzymatic reduction of ketones [53]. Badalyan et al. used the negatively charged viologen derivative [$(SPr)_2V^•$]$^-$ as an effective mediator for nitrogenase electrocatalysis [54]. Tosstorff et al. used three different mediators, cobalt sepulchrate, safranin T, and [$Cp^*Rh(bpy)(H_2O)$]$^{2+}$ in an NADH regeneration system that was combined with old yellow enzyme for the asymmetric reduction of C=C [55].

Redox polymers are used to wire enzymes and also act as an immobilisation matrix facilitating electron transfer between the electrode and enzyme by transferring electrons within the polymer matrix. Redox polymers possess a polymeric backbone with attached redox mediators [56,57]. Early reports focused on redox polymers based on osmium complexes with polymer backbones comprised of poly(vinyl imidazole)s and poly(vinyl pyridine)s [58,59]. A range of other redox polymers based on poly(vinylalcohol), poly(vinyl imidazole), poly(vinyl pyridine) and poly(ethylenimine) modified with redox species such as ferrocene [60], cobaltocene [61], viologen [62] and quinone [44] have been described. For example, Alkotaini et al. used a redox polymer N-benzyl-N'-propyl-4,4'-bipyridinium-modified linear polyethylenimine benzylpropylviologen (BPV-LPEI) and diaphorase for effective cofactor regeneration for the bioelectrosynthesis of the bioplastic, polyhydroxybutyrate (PHB) [62]. Szczesny et al. used a viologen-modified redox polymer for the electrical wiring of W-dependent formate dehydrogenase to reduce CO_2 gas to formate [63], whereas Yuan et al. described the electrochemical regeneration of NADH using a cobaltocene-modified poly(allylamine) redox polymer and diaphorase. The system produced 1,4-NADH with high yields close to 100% and turnover frequencies between 2091 and 3680 h^{-1} at different temperatures. The system was coupled with ADH to produce methanol and propanol (Figure 2) [57].

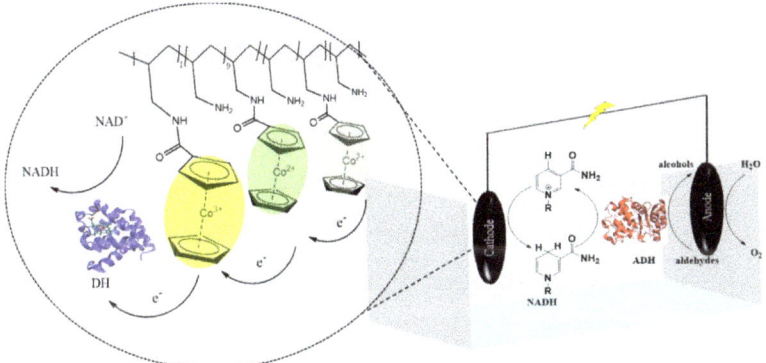

Figure 2. A schematic of the designed system for the electrochemical regeneration of NADH [57].

2.2. Direct Electron Transfer

Direct electron transfer has a significant advantage in comparison to mediated electron transfer [64] as the use of mediators can cause voltage losses stemming from the difference between the redox potentials of the enzyme and the mediator. In addition, the development and miniaturization of biofuel cells can be easier as no membrane or compartment is required [51,64,65]. Some oxidoreductases, such as multi copper oxidases, can undergo direct electron transfer with the electrode. In direct electron transfer, the distance (d) between the electrode and cofactor plays a key role in the rate of electron transfer. In an electron transfer reaction, the heterogeneous rate constant k° depends on the distance between the mediator and the electrode, (Equation (1)) [66,67]:

$$k° = k°_{maximum} \exp(-\beta d) \qquad (1)$$

where β is the decay coefficient, d is the distance between the redox centre of the enzyme and the electrode and $k°_{maximum}$ is the rate constant at the closest approach. The orientation of enzymes on the electrode surface can affect d so it is crucial that enzymes have the optimal orientation on the surface for direct electron transfer [67]. In general, the distance should be less than 2 nm. At longer distances, the rate of electron transfer between the electrode and cofactor is too low [35]. DET studies of multi-copper oxidases such as laccase, cellobiose dehydrogenase and pyrroloquinoline quinone-dependent glucose dehydrogenases have been widely reported [34]. Interprotein electron transfer occurs readily in multi-copper oxidases, minimizing the distance for electron transfer and increasing the rate of electron transfer [68]. For example; the copper T_1 site is the primary electron acceptor site and then transfers electrons via an intermolecular mechanism to other copper sites, where O_2 is reduced [69]. The greatest challenge in direct electron transfer is to enable electrical communication between the electrode and enzyme. Some methods of enzyme immobilisation can improve DET [35]. For example, Lee et al. used a gold-binding peptide (GBP) to bind glucose dehydrogenase which facilitates enzyme orientation on the surface of the gold electrode, decreasing the distance between flavin adenine dinucleotide (FAD) and the electrode surface [70]. Cross-linked hydrogels can be used for enzyme immobilisation on the surface of an electrode. Liu et al. used an agarose hydrogel to immobilise haemoglobin, myoglobin and horseradish peroxidase on the electrode surface whereby the proteins could undergo direct electron transfer [71]. Kuk et al. reported the electroenzymatic reduction of CO_2 using NADH-free formate dehydrogenase immobilised on a conductive polyaniline hydrogel. The hydrogel amplified the electron transfer between the electrode and enzyme and moreover, it increased enzyme loading [72]. Hickey et al. reported successful direct electron transfer between enzymes (laccase and nitrogenase) and an electrode using carbon electrodes modified with multi-walled carbon nanotubes and poly(ethylenimine) attached to pyrene moieties. The system was used for the catalytic reduction of O_2 and the production of NH_3 from N_2 [73]. Much of the work on the direct electron transfer of redox enzymes has focused on biosensors and biofuel cells as DET offers a number of advantages. DET enables the use of reagentless biosensors, an important advantage for such devices [74] while with biofuel cells, the absence of a mediator reduces voltage losses arising from differences in the redox potentials of the mediator and the redox enzyme.

3. Immobilisation Strategies

There are five general methods of immobilising enzymes; physical adsorption, covalent binding, immobilisation via ionic interactions, cross linking and entrapment in a polymeric gel or capsule [75]. Physical adsorption is the easiest method of immobilisation [33,34]. Sakai et al. used a glassy carbon electrode modified with 4-mercaptopyridine and gold nanoparticles to adsorb formate dehydrogenase for the oxidation of $HCOO^-$ oxidation and the reduction of CO_2 [76]. However, due to weak interactions, physisorbed enzymes can leach from the electrode surface. Covalent binding onto electrode surfaces provides very stable enzyme immobilisation [75,77]. The most commonly used electrodes in this method are gold and carbon electrodes [34]. The immobilisation of bilirubin oxidases on nanoporous gold electrodes was performed by attaching -COOH via diazonium surface coupling to the electrode surface which was then used for the covalent coupling of bilirubin oxidases [78]. Immobilisation via ionic interactions is commonly used and is dependent on parameters such as pH and the concentration of salt [75]. The crosslinking of enzymes occurs via a bifunctional agent such as glutaraldehyde which results in enzyme aggregation, with the enzymes acting as their own carrier [75]. Encapsulation techniques entail the entrapment of enzymes in the pores of polymers, hydrogels and sol–gels. Polymers are crossed linked in the presence of enzymes, encapsulating the enzymes in the polymers [34,79]. Redox polymers (Section 2.1) were widely used for the immobilisation of enzymes, with immobilisation occurring via electrostatic interactions, cross linking or/and encapsulation. These polymers are used for the construction of cathodes and anodes in biofuel cells as they enable the successful electrical connection of enzymes, providing a stable means of attachment and can be miniaturized [80,81]. On account of these advantages, redox polymers are successfully and widely used for the construction of biofuel cells [82], biosensors [83], biosupercapacitors [84] and bioelectrosynthesis [85]. As an example of a biocatalytic system, Alsaoub et al. constructed a biosupercapacitor using an Os complex modified polymer to immobilize glucose oxidase and flavin adenine dinucleotide (FAD)-dependent glucose dehydrogenase at the anode and bilirubin oxidase at the cathode [86].

4. Biocatalytic Reactors

Biocatalytic reactors consist of one of two types of reactor, batch and flow reactors [87,88].

4.1. Batch Reactors

Batch reactors are simple and flexible in terms of manufacture. They can vary in size from small (mL) to large (m^3). Immobilised enzymes can be recovered and reused in batch reactors, while additional amounts of enzyme can be added if required [87,89]. Markle et al. developed a batch reactor in which deracemization, stereoinversion and the asymmetric synthesis of l-leucine were carried out by combining enzymatic oxidation using d-amino acid oxidase with an electrochemical reduction step [90]. Mazurenko et al. used membrane-bound (S)-mandelate dehydrogenase encapsulated in silica film on an electrode surface to prepare phenylglyoxylic acid from an (S)-mandelic acid in a batch reactor [91]. Ali et al. developed an electrochemical batch reactor for the regeneration of 1,4-NADH via platinum and nickel nano-particles deposited on the electrode surface [92]. While batch systems are in widespread use in industrial applications, the development of flow reactors has been the subject of significant research as outlined below [93].

4.2. Flow Reactors

The use of flow can increase rates of mass transfer with subsequent increases in the rates of reaction, decreasing the reaction time and making it feasible to perform reactions at a large-scale using relatively small scale equipment [89]. Kundu et al. constructed a packed microreactor to investigate the polymerization of polycaprolactone from ε-caprolactone in batch and continuous flow modes. Polymerization in continuous flow mode was faster than in batch mode and higher molecular mass polymers were obtained in continuous flow mode [94]. Using lipase, the time for the resolution of (±)-1,2-propanediol decreased significantly from 6 h (batch) to 7 min (flow) [95]. Using enzymes as a catalyst for the production of chemicals on a large scale is limited due to the requirement for high concentrations of enzyme, a limitation that does not necessarily apply to flow reactors [96]. For example, Cosgrove et al. used galactose oxidase to investigate the oxidation of lactose on a large scale in batch and flow systems. In the batch reactor rate of reaction of 0.74 mmol L^{-1} h^{-1} was reported, a rate that increased 224-fold in the flow reactor (167 mmol L^{-1} h^{-1}). Moreover, the concentration of enzyme decreased considerably from 2.5 mg mL^{-1} (batch reactor) to 0.5 mg mL^{-1} (flow reactor) [96]. Flow reactors possess additional advantages over batch reactors such as better mixing and thermal control, improved stability and life time [23,97]. A range of studies have described the use of flow reactors to prepare intermediates for the preparation of pharmaceutical drugs [98], agrochemicals such as fluorinated and chlorinated organic compounds [99] and electronic materials such as iron silicide–carbon nanotubes used in Li ion batteries [100,101] A flow reactor is based on the principle of using a pump to pump the substrate into a reactor where a product is formed and is then pumped out of the reactor. Pumps can be divided into two groups, continuous and semi-continuous systems. Semi-continuous systems need to be refilled while continuous flow systems do not require refilling. Generally, reactors are fabricated from glass, polymers (e.g., polytetrafluoroethylene, polyfluoroacetate and polyether ether ketone), stainless steel and metals [23,102].

In general, enzyme immobilisation brings about a number of benefits. Lower amounts of enzyme are required, making the process more cost effective [103], while the removal of product from the reactor is simple and easy. In addition, immobilised enzymes show high stability, selectivity in comparison with free enzymes [23]. As described earlier, there are five general methods of immobilising enzymes. A number of challenges can arise when enzymes are immobilised in a microreactor. The enzymes need to be stable under conditions of fluid flow. In addition, during the immobilisation process, clogging and solid formation should be minimized [104]. Enzymes can be immobilised in flow reactors in a number of ways. Enzymes can be attached to beads, loaded into the reactor. This approach can result in high enzyme loadings but requires high back pressures [105]. The enzyme can be immobilised on the walls of reactors [89,106,107], attached to inorganic and polymeric monoliths in the channels [108–110] immobilised on a membrane (a method that is outside the scope of this paper) [111–114].

In order to select the optimal conditions for the immobilisation of an enzyme, the support needs to be selected. The type of support chosen will determine the immobilisation procedure [106]. The process of enzyme immobilisation should be cost-effective and technologically efficient. Flow reactors can be divided into three groups, packed-bed microreactors, monolithic microreactors and wall-coated microreactors (Figure 3).

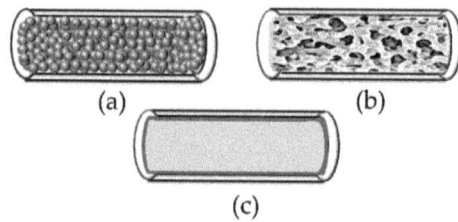

Figure 3. A schematic of (**a**) packed-bed; (**b**) monolithic; and (**c**) wall-coated microreactors.

4.2.1. Packed-Bed Reactors

Conventional packed-bed reactors suffer from disadvantages such as inefficient heat and mass transfer due to the use of large particles (mm in size) in the channels [97]. The diameter of the particles should be less than 1/20th of the channel diameter in order to prevent channelling and ensure even flow solution. For microreactors, the diameters of the particles should be less than 50 nm [115]. Enzymes can be immobilised on porous beads, streptavidin-coated magnetic microbeads, porous resins or various hydrogels. Oglio et al. designed a pack-bed flow reactor using immobilised ketoreductase and glucose dehydrogenase on aldehyde agarose for different ketones reduction. Both enzymes showed good stability in DMSO and the system was used continuously for a number of weeks [116]. A packed bed bioreactor with high selectivity for the continuous production of a range of chiral alcohols was prepared by immobilising alcohol dehydrogenase fused with HaloTag^{TM} on a resin containing sepharose beads [117]. A cascade reaction for the synthesis of (1S,2S)-1-phenylpropane-1,2-diol using a packed bed reactor containing immobilised fusion alcohol dehydrogenase and benzoylformate decarboxylase was described [118]. Peschke et al. immobilised (R)-stereoselective ketoreductase LbADH (alcohol dehydrogenase from Lactobacillus brevis) and glucose dehydrogenase (GDH) on magnetic beads for the preparation of (R)-alcohols. Glucose dehydrogenase was used to recycle the cofactor. The packed-bed reactor operated continuously for four days [119]. However, using beads in the reactors has disadvantages such as the need for high backpressures to achieve adequate flow rates. Moreover, it is difficult to control a multiphase flow inside the microreactor [120].

4.2.2. Monolithic Reactors

To address the disadvantages of packed-bed microreactors, monolithic microreactors have been prepared [93,121]. Monolithic reactors include a network of micro- or mesoporous materials that can be divided into two groups: inorganic monoliths (such as silica-based monoliths [122]) and polymeric monoliths (such as acrylamide copolymers [123] and glycidyl methacrylate copolymers [124]). The porosity of monolithic microreactors enables rapid rates of mass transfer with high flow rates possible at low pressures, resulting in improved performance when compared to conventional packed beds [93]. Biggelaar et al. immobilised ω-transaminases on silica monoliths for the enantioselective transamination reaction [125]. Szymańska et al. developed a silica monolithic flow reactor for the esterification of a primary diol by immobilising acyltransferase on silica monoliths [126]. Sandig et al. used novel hybrid monolith materials (monolithic polyurethane matrix containing cellulose beads) for the immobilisation of lipase. The use of the continuous flow resulted in an efficient reactor with a high turnover number of 5.1×10^6 [127]. Logan et al. immobilized three enzymes (invertase, glucose oxidase, and horseradish peroxidase) on polymer monoliths in different regions of a microfluidic device [128]. Qiao et al. prepared a monolithic and a wall-coated reactor, using L-asparaginase immobilised on poly(GMA-co-EDMA). A lower value of K_M was obtained with the monolith reactor

(3.8 µM) in comparison to the wall-coated microreactor (7.7 µM) with a lower value for V_{max} also reported (106.2 vs. 157 µMmin^{-1}, respectively), indicating that at a lower substrate concentration, a higher reaction rate was achieved in a monolithic reactor. The monolithic reactor showed a better affinity between substrate and enzyme than wall-coated reactors [129].

4.2.3. Wall-Coated Reactors

In wall-coated microreactors, the mass transfer resistance was decreased and fluid flow through the channel can occur without difficulties such as pressure drop and channel blocking. Fluid dynamics as well as heat transfer can be controlled more easily than in other microreactors [130]. A disadvantage of wall-coated microreactors is that lower enzyme loadings occur in comparison with other microreactors. In order to increase enzyme loadings, different strategies can be used, including the deposition of nanostructured materials such as gold nanoparticles [107], nanosprings [131], graphene oxide [132] and dopamine [133,134] on the wall as well as using multiple layers immobilised with enzymes attached to the surface of the wall [106,135,136]. For example, Valikhani et al. constructed a wall-coated microreactor with the use of silica nanosprings, comprised of helical silicon dioxide (SiO$_2$) structures grown via a chemical deposition process. The nanosprings were attached on to the channel wall and used to immobilise sucrose phosphorylase. In comparison with the unmodified surface, the loading of enzyme was significantly increased [137]. Bi et al. developed a wall-coated micro reactor in which polyethyleneimine (PEI) and Candida Antarctica lipase B were alternatively absorbed. The loading of the enzyme increased with the increasing number of layers, showing good stability and performance for the synthesis of wax ester [138]. An interesting study was carried out by Britton et al. in which enzymes were attached on the wall in specific separated zones. This type of reactor can be used for multi-step reactions for the synthesis of products such as alpha-d-glucose1-phosphate [139]. For example, Valikhani et al. described the use of sucrose phosphorylase to attach enzymes on the walls of glass microchannels for the synthesis of alpha-d-glucose1-phosphate [140].

4.3. Electrochemical Reactors

The coenzyme nicotinamide adenine dinucleotide is required as a co-substrate for over 300 dehydrogenases [141]. A number of studies have been carried out to use electrochemical methods to regenerate the cofactor. Yoon et al. prepared an electrochemical laminar flow microreactor to regenerate NADH in the synthesis of chiral L-lactate from the achiral substrate pyruvate (Figure 4). For this purpose, gold electrodes were deposited on the inside of a Y-shaped reactor, with two separate streams, one with buffer and the second with reagents (FAD, NAD$^+$, enzymes and substrate), with the flow directed to the cathode. The reduced cofactor, FADH$_2$ was produced at the cathode and used for the reduction of NAD$^+$ to NADH, resulting in a 41% yield of L-lactate (theoretical yield of 50%) [142].

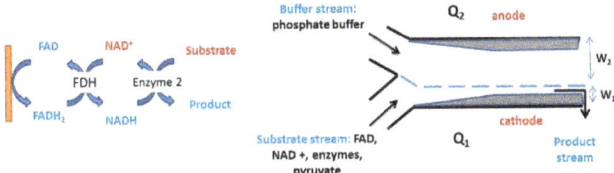

Figure 4. Schematic diagram of a laminar Y-shaped microreactor used for the electrochemical regeneration of NAD$^+$ using formate dehydrogenase (FDH) [142].

A filter-press microreactor with semi-cylindrical channels on the electrode surface resulting in high surface areas was used to electrochemically regenerate FADH$_2$ (Figure 5) [143].

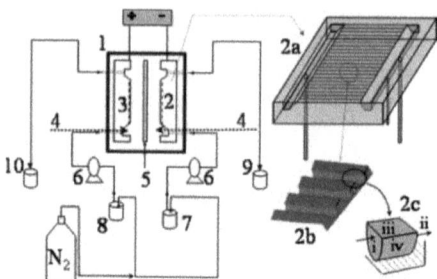

Figure 5. A schematic of a filter-press microreactor. Reprinted from [143] Copyright (2020), with permission from John Wiley and Sons.

As electrochemical cofactor regeneration has, by definition, to occur at the surface of the electrode, scaling up reactions is likely to cause diffusion limitations resulting in reduced rates of reaction [23]. This challenge can be addressed by using three-dimensional electrodes in continuous flow reactors. Kochius et al. designed a system for the efficient electrochemical regeneration of NAD^+, based on 3D electrodes with a high working surface area 24 m^2. The anode, comprised of a packed bed of glassy carbon particles, was surrounded by two cathodes (titanium net) (Figure 6). The oxidation of NADH was carried out using ABTS as a mediator, with turnover numbers of 1860 and 93 h^{-1} for the mediator and the cofactor, respectively, rates that were significantly higher than previously reported. A space time yield 1.4 g L^{-1} h^{-1} was achieved for the three-dimensional electrochemical reactor, a value higher than that at a two-dimensional cell [48].

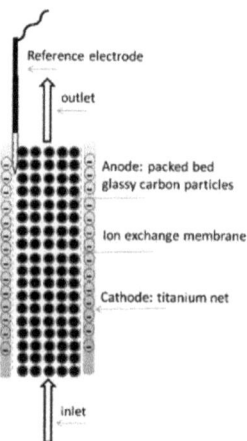

Figure 6. Schematic diagram of the three-dimensional electrode [48].

Ruinatscha et al. fabricated a reactor with three-dimensional porous carbon electrodes with a surface area to volume ratio of 19,685 $m^2 m^{-3}$ (Figure 7).

The reactor was used to produce $FADH_2$, which was then coupled with styrene monooxygenases for the synthesis of styrene oxide. The rate of mass transfer increased in this system and FAD was reduced at a rate of 93 mM h^{-1} producing styrene oxide at a rate of 1.3 mM h^{-1} [144]. Due to the benefits of using microreactors for cofactor regeneration, the direct regeneration of NAD^+ was examined. Rodríguez-Hinestroza et al. designed an electrochemical filter press microreactor for the direct anodic regeneration of NAD^+, via the horse-liver alcohol dehydrogenase-catalysed synthesis of β-alanine (Figure 8). The platinum and gold electrodes used had 150 microchannels, with a high surface area of 250 cm^{-2}. NADH was directly oxidized at the gold electrode, with a 92% conversion of NADH to NAD$^+$. The produced NAD$^+$ was used for the enzymatic oxidation of carboxybenzyl-β-amino propanol [145].

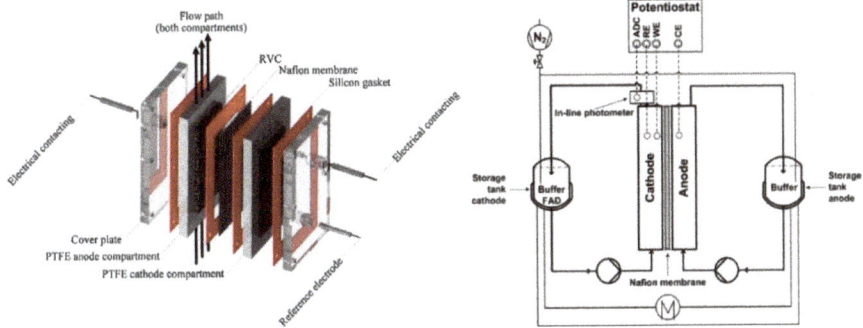

Figure 7. Schematic diagram of system for a reactor for the regeneration of FAD. Reprinted from [144] Copyright (2020), with permission from Elsevier.

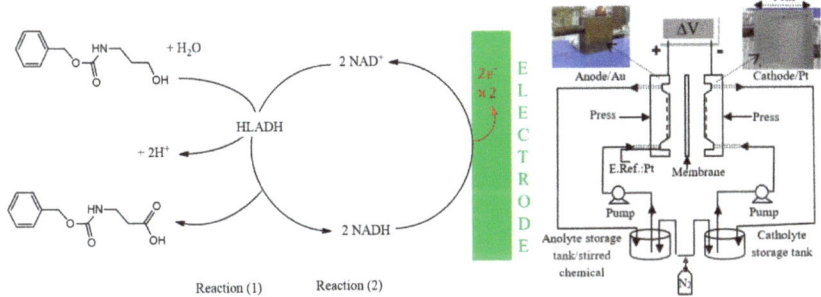

Figure 8. Schematic diagram of an electrochemical microreactor system and the electroenzymatic oxidation of carboxybenzyl-β-amino propanol. Reprinted from [145]. Copyright (2020), with permission from Elsevier.

Fisher et al. prepared a multichannel segmented flow bioreactor in which the oxidoreductase pentaerythritol tetranitrate reductase (PETNR) was regenerated electrochemically with the use of methyl viologen as a mediator (Figure 9), replacing NADPH as cofactor. The use of methyl viologen resulted in rates of substrate reduction that were 15–70% of those observed with NADPH [146].

Mazurenko et al. carried out mediated regeneration NAD$^+$ via poly(methylene green) in a flow reactor for the synthesis of D-fructose from D-sorbitol using D-sorbitol dehydrogenase [142]. Electrochemical methods can also be used to immobilise enzymes on the surface of electrodes for subsequent use in a flow reactor. For example; Xiao et al. designed a continuous flow bioreactor based on lipase immobilised by the electrochemical generation of a silica film nanoporous gold-modified glassy carbon electrodes (Figure 10a). In addition, −1.1 V vs. Ag/AgCl was applied on the electrode surface immersed in tetraethoxysilane and lipase solution. This triggered hydroxyl ion production that caused tetraethoxysilane condensation on the electrode surface and lipase was entrapped into silica structure.

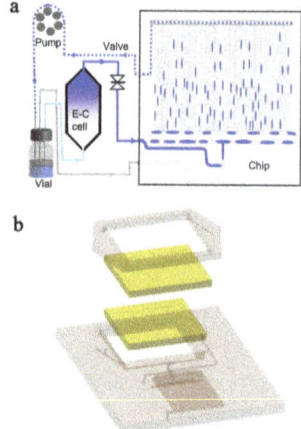

Figure 9. (a) Schematic diagram of a segmented flow bioreactor system, (b) 3D representation of the bioreactor. Reprinted from [146] Copyright (2020), with permission from The Royal Society of Chemistry.

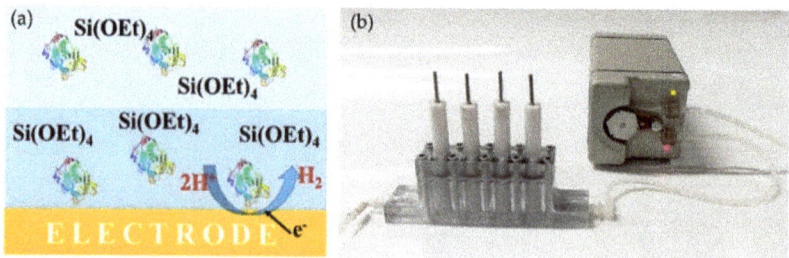

Figure 10. (a) Schematic diagram of the electrodeposition of a silica layer with entrapped enzyme on an electrode and (b) an image of the continuous flow reactor [107].

The modified electrodes were inserted into the flow reactor (Figure 10b) and used to prepare p-nitrophenol from 4-nitrophenyl butyrate (4-NPB), with the full conversion of the substrate (0.075 mM in 2 mL) after eight cycles [107].

5. Strategies for Improving Bioelectrocatalysts

5.1. High Surface Area Electrodes

One method of improving the rate of reaction in electrochemical bioreactors is to use electrodes with high surface areas to enable high enzyme loadings and consequently higher current densities. Electrodes can be designed with a range of three-dimensional (3D) architectures. Parameters such as the method of enzyme immobilisation, the orientation of the enzyme on the support, the porosity of the support, and diffusional limitations within the pores need to be taken into account [147].

Due to their ease of preparation, porous metal electrodes have been used for a range of applications. For example, dealloyed nanoporous gold (NPG) can be prepared on the electrode surface by electro/chemically dissolving Ag from Au/Ag alloys. The resultant pores have diameters that are sufficiently large to accommodate the enzymes and to enable substrates and products to readily diffuse into and out of the pores. Through changing parameters such as the alloy composition and the dealloying conditions, the pore diameters can vary over the range of 5–700 nm [148]. For example, dealloyed nanoporous gold (NPG) electrodes modified with osmium polymer as a mediator were used for the determination of glucose and lactose [149]. Dealloyed nanoporous gold (NPG) electrodes have also been used for the direct electron transfer of enzymes such as laccase [69]. The immobilisation of FAD-dependent glucose dehydrogenase, bilirubin oxidases [78] and fructose dehydrogenase [150] were also investigated on porous gold electrodes. Nanoporous gold electrodes are widely used as a biofuel cell [78,151] and biosensor [152]. In the construction of biofuel cells, gold nanoparticles have been attached to the electrode surface to enhance the efficiency of the cells [153,154].

Carbon nanotubes have been widely used to modify electrodes [155–158]. Carbon nanotubes possess properties such as high conductivity and surface areas. A simple method for the modification of electrodes with carbon nanotubes and enzymes is to place a drop of a suspension of enzyme and carbon nanotubes onto the surface of the electrode surface [77]. Electrode modification with carbon nanotubes can facilitate direct electron transfer between the electrode and enzymes such as fructose dehydrogenase [159], or horseradish peroxidase [160], resulting in the preparation of efficient biofuel cells [161,162]. Jourdin et al. used multiwalled carbon nanotubes to improve the performance of a microbial system for the bioreduction of carbon dioxide. Multiwalled carbon nanotube electrodes considerably increased the rate of electron transfer between electrode and microorganisms by 1.65-fold and the acetate production rate by 2.6-fold in comparison with the graphite plate electrodes [163]. The microbial electrosynthesis of acetate from CO_2 was described using a vitreous carbon electrode modified with multi-walled carbon nanotubes [164]. Bulutoglu et al. immobilised fused alcohol dehydrogenase on multi-walled carbon nanotubes for the electrocatalytic oxidation of 2,3-butanediol [165]. Bucky papers are flexible, light materials prepared from carbon nanotubes [147,166]. Zhang et al. developed a bioelectrode for electroenzymatic synthesis using a bucky paper electrode on which [Cp*Rh(bpy)Cl]$^+$ was immobilized as a mediator for the regeneration of NADH. A turnover frequency of 1.3 s^{-1} was achieved for the regeneration of NADH and the system was used for the preparation of D-sorbitol from D-fructose using immobilized D-sorbitol dehydrogenase.

The high porosity of 3D graphene materials can enable higher enzyme loadings, increasing the performance of the electrodes. Choi et al. used a combination of graphitic carbon nitride and reduced graphene oxide as an efficient cathode for the reduction of O_2. The H_2O_2 produced at the surface of the cathode surface was used for the peroxygenase-catalysed selective hydroxylation of ethylbenzene to (R)-1-phenylethanol [167]. Enzymes can be covalently attached on the graphene hybrid electrodes through the controlled functionalization of graphene causing high stability. Effective enzyme–graphene conjugation can pave the way for direct electron transfer on the electrode surface. Seelajaroen et al. used enzyme–graphene hybrids for the electrochemical preparation of methanol from CO_2 using NAD(P)-linked enzymes including formate dehydrogenase, formaldehyde dehydrogenase and alcohol dehydrogenase (Figure 11). Enzymes were covalently bound on to carboxylate-modified graphene surfaces.

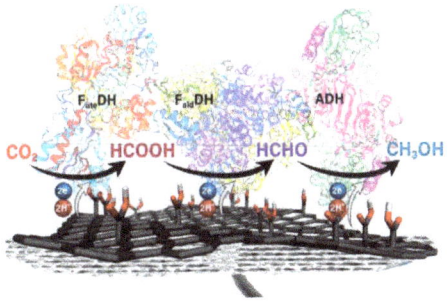

Figure 11. The cofactor-free biosynthesis of methanol using formate dehydrogenase, formaldehyde dehydrogenase and alcohol dehydrogenase. Reprinted from [168] Copyright (2020), with permission from the American Chemical Society.

Due to the effective rates of electron transfer from the graphene-modified electrode to the enzymes, considerable improvement was observed with a high faradaic efficiency of 12%. A higher production rate of 0.6 µmol·h^{-1} was achieved and the current was stable for 20 h. This shows the advantage of using conductive graphene carboxylic acid as support [168].

5.2. Enzyme Engineering

Protein engineering can be used to improve the catalytic performance of enzymes [169,170]. Rational design, directed evolution and combinations of these approaches are widely used to prepare more active and stable enzymes [170]. In rational design, mutations are inserted into specific locations in the protein through site-directed mutagenesis. The lack of a full understanding of the relationship between the protein structure and the rate of electron transfer makes it difficult to improve the catalytic activity using rational design. The structural knowledge of proteins is not required for directed evolution and catalytic activity can be improved in the absence of a detailed crystal structure of the enzyme by mimicking Darwinian evolution [170]. Modifying the heme centre of myoglobin increased its dehalogenation activity to over 1000 times that of a native dehaloperoxidase [171]. Recently, Chen et al. prepared cytochrome P450 enzymes through the directed evolution of serine-ligated P450 variants for the preparation of cyclopropene with high efficiency (with a total turnover number (TTN) of up to 5760) and high selectivity (>99.9% ee) [172]. Brandenberg et al. reported a cytochrome P450 variant that was able to catalyse the C2-amidation of indole. Both the heme and reductase domains were modified, improving the catalytic activity with the total turnover number increasing from 100 to 8400 and the product yield from 2.1 to 90% [173]. Mateljak et al. used computational design with directed evolution to design fungal high-redox-potential laccases that exhibited high stability and activity toward redox mediators [174]. Further work was performed to investigate the use of these laccases for the reduction of O_2 in the presence and absence of ABTS. The designed laccases could reduce O_2 at low overpotentials [175]. Protein engineering has been used to adjust the properties of NAD(P)H-dependent oxidoreductases [176]. For example, Liu et al. used directed evolution to prepare an NADH-dependent alcohol dehydrogenase from *Lactococcus lactis* for the production of isobutanol. The catalytic efficiency of the engineered enzyme increased by a factor of 160 in comparison with the wild-type enzyme [177]. Li et al. used directed evolution to improve the catalytic activity of puritative oxidreductase in the production of 1,3-propanediol [178].

5.3. Enzyme Cascades

Typically, a biocatalytic cascade reaction is a reaction system in which two or more transformations are performed simultaneously [179]. Through biocatalytic cascades, the isolation of reaction intermediates is circumvented, saving time and the use of reagents. The synthesis of chemicals with unstable intermediates is feasible as the isolation of the intermediates is not necessary [180]. Enzyme cascades can enhance the current density as the oxidation of the substrate in a sequential manner enables more electrons to be extracted [34]. Dong et al. designed a biphasic bioelectrocatalytic system that used a reaction cascade to prepare (R)-ethyl-4-cyano-3-hydroxybutyrate from ethyl 4-chloroacetoacetate using alcohol dehydrogenase and halohydrin dehalogenase (Figure 12). Ethyl 4-chloroacetoacetate was reduced to (S)-4-chloro-3-hydroxybutanoate using alcohol dehydrogenase which was then converted to its R enantiomer using halohydrin dehalogenase. Efficient NADH regeneration was performed using diaphorase with the redox polymer cobaltocene-modified poly-(allylamine) as a mediator on the electrode surface [181]. In total, 85% of substrate was converted into (R)-ethyl-4-cyano-3-hydroxybutyrate, an 8.8 higher yield than that achieved with a single-phase system.

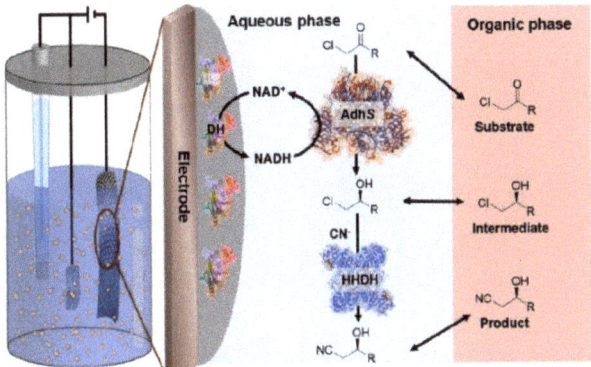

Figure 12. Schematic diagram of the biphasic bioelectrocatalytic reaction cascade for the synthesis of (R)-ethyl-4-cyano-3-hydroxybutyrate. Reprinted from [181] Copyright (2020) with permission from American Chemical Society.

Abdellaoui et al. combined aldehyde deformylating oxygenase, immobilised on the surface of an electrode, with NAD^+-dependent alcohol dehydrogenase to create an enzymatic cascade reaction for the conversion of fatty alcohols to aldehydes followed by the decarbonylation to produce alkenes (Figure 13) [182].

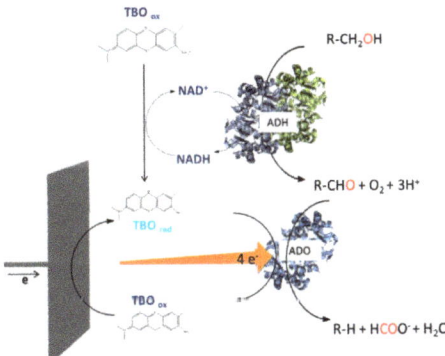

Figure 13. Schematic diagram of a bioelectrochemical enzymatic cascade for the preparation of alkenes [182].

Kuk et al. constructed a photoelectrochemical system for the production of methanol by a multienzyme cascade (a three-dehydrogenase cascade system) and the efficient regeneration of the NADH via visible-light assistance (Figure 14) [183].

Chen et al. developed an enzyme cascade to produce chiral amines from nitrogen. NH_3 was produced from N_2 via nitrogenase and then used, with L-alanine dehydrogenase, to produce alanine from pyruvate. The desired chiral amines were produced via the ω-transaminase transfer of an amino group from alanine. Pyruvate was generated as a by-product and converted to alanine using L-alanine dehydrogenase. The system was used for the successful amination of the substrates, 4-phenyl-2 butanone, 4-methyl methoxy phenyl acetone, phenoxyacetone, 2-pentanone, methoxyacetone and 2-octanone with an enantiomeric excess >99% (Figure 15) [52].

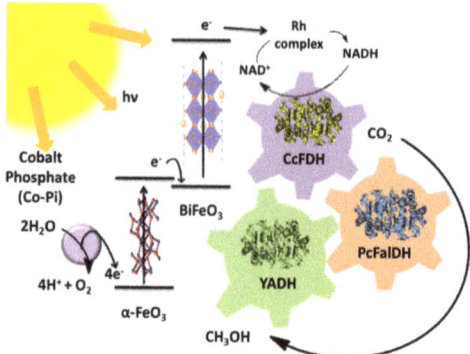

Figure 14. Schematic diagram of the light-assisted synthesis of methanol from CO_2 via an enzyme cascade (formaldehyde dehydrogenase, formate dehydrogenase, and alcohol dehydrogenase) [183].

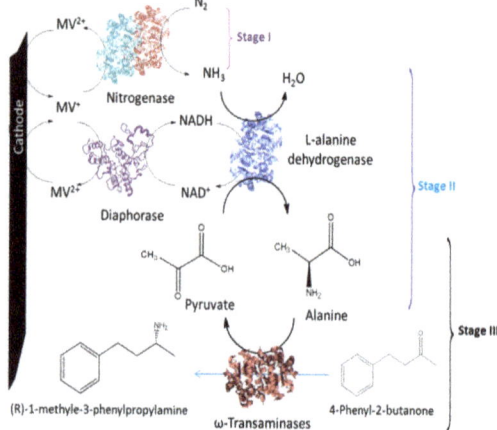

Figure 15. A schematic of bioelectrochemical asymmetric amination [52].

In a sequential catalytic cascade, noble metal catalysts can also be used. For example, the biocatalytic reduction of nitrite to ammonia is slow in comparison with metal catalysts [184]. Duca et al. fabricated a hybrid electrochemical cascade that combined nitrate reductase and Pt nanoparticles on a carbon electrode surface. The reduction of nitrate to nitrite and nitrite to ammonia was carried out by the Pt nanoparticles and nitrate reductase, respectively [185].

6. Bioelectrosynthesis

A variety of oxidoreductases have been used for the asymmetric synthesis of amines, amino acids and alcohols [186–188]. Dong et al. developed a biphasic bioelectrocatalysis system for the synthesis of (R)-CHCN (81.2%), (R)-3-hydroxy-3-Phenylpropanenitrile (96.8% ee), (S)-3-hydroxy-4-phenylbutanenitrile (94.6% ee) using alcohol dehydrogenase and halohydrin dehalogenase. The system showed high selectivity for the synthesis of chiral β-hydroxy nitriles. The NADH cofactor was efficiently regenerated using diaphorase and a cobaltocene-modified poly(allylamine) redox polymer [181]. Combining enzymatic electrosynthesis with biofuel cells can pave the way for the self-powered bioelectrosynthesis of valuable chemicals. Wu et al. designed an electroenzymatic system by combining a bioelectrosynthesis cell and an enzymatic fuel cell to prepare L-3,4-dDihydroxyphenylalanine with a coulombic efficiency of 90% [189]. Chen et al. combined a H_2/α-keto acid enzymatic fuel cells with bioelectrosynthesis to synthesis chiral amino acids with high efficiency. An enzymatic cascade consisting of nitrogenase, diaphorase and L-leucine dehydrogenase was used at the cathode to convert nitrogen to chiral amino acids [190]. The NH_3 and NADH prepared in the reaction were used for the synthesis of L-norleucine from 2-ketohexanoic acid via leucine dehydrogenase (LeuDH). A high NH_3 conversion ratio (92%) and a high faradaic efficiency (87.1%) was observed.

Carbon dioxide emissions are increasing, dramatically bringing about an environmental crisis. Therefore, electrochemical CO_2 reduction to useful chemicals has gained great interest to reduce CO_2 emissions. Cai et al. reduced CO_2 to ethylene and propene via the VFe protein of vanadium nitrogenase that was contacted with the electrode mediated by cobaltocene/cobaltocenium (Figure 16). A study represented a new approach for preparing C–C bonds through a single metalloenzyme [191].

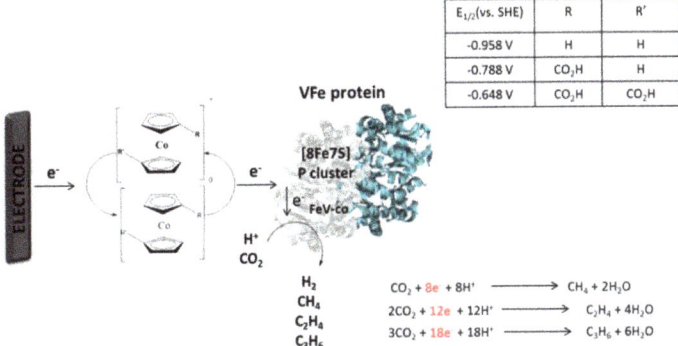

Figure 16. A schematic of C–C bond preparation via bioelectrocatalysts, Amperometric i–t trace carried out for enzymatic electroreduction of CO_2 via VFe nitrogenase [191].

Yuan et al. reported CO_2 reduction producing formate with a high faradaic efficiency of 99% at a low potential −0.66 V vs. The standard hydrogen electrode (SHE). They used immobilised molybdenum-dependent formate mediated by cobaltocene on the electrode surface [192]. In some studies, CO_2 reduction was carried out via enzymatic cascade, using formate dehydrogenases, formaldehyde dehydrogenase and alcohol dehydrogenase to produce CH_3OH [193]. Bioplastics can be prepared from CO_2 reduction via bioelectrocatalysts. For example, Sciarria et al. synthesized polyhydroxyalkanoates (PHA) from CO_2 through a bioelectrochemical reactor. CO_2 was converted into bioplastics successfully in a way that 0.41 kg of carbon as PHA was achieved per 1 kg of C_{CO2} in the whole system [194].

Nitrogenase catalyses the reduction of N_2 to ammonia. Nitrogenase can also be used for the reduction of CO, nitrite, azide and cyanide that can then be used for the synthesis of a range of products [195]. Lee et al. wired nitrogenase to a redox-polymer (neutral red-modified poly(glycidyl methacrylate-co-methylmethacrylate-co-poly(ethyleneglycol)methacrylate)) for the conversion of nitrogen to ammonia. [196]. Milton et al. reported on the use of nitrogenase to prepare ammonia from azide and nitrite using cobaltocene as a mediator. The system operated at a potential of −1.25 V with 70 and 234 nmol of NH_3 produced from the reduction of N_3^- and NO_2^-, respectively [197]. An ATP-free-mediated electron-transfer system operating at a potential of −0.58 V was developed by Lee et al. for the synthesis of ammonia using cobaltocene-functionalized poly(allylamine) (Cc-PAA) as a mediator. The amount of ammonia prepared was 30 ± 5 and 7 ± 2 nmol from the reduction of N_3^- and NO_2^-, respectively [198].

7. Conclusions

Oxidoreductases catalyse redox reactions and have been successfully used to prepare biofuel cells and biosensors. The enzymes undergo electron transfer with the surface of the electrode via mediated or direct electron transfer mechanisms. However, relatively few oxidoreductases can undergo direct electron transfer because their redox active centres are placed deep within their structures. While enzymes have been widely used in batch and flow reactors, the use of oxidoreductases in bioreactors has been mainly confined to cofactor regeneration. A number of strategies that entail the use of high surface area electrodes, enzyme engineering and enzyme cascades have been used to improve the performance of bioelectrocatalysts. As an example of the scope of bioelectrocatalysts, a variety of oxidoreductases has been used in asymmetric synthesis reactions. While the scope of bioelectrocatalysts can be broadened, the successful use of bioelectrocatalysts will lie initially in the development of small-scale reaction systems that focus on high-value fine chemicals. The scale up of the reaction systems to the size required for bulk manufacture will be a significant challenge. Due to this challenge, the development of electrocatalytic bioreactors will likely focus on the preparation of products at a small scale taking advantage of the relatively low-cost systems that can be assembled in a relatively simple and rapid manner. Such systems are adaptable and are particularly suited for systems that may be required in process optimization or in the preparation of products or intermediates on a small (g) scale. Screening the performance of mutant enzymes is also more feasible given the relatively small amounts of enzymes required in such reactors. A particular area of interest lies in the use of multiple enzymes for cascade reactions, where rapid prototyping is particularly useful in screening and selecting the optimal operational conditions.

Author Contributions: Writing, S.A.; editing and review, M.N.-A.; writing—review and editing, E.M. All authors have read and agreed to the published version of the manuscript.

Funding: This research was funded by the Science Foundation Ireland (SFI) Research Centre for Pharmaceuticals under Grant Number 12/RC/2275.

Conflicts of Interest: The authors declare no conflict of interest.

References

1. Jemli, S.; Ayadi-Zouari, D.; Hlima, H.B.; Bejar, S. Biocatalysts: Application and engineering for industrial purposes. *Crit. Rev. Biotechnol.* **2016**, *36*, 246–258. [CrossRef] [PubMed]
2. van Beilen, J.B.; Li, Z. Enzyme technology: An overview. *Curr. Opin. Biotechnol.* **2002**, *13*, 338–344. [CrossRef]
3. Otten, L.G.; Quax, W.J. Directed evolution: Selecting today's biocatalysts. *Biomol. Eng.* **2005**, *22*, 1–9. [CrossRef] [PubMed]
4. Hartmann, M.; Kostrov, X. Immobilization of enzymes on porous silicas–benefits and challenges. *Chem. Soc. Rev.* **2013**, *42*, 6277–6289. [CrossRef]
5. Schulze, B.; Wubbolts, M.G. Biocatalysis for industrial production of fine chemicals. *Curr. Opin. Biotechnol.* **1999**, *10*, 609–615. [CrossRef]
6. Davis, B.G.; Boyer, V. Biocatalysis and enzymes in organic synthesis. *Nat. Prod. Rep.* **2001**, *18*, 618–640.
7. Martinez, A.T.; Ruiz-Dueñas, F.J.; Camarero, S.; Serrano, A.; Linde, D.; Lund, H.; Vind, J.; Tovborg, M.; Herold-Majumdar, O.M.; Hofrichter, M. Oxidoreductases on their way to industrial biotransformations. *Biotechnol. Adv.* **2017**, *35*, 815–831. [CrossRef]
8. Schmitz, L.M.; Rosenthal, K.; Lütz, S. Enzyme-based electrobiotechnological synthesis. In *Bioelectrosynthesis*; Springer: Cham, Switzerland, 2017; pp. 87–134.
9. Giroud, F.; Sawada, K.; Taya, M.; Cosnier, S. 5, 5-Dithiobis (2-nitrobenzoic acid) pyrene derivative-carbon nanotube electrodes for NADH electrooxidation and oriented immobilization of multicopper oxidases for the development of glucose/O_2 biofuel cells. *Biosens. Bioelectron.* **2017**, *87*, 957–963. [CrossRef]
10. Sakai, K.; Kitazumi, Y.; Shirai, O.; Takagi, K.; Kano, K. High-power formate/dioxygen biofuel cell based on mediated electron transfer type bioelectrocatalysis. *ACS Catal.* **2017**, *7*, 5668–5673. [CrossRef]
11. Kang, Z.; Jiao, K.; Yu, C.; Dong, J.; Peng, R.; Hu, Z.; Jiao, S. Direct electrochemistry and bioelectrocatalysis of glucose oxidase in CS/CNC film and its application in glucose biosensing and biofuel cells. *RSC Adv.* **2017**, *7*, 4572–4579. [CrossRef]
12. Ruff, A.; Conzuelo, F.; Schuhmann, W. Bioelectrocatalysis as the basis for the design of enzyme-based biofuel cells and semi-artificial biophotoelectrodes. *Nat. Catal.* **2019**, 1–11. [CrossRef]
13. Sekretaryova, A.N.; Eriksson, M.; Turner, A.P. Bioelectrocatalytic systems for health applications. *Biotechnol. Adv.* **2016**, *34*, 177–197. [CrossRef] [PubMed]

14. Gomes, F.O.; Maia, L.B.; Loureiro, J.A.; Pereira, M.C.; Delerue-Matos, C.; Moura, I.; Moura, J.J.; Morais, S. Biosensor for direct bioelectrocatalysis detection of nitric oxide using nitric oxide reductase incorporated in carboxylated single-walled carbon nanotubes/lipidic 3 bilayer nanocomposite. *Bioelectrochemistry* **2019**, *127*, 76–86. [CrossRef] [PubMed]
15. Voitechovič, E.; Vektarienė, A.; Vektaris, G.; Jančienė, R.; Razumienė, J.; Gurevičienė, V. 1, 4-Benzoquinone derivatives for enhanced bioelectrocatalysis by fructose dehydrogenase from Gluconobacter japonicus: Towards promising D-fructose biosensor development. *Electroanalysis* **2020**. [CrossRef]
16. Miyata, M.; Kitazumi, Y.; Shirai, O.; Kataoka, K.; Kano, K. Diffusion-limited biosensing of dissolved oxygen by direct electron transfer-type bioelectrocatalysis of multi-copper oxidases immobilized on porous gold microelectrodes. *J. Electroanal. Chem.* **2020**, *860*, 113895. [CrossRef]
17. Höllrigl, V.; Otto, K.; Schmid, A. Electroenzymatic asymmetric reduction of rac-3-methylcyclohexanone to (1S, 3S)-3-methylcyclohexanol in organic/aqueous media catalyzed by a thermophilic alcohol dehydrogenase. *Adv. Synth. Catal.* **2007**, *349*, 1337–1340. [CrossRef]
18. Kohlmann, C.; Lütz, S. Electroenzymatic synthesis of chiral sulfoxides. *Eng. Life Sci.* **2006**, *6*, 170–174. [CrossRef]
19. Lütz, S.; Steckhan, E.; Liese, A. First asymmetric electroenzymatic oxidation catalyzed by a peroxidase. *Electrochem. Commun.* **2004**, *6*, 583–587. [CrossRef]
20. Atsumi, S.; Hanai, T.; Liao, J.C. Non-fermentative pathways for synthesis of branched-chain higher alcohols as biofuels. *Nature* **2008**, *451*, 86–89. [CrossRef]
21. Liu, C.; Sakimoto, K.K.; Colón, B.C.; Silver, P.A.; Nocera, D.G. Ambient nitrogen reduction cycle using a hybrid inorganic–biological system. *Proc. Natl. Acad. Sci. USA* **2017**, *114*, 6450–6455. [CrossRef]
22. Cooney, C.L. Bioreactors: Design and operation. *Science* **1983**, *219*, 728–733. [CrossRef]
23. Britton, J.; Majumdar, S.; Weiss, G.A. Continuous flow biocatalysis. *Chem. Soc. Rev.* **2018**, *47*, 5891–5918. [CrossRef] [PubMed]
24. Milton, R.D.; Minteer, S.D. Direct enzymatic bioelectrocatalysis: Differentiating between myth and reality. *J. R. Soc. Interface* **2017**, *14*, 20170253. [CrossRef] [PubMed]
25. Nguyen, K.V.; Holade, Y.; Minteer, S.D. DNA redox hydrogels: Improving mediated enzymatic bioelectrocatalysis. *ACS Catal.* **2016**, *6*, 2603–2607. [CrossRef]
26. Shiraiwa, S.; So, K.; Sugimoto, Y.; Kitazumi, Y.; Shirai, O.; Nishikawa, K.; Higuchi, Y.; Kano, K. Reactivation of standard [NiFe]-hydrogenase and bioelectrochemical catalysis of proton reduction and hydrogen oxidation in a mediated-electron-transfer system. *Bioelectrochemistry* **2018**, *123*, 156–161. [CrossRef] [PubMed]
27. Gross, A.J.; Chen, X.; Giroud, F.; Travelet, C.; Borsali, R.; Cosnier, S. Redox-active glyconanoparticles as electron shuttles for mediated electron transfer with bilirubin oxidase in solution. *J. Am. Chem. Soc.* **2017**, *139*, 16076–16079. [CrossRef] [PubMed]
28. Elouarzaki, K.; Cheng, D.; Fisher, A.C.; Lee, J.-M. Coupling orientation and mediation strategies for efficient electron transfer in hybrid biofuel cells. *Nat. Energy* **2018**, *3*, 574–581. [CrossRef]
29. Sakai, K.; Sugimoto, Y.; Kitazumi, Y.; Shirai, O.; Takagi, K.; Kano, K. Direct electron transfer-type bioelectrocatalytic interconversion of carbon dioxide/formate and NAD+/NADH redox couples with tungsten-containing formate dehydrogenase. *Electrochim. Acta* **2017**, *228*, 537–544. [CrossRef]
30. Sakai, K.; Xia, H.-q.; Kitazumi, Y.; Shirai, O.; Kano, K. Assembly of direct-electron-transfer-type bioelectrodes with high performance. *Electrochim. Acta* **2018**, *271*, 305–311. [CrossRef]
31. Kaida, Y.; Hibino, Y.; Kitazumi, Y.; Shirai, O.; Kano, K. Ultimate downsizing of D-fructose dehydrogenase for improving the performance of direct electron transfer-type bioelectrocatalysis. *Electrochem. Commun.* **2019**, *98*, 101–105. [CrossRef]
32. Guzik, U.; Hupert-Kocurek, K.; Wojcieszyńska, D. Immobilization as a strategy for improving enzyme properties-application to oxidoreductases. *Molecules* **2014**, *19*, 8995–9018. [CrossRef]
33. Cooney, M.; Svoboda, V.; Lau, C.; Martin, G.; Minteer, S.D. Enzyme catalysed biofuel cells. *Energy Environ. Sci.* **2008**, *1*, 320–337. [CrossRef]
34. Hickey, D.P.; Milton, R.D.; Rasmussen, M.; Abdellaoui, S.; Nguyen, K.; Minteer, S.D. Fundamentals and applications of bioelectrocatalysis. *Electrochemistry* **2015**, *13*, 97.
35. Chen, H.; Dong, F.; Minteer, S.D. The progress and outlook of bioelectrocatalysis for the production of chemicals, fuels and materials. *Nat. Catal.* **2020**, *3*, 1–20. [CrossRef]

36. Hilt, G.; Lewall, B.; Montero, G.; Utley, J.H.; Steckhan, E. Efficient In-Situ Redox Catalytic NAD (P)+ Regeneration in Enzymatic Synthesis Using Transition-Metal Complexes of 1, 10-Phenanthroline-5, 6-dione and Its N-Monomethylated Derivative as Catalysts. *Liebigs Annalen* **1997**, *1997*, 2289–2296. [CrossRef]
37. Kroutil, W.; Mang, H.; Edegger, K.; Faber, K. Biocatalytic oxidation of primary and secondary alcohols. *Adv. Synth. Catal.* **2004**, *346*, 125–142. [CrossRef]
38. Findrik, Z.; Šimunović, I.; Vasić-Rački, Đ. Coenzyme regeneration catalyzed by NADH oxidase from Lactobacillus brevis in the reaction of L-amino acid oxidation. *Biochem. Eng. J.* **2008**, *39*, 319–327. [CrossRef]
39. Tahar, A.B.; Szymczyk, A.; Tingry, S.; Vadgama, P.; Zelsmanne, M.; Tsujumura, S.; Cinquin, P.; Martin, D.; Zebda, A. One-year stability of glucose dehydrogenase confined in a 3D carbon nanotube electrode with coated poly-methylene green: Application as bioanode for a glucose biofuel cell. *J. Electroanal. Chem.* **2019**, *847*, 113069. [CrossRef]
40. Bonfin, C.S.; Franco, J.H.; de Andrade, A.R. Ethanol bioelectrooxidation in a robust poly (methylene green-pyrrole)-mediated enzymatic biofuel cell. *J. Electroanal. Chem.* **2019**, *844*, 43–48. [CrossRef]
41. Karyakin, A.A.; Karyakina, E.E.; Schuhmann, W.; Schmidt, H.L.; Varfolomeyev, S.D. New amperometric dehydrogenase electrodes based on electrocatalytic NADH-oxidation at poly (methylene blue)-modified electrodes. *Electroanalysis* **1994**, *6*, 821–829. [CrossRef]
42. Yuan, M.; Minteer, S.D. Redox polymers in electrochemical systems: From methods of mediation to energy storage. *Curr. Opin. Electrochem.* **2019**. [CrossRef]
43. Carucci, C.; Salis, A.; Magner, E. Specific Ion Effects on the Mediated Oxidation of NADH. *ChemElectroChem* **2017**, *4*, 3075–3080. [CrossRef]
44. Abdellaoui, S.; Milton, R.D.; Quah, T.; Minteer, S.D. NAD-dependent dehydrogenase bioelectrocatalysis: The ability of a naphthoquinone redox polymer to regenerate NAD. *Chem. Commun.* **2016**, *52*, 1147–1150. [CrossRef] [PubMed]
45. Zhang, W.; Hollmann, F. Nonconventional regeneration of redox enzymes–a practical approach for organic synthesis? *Chem. Commun.* **2018**, *54*, 7281–7289. [CrossRef] [PubMed]
46. Kwon, S.J.; Yang, H.; Jo, K.; Kwak, J. An electrochemical immunosensor using p-aminophenol redox cycling by NADH on a self-assembled monolayer and ferrocene-modified Au electrodes. *Analyst* **2008**, *133*, 1599–1604. [CrossRef] [PubMed]
47. Kim, S.; Yun, S.-E.; Kang, C. Electrochemical evaluation of the reaction rate between methyl viologen mediator and diaphorase enzyme for the electrocatalytic reduction of NAD+ and digital simulation for its voltammetric responses. *J. Electroanal. Chem.* **1999**, *465*, 153–159. [CrossRef]
48. Kochius, S.; Park, J.B.; Ley, C.; Könst, P.; Hollmann, F.; Schrader, J.; Holtmann, D. Electrochemical regeneration of oxidised nicotinamide cofactors in a scalable reactor. *J. Mol. Catal. B Enzym.* **2014**, *103*, 94–99. [CrossRef]
49. Hollmann, F.; Arends, I.W.; Buehler, K. Biocatalytic redox reactions for organic synthesis: Nonconventional regeneration methods. *ChemCatChem* **2010**, *2*, 762–782. [CrossRef]
50. Weckbecker, A.; Gröger, H.; Hummel, W. Regeneration of nicotinamide coenzymes: Principles and applications for the synthesis of chiral compounds. In *Biosystems Engineering I*; Springer: Cham, Switzerland, 2010; pp. 195–242.
51. Kochius, S.; Magnusson, A.O.; Hollmann, F.; Schrader, J.; Holtmann, D. Immobilized redox mediators for electrochemical NAD (P)+ regeneration. *Appl. Microbiol. Biotechnol.* **2012**, *93*, 2251–2264. [CrossRef]
52. Chen, H.; Cai, R.; Patel, J.; Dong, F.; Chen, H.; Minteer, S.D. Upgraded bioelectrocatalytic N2 fixation: From N2 to chiral amine intermediates. *J. Am. Chem. Soc.* **2019**, *141*, 4963–4971. [CrossRef]
53. Kashiwagi, Y.; Yanagisawa, Y.; Shibayama, N.; Nakahara, K.; Kurashima, F.; Anzai, J.; Osa, T. Preparative, electroenzymatic reduction of ketones on an all components-immobilized graphite felt electrode. *Electrochim. Acta* **1997**, *42*, 2267–2270. [CrossRef]
54. Badalyan, A.; Yang, Z.-Y.; Hu, B.; Luo, J.; Hu, M.; Liu, T.L.; Seefeldt, L.C. An Efficient Viologen-Based Electron Donor to Nitrogenase. *Biochemistry* **2019**, *58*, 4590–4595. [CrossRef] [PubMed]
55. Tosstorff, A.; Kroner, C.; Opperman, D.J.; Hollmann, F.; Holtmann, D. Towards electroenzymatic processes involving old yellow enzymes and mediated cofactor regeneration. *Eng. Life Sci.* **2017**, *17*, 71–76. [CrossRef] [PubMed]
56. Ruff, A. Redox polymers in bioelectrochemistry: Common playgrounds and novel concepts. *Curr. Opin. Electrochem.* **2017**, *5*, 66–73. [CrossRef]

57. Yuan, M.; Kummer, M.J.; Milton, R.D.; Quah, T.; Minteer, S.D. Efficient NADH regeneration by a redox polymer-immobilized enzymatic system. *ACS Catal.* **2019**, *9*, 5486–5495. [CrossRef]
58. Heller, A. Electrical wiring of redox enzymes. *Acc. Chem. Res.* **1990**, *23*, 128–134. [CrossRef]
59. Heller, A. Electrical connection of enzyme redox centers to electrodes. *J. Phys. Chem.* **1992**, *96*, 3579–3587. [CrossRef]
60. Campbell, A.S.; Murata, H.; Carmali, S.; Matyjaszewski, K.; Islam, M.F.; Russell, A.J. Polymer-based protein engineering grown ferrocene-containing redox polymers improve current generation in an enzymatic biofuel cell. *Biosens. Bioelectron.* **2016**, *86*, 446–453. [CrossRef]
61. Tapia, C.; Milton, R.D.; Pankratova, G.; Minteer, S.D.; Åkerlund, H.E.; Leech, D.; De Lacey, A.L.; Pita, M.; Gorton, L. Wiring of photosystem I and hydrogenase on an electrode for photoelectrochemical H2 production by using redox polymers for relatively positive onset potential. *ChemElectroChem* **2017**, *4*, 90–95. [CrossRef]
62. Alkotaini, B.; Abdellaoui, S.; Hasan, K.; Grattieri, M.; Quah, T.; Cai, R.; Yuan, M.; Minteer, S.D. Sustainable bioelectrosynthesis of the bioplastic polyhydroxybutyrate: Overcoming substrate requirement for NADH regeneration. *ACS Sustain. Chem. Eng.* **2018**, *6*, 4909–4915. [CrossRef]
63. Szczesny, J.; Ruff, A.; Oliveira, A.R.; Pita, M.; Pereira, I.A.; De Lacey, A.L.; Schuhmann, W. Electroenzymatic CO_2 Fixation Using Redox Polymer/Enzyme-Modified Gas Diffusion Electrodes. *ACS Energy Lett.* **2020**, *5*, 321–327. [CrossRef]
64. Falk, M.; Blum, Z.; Shleev, S. Direct electron transfer based enzymatic fuel cells. *Electrochim. Acta* **2012**, *82*, 191–202. [CrossRef]
65. Newman, J.D.; Turner, A.P. Home blood glucose biosensors: A commercial perspective. *Biosens. Bioelectron.* **2005**, *20*, 2435–2453. [CrossRef] [PubMed]
66. Allen, J.B.; Larry, R.F. *Electrochemical Methods Fundamentals and Applications*; John Wiley & Sons: New York, NY, USA, 2001.
67. Adachi, T.; Kitazumi, Y.; Shirai, O.; Kano, K. Direct Electron Transfer-Type Bioelectrocatalysis of Redox Enzymes at Nanostructured Electrodes. *Catalysts* **2020**, *10*, 236. [CrossRef]
68. Shleev, S.; Tkac, J.; Christenson, A.; Ruzgas, T.; Yaropolov, A.I.; Whittaker, J.W.; Gorton, L. Direct electron transfer between copper-containing proteins and electrodes. *Biosens. Bioelectron.* **2005**, *20*, 2517–2554. [CrossRef]
69. Salaj-Kosla, U.; Pöller, S.; Schuhmann, W.; Shleev, S.; Magner, E. Direct electron transfer of Trametes hirsuta laccase adsorbed at unmodified nanoporous gold electrodes. *Bioelectrochemistry* **2013**, *91*, 15–20. [CrossRef]
70. Lee, Y.S.; Baek, S.; Lee, H.; Reginald, S.S.; Kim, Y.; Kang, H.; Choi, I.-G.; Chang, I.S. Construction of Uniform Monolayer-and Orientation-Tunable Enzyme Electrode by a Synthetic Glucose Dehydrogenase without Electron-Transfer Subunit via Optimized Site-Specific Gold-Binding Peptide Capable of Direct Electron Transfer. *ACS Appl. Mater. Interfaces* **2018**, *10*, 28615–28626. [CrossRef] [PubMed]
71. Feng, J.-J.; Zhao, G.; Xu, J.-J.; Chen, H.-Y. Direct electrochemistry and electrocatalysis of heme proteins immobilized on gold nanoparticles stabilized by chitosan. *Anal. Biochem.* **2005**, *342*, 280–286. [CrossRef]
72. Kuk, S.K.; Gopinath, K.; Singh, R.K.; Kim, T.-D.; Lee, Y.; Choi, W.S.; Lee, J.-K.; Park, C.B. NADH-Free Electroenzymatic Reduction of CO_2 by Conductive Hydrogel-Conjugated Formate Dehydrogenase. *ACS Catal.* **2019**, *9*, 5584–5589. [CrossRef]
73. Hickey, D.P.; Lim, K.; Cai, R.; Patterson, A.R.; Yuan, M.; Sahin, S.; Abdellaoui, S.; Minteer, S.D. Pyrene hydrogel for promoting direct bioelectrochemistry: ATP-independent electroenzymatic reduction of N 2. *Chem. Sci.* **2018**, *9*, 5172–5177. [CrossRef]
74. Bollella, P.; Gorton, L.; Antiochia, R. Direct electron transfer of dehydrogenases for development of 3rd generation biosensors and enzymatic fuel cells. *Sensors* **2018**, *18*, 1319. [CrossRef] [PubMed]
75. Hanefeld, U.; Gardossi, L.; Magner, E. Understanding enzyme immobilisation. *Chem. Soc. Rev.* **2009**, *38*, 453–468. [CrossRef] [PubMed]
76. Sakai, K.; Kitazumi, Y.; Shirai, O.; Takagi, K.; Kano, K. Direct electron transfer-type four-way bioelectrocatalysis of CO_2/formate and NAD+/NADH redox couples by tungsten-containing formate dehydrogenase adsorbed on gold nanoparticle-embedded mesoporous carbon electrodes modified with 4-mercaptopyridine. *Electrochem. Commun.* **2017**, *84*, 75–79. [CrossRef]
77. Yates, N.D.; Fascione, M.A.; Parkin, A. Methodologies for "wiring" redox proteins/enzymes to electrode surfaces. *Chem. Eur. J.* **2018**, *24*, 12164–12182. [CrossRef] [PubMed]

78. Siepenkoetter, T.; Salaj-Kosla, U.; Xiao, X.; Conghaile, P.Ó.; Pita, M.; Ludwig, R.; Magner, E. Immobilization of redox enzymes on nanoporous gold electrodes: Applications in biofuel cells. *ChemPlusChem* **2017**, *82*, 553–560. [CrossRef]
79. Besic, S.; Minteer, S.D. Micellar Polymer Encapsulation of Enzymes. In *Enzyme Stabilization and Immobilization*; Springer: Cham, Swtzerland, 2017; pp. 93–108.
80. Xiao, X.; Xia, H.-Q.; Wu, R.; Bai, L.; Yan, L.; Magner, E.; Cosnier, S.; Lojou, E.; Zhu, Z.; Liu, A. Tackling the challenges of enzymatic (bio) fuel cells. *Chem. Rev.* **2019**, *119*, 9509–9558. [CrossRef]
81. Leech, D.; Kavanagh, P.; Schuhmann, W. Enzymatic fuel cells: Recent progress. *Electrochim. Acta* **2012**, *84*, 223–234. [CrossRef]
82. Milton, R.D.; Hickey, D.P.; Abdellaoui, S.; Lim, K.; Wu, F.; Tan, B.; Minteer, S.D. Rational design of quinones for high power density biofuel cells. *Chem. Sci.* **2015**, *6*, 4867–4875. [CrossRef]
83. Prévoteau, A.; Mano, N. Oxygen reduction on redox mediators may affect glucose biosensors based on "wired" enzymes. *Electrochim. Acta* **2012**, *68*, 128–133. [CrossRef]
84. Pankratov, D.; Conzuelo, F.; Pinyou, P.; Alsaoub, S.; Schuhmann, W.; Shleev, S. A Nernstian biosupercapacitor. *Angew. Chem. Int. Ed.* **2016**, *55*, 15434–15438. [CrossRef]
85. Ruff, A.; Szczesny, J.; Zacarias, S.N.; Pereira, I.S.A.; Plumeré, N.; Schuhmann, W. Protection and reactivation of the [NiFeSe] hydrogenase from Desulfovibrio vulgaris Hildenborough under oxidative conditions. *ACS Energy Lett.* **2017**, *2*, 964–968. [CrossRef] [PubMed]
86. Alsaoub, S.; Ruff, A.; Conzuelo, F.; Ventosa, E.; Ludwig, R.; Shleev, S.; Schuhmann, W. An Intrinsic Self-Charging Biosupercapacitor Comprised of a High-Potential Bioanode and a Low-Potential Biocathode. *ChemPlusChem* **2017**, *82*, 576–583. [CrossRef] [PubMed]
87. Spahn, C.; Minteer, S.D. Enzyme immobilization in biotechnology. *Recent Pat. Eng.* **2008**, *2*, 195–200. [CrossRef]
88. Jankowska, K.; Bachosz, K.; Zdarta, J.; Jesionowski, T. Application of Enzymatic-Based Bioreactors. In *Seminar on Practical Aspects of Chemical Engineering*; Springer: Cham, Switzerland, 2019; pp. 110–121.
89. Tamborini, L.; Fernandes, P.; Paradisi, F.; Molinari, F. Flow bioreactors as complementary tools for biocatalytic process intensification. *Trends Biotechnol.* **2018**, *36*, 73–88. [CrossRef]
90. Märkle, W.; Lütz, S. Electroenzymatic strategies for deracemization, stereoinversion and asymmetric synthesis of amino acids. *Electrochim. Acta* **2008**, *53*, 3175–3180. [CrossRef]
91. Mazurenko, I.; Ghach, W.; Kohring, G.-W.; Despas, C.; Walcarius, A.; Etienne, M. Immobilization of membrane-bounded (S)-mandelate dehydrogenase in sol–gel matrix for electroenzymatic synthesis. *Bioelectrochemistry* **2015**, *104*, 65–70. [CrossRef]
92. Ali, I.; Gill, A.; Omanovic, S. Direct electrochemical regeneration of the enzymatic cofactor 1, 4-NADH employing nano-patterned glassy carbon/Pt and glassy carbon/Ni electrodes. *Chem. Eng. J.* **2012**, *188*, 173–180. [CrossRef]
93. Zhu, Y.; Chen, Q.; Shao, L.; Jia, Y.; Zhang, X. Microfluidic immobilized enzyme reactors for continuous biocatalysis. *React. Chem. Eng.* **2020**, *5*, 9–32. [CrossRef]
94. Kundu, S.; Bhangale, A.S.; Wallace, W.E.; Flynn, K.M.; Guttman, C.M.; Gross, R.A.; Beers, K.L. Continuous flow enzyme-catalyzed polymerization in a microreactor. *J. Am. Chem. Soc.* **2011**, *133*, 6006–6011. [CrossRef]
95. Aguillón, A.R.; Avelar, M.N.; Gotardo, L.E.; de Souza, S.P.; Leão, R.A.; Itabaiana Jr, I.; Miranda, L.S.; de Souza, R.O. Immobilized lipase screening towards continuous-flow kinetic resolution of (±)-1, 2-propanediol. *Mol. Catal.* **2019**, *467*, 128–134. [CrossRef]
96. Cosgrove, S.C.; Mattey, A.P.; Riese, M.; Chapman, M.R.; Birmingham, W.R.; Blacker, A.J.; Kapur, N.; Turner, N.J.; Flitsch, S.L. Biocatalytic oxidation in continuous flow for the generation of carbohydrate dialdehydes. *ACS Catal.* **2019**, *9*, 11658–11662. [CrossRef]
97. Yue, J. Multiphase flow processing in microreactors combined with heterogeneous catalysis for efficient and sustainable chemical synthesis. *Catal. Today* **2018**, *308*, 3–19. [CrossRef]
98. Zhang, P.; Russell, M.G.; Jamison, T.F. Continuous flow total synthesis of rufinamide. *Org. Process Res. Dev.* **2014**, *18*, 1567–1570. [CrossRef]
99. Cantillo, D.; Kappe, C.O. Halogenation of organic compounds using continuous flow and microreactor technology. *React. Chem. Eng.* **2017**, *2*, 7–19. [CrossRef]

100. Jo, C.; Groombridge, A.S.; De La Verpilliere, J.; Lee, J.T.; Son, Y.; Liang, H.-L.; Boies, A.M.; De Volder, M. Continuous Flow Synthesis of Carbon Coated Silicon/Iron Silicide Secondary Particles for Li-Ion Batteries. *ACS Nano* **2019**, *14*, 698–707. [CrossRef]
101. Rubio-Martinez, M.; Hadley, T.D.; Batten, M.P.; Constanti-Carey, K.; Barton, T.; Marley, D.; Mönch, A.; Lim, K.S.; Hill, M.R. Scalability of Continuous Flow Production of Metal–Organic Frameworks. *ChemSusChem* **2016**, *9*, 938–941. [CrossRef]
102. Pastre, J.C.; Browne, D.L.; Ley, S.V. Flow chemistry syntheses of natural products. *Chem. Soc. Rev.* **2013**, *42*, 8849–8869. [CrossRef]
103. Sheldon, R.A.; van Pelt, S. Enzyme immobilisation in biocatalysis: Why, what and how. *Chem. Soc. Rev.* **2013**, *42*, 6223–6235. [CrossRef]
104. Valikhani, D.; Bolivar, J.M.; Nidetzky, B. Enzyme Immobilization in Wall-Coated Flow Microreactors. In *Immobilization of Enzymes and Cells*; Springer: Cham, Switzerland, 2020; pp. 243–257.
105. Thompson, M.P.; Peñafiel, I.; Cosgrove, S.C.; Turner, N.J. Biocatalysis using immobilized enzymes in continuous flow for the synthesis of fine chemicals. *Org. Process Res. Dev.* **2018**, *23*, 9–18. [CrossRef]
106. Bolivar, J.M.; Nidetzky, B. Smart enzyme immobilization in microstructured reactors. *Chim Oggi* **2013**, *31*, 50–54.
107. Xiao, X.; Siepenkoetter, T.; Whelan, R.; Salaj-Kosla, U.; Magner, E. A continuous fluidic bioreactor utilising electrodeposited silica for lipase immobilisation onto nanoporous gold. *J. Electroanal. Chem.* **2018**, *812*, 180–185. [CrossRef]
108. Svec, F. Less common applications of monoliths: I. Microscale protein mapping with proteolytic enzymes immobilized on monolithic supports. *Electrophoresis* **2006**, *27*, 947–961. [CrossRef] [PubMed]
109. Wohlgemuth, R.; Plazl, I.; Žnidaršič-Plazl, P.; Gernaey, K.V.; Woodley, J.M. Microscale technology and biocatalytic processes: Opportunities and challenges for synthesis. *Trends Biotechnol.* **2015**, *33*, 302–314. [CrossRef] [PubMed]
110. Szymańska, K.; Pietrowska, M.; Kocurek, J.; Maresz, K.; Koreniuk, A.; Mrowiec-Białoń, J.; Widłak, P.; Magner, E.; Jarzębski, A. Low back-pressure hierarchically structured multichannel microfluidic bioreactors for rapid protein digestion–Proof of concept. *Chem. Eng. J.* **2016**, *287*, 148–154. [CrossRef]
111. Hong, J.; Tsao, G.; Wankat, P. Membrane reactor for enzymatic hydrolysis of cellobiose. *Biotechnol. Bioeng.* **1981**, *23*, 1501–1516. [CrossRef]
112. Rios, G.; Belleville, M.; Paolucci, D.; Sanchez, J. Progress in enzymatic membrane reactors—A review. *J. Membr. Sci.* **2004**, *242*, 189–196. [CrossRef]
113. Lopez, C.; Mielgo, I.; Moreira, M.; Feijoo, G.; Lema, J. Enzymatic membrane reactors for biodegradation of recalcitrant compounds. Application to dye decolourisation. *J. Biotechnol.* **2002**, *99*, 249–257. [CrossRef]
114. Prazeres, D.; Cabral, J. Enzymatic membrane bioreactors and their applications. *Enzyme Microb. Technol.* **1994**, *16*, 738–750. [CrossRef]
115. Tanimu, A.; Jaenicke, S.; Alhooshani, K. Heterogeneous catalysis in continuous flow microreactors: A review of methods and applications. *Chem. Eng. J.* **2017**, *327*, 792–821. [CrossRef]
116. Dall'Oglio, F.; Contente, M.L.; Conti, P.; Molinari, F.; Monfredi, D.; Pinto, A.; Romano, D.; Ubiali, D.; Tamborini, L.; Serra, I. Flow-based stereoselective reduction of ketones using an immobilized ketoreductase/glucose dehydrogenase mixed bed system. *Catal. Commun.* **2017**, *93*, 29–32. [CrossRef]
117. Döbber, J.; Pohl, M.; Ley, S.; Musio, B. Rapid, selective and stable HaloTag-Lb ADH immobilization directly from crude cell extract for the continuous biocatalytic production of chiral alcohols and epoxides. *Reaction Chem. Eng.* **2018**, *3*, 8–12. [CrossRef]
118. Döbber, J.; Gerlach, T.; Offermann, H.; Rother, D.; Pohl, M. Closing the gap for efficient immobilization of biocatalysts in continuous processes: HaloTag™ fusion enzymes for a continuous enzymatic cascade towards a vicinal chiral diol. *Green Chem.* **2018**, *20*, 544–552. [CrossRef]
119. Peschke, T.; Bitterwolf, P.; Rabe, K.S.; Niemeyer, C.M. Self-Immobilizing Oxidoreductases for Flow Biocatalysis in Miniaturized Packed-Bed Reactors. *Chem. Eng. Technol.* **2019**, *42*, 2009–2017. [CrossRef]
120. Bolivar, J.M.; Wiesbauer, J.; Nidetzky, B. Biotransformations in microstructured reactors: More than flowing with the stream? *Trends Biotechnol.* **2011**, *29*, 333–342. [CrossRef]
121. Russell, M.G.; Veryser, C.; Hunter, J.F.; Beingessner, R.L.; Jamison, T.F. Monolithic Silica Support for Immobilized Catalysis in Continuous Flow. *Adv. Synth. Catal.* **2020**, *362*, 314–319. [CrossRef]

122. Liu, X.; Zhu, X.; Camara, M.A.; Qu, Q.; Shan, Y.; Yang, L. Surface modification with highly-homogeneous porous silica layer for enzyme immobilization in capillary enzyme microreactors. *Talanta* **2019**, *197*, 539–547. [CrossRef]
123. Zhao, X.; Fan, P.-R.; Mo, C.-E.; Huang, Y.-P.; Liu, Z.-S. Green synthesis of monolithic enzyme microreactor based on thiol-ene click reaction for enzymatic hydrolysis of protein. *J. Chromatogr. A* **2020**, *1611*, 460618. [CrossRef]
124. Meller, K.; Pomastowski, P.; Szumski, M.; Buszewski, B. Preparation of an improved hydrophilic monolith to make trypsin-immobilized microreactors. *J. Chromatogr. B* **2017**, *1043*, 128–137. [CrossRef] [PubMed]
125. Van den Biggelaar, L.; Soumillion, P.; Debecker, D.P. Enantioselective transamination in continuous flow mode with transaminase immobilized in a macrocellular silica monolith. *Catalysts* **2017**, *7*, 54. [CrossRef]
126. Szymańska, K.; Odrozek, K.; Zniszczoł, A.; Torrelo, G.; Resch, V.; Hanefeld, U.; Jarzębski, A.B. MsAcT in siliceous monolithic microreactors enables quantitative ester synthesis in water. *Catal. Sci. Technol.* **2016**, *6*, 4882–4888. [CrossRef]
127. Sandig, B.; Buchmeiser, M.R. Highly productive and enantioselective enzyme catalysis under continuous supported liquid–liquid conditions using a hybrid monolithic bioreactor. *ChemSusChem* **2016**, *9*, 2917–2921. [CrossRef] [PubMed]
128. Logan, T.C.; Clark, D.S.; Stachowiak, T.B.; Svec, F.; Frechet, J.M. Photopatterning enzymes on polymer monoliths in microfluidic devices for steady-state kinetic analysis and spatially separated multi-enzyme reactions. *Anal. Chem.* **2007**, *79*, 6592–6598. [CrossRef] [PubMed]
129. Qiao, J.; Qi, L.; Mu, X.; Chen, Y. Monolith and coating enzymatic microreactors of L-asparaginase: Kinetics study by MCE–LIF for potential application in acute lymphoblastic leukemia (ALL) treatment. *Analyst* **2011**, *136*, 2077–2083. [CrossRef]
130. Irfan, M.; Glasnov, T.N.; Kappe, C.O. Heterogeneous catalytic hydrogenation reactions in continuous-flow reactors. *ChemSusChem* **2011**, *4*, 300–316. [CrossRef] [PubMed]
131. Schilke, K.F.; Wilson, K.L.; Cantrell, T.; Corti, G.; McIlroy, D.N.; Kelly, C. A novel enzymatic microreactor with Aspergillus oryzae β-galactosidase immobilized on silicon dioxide nanosprings. *Biotechnol. Progr.* **2010**, *26*, 1597–1605. [CrossRef]
132. Li, X.; Yin, Z.; Cui, X.; Yang, L. Capillary electrophoresis-integrated immobilized enzyme microreactor with graphene oxide as support: Immobilization of negatively charged l-lactate dehydrogenase via hydrophobic interactions. *Electrophoresis* **2020**, *41*, 175–182. [CrossRef] [PubMed]
133. Feng, H.; Zhang, B.; Zhu, X.; Chen, R.; Liao, Q.; Ye, D.-d.; Liu, J.; Liu, M.; Chen, G. Multilayered Pd nanocatalysts with nano-bulge structure in a microreactor for multiphase catalytic reaction. *Chem. Eng. Res. Des.* **2018**, *138*, 190–199. [CrossRef]
134. Liu, J.; Zhu, X.; Liao, Q.; Chen, R.; Ye, D.; Feng, H.; Liu, M.; Chen, G. Layer-by-layer self-assembly of palladium nanocatalysts with polyelectrolytes grafted on the polydopamine functionalized gas-liquid-solid microreactor. *Chem. Eng. J.* **2018**, *332*, 174–182. [CrossRef]
135. Munirathinam, R.; Huskens, J.; Verboom, W. Supported catalysis in continuous-flow microreactors. *Adv. Synth. Catal.* **2015**, *357*, 1093–1123. [CrossRef]
136. Bi, Y.; Zhou, H.; Jia, H.; Wei, P. A flow-through enzymatic microreactor immobilizing lipase based on layer-by-layer method for biosynthetic process: Catalyzing the transesterification of soybean oil for fatty acid methyl ester production. *Process Biochem.* **2017**, *54*, 73–80. [CrossRef]
137. Valikhani, D.; Bolivar, J.M.; Viefhues, M.; McIlroy, D.N.; Vrouwe, E.X.; Nidetzky, B. A spring in performance: Silica nanosprings boost enzyme immobilization in microfluidic channels. *ACS Appl. Mater. Interfaces* **2017**, *9*, 34641–34649. [CrossRef] [PubMed]
138. Bi, Y.; Zhou, H.; Jia, H.; Wei, P. Polydopamine-mediated preparation of an enzyme-immobilized microreactor for the rapid production of wax ester. *RSC Adv.* **2017**, *7*, 12283–12291. [CrossRef]
139. Britton, J.; Dyer, R.P.; Majumdar, S.; Raston, C.L.; Weiss, G.A. Ten-Minute Protein Purification and Surface Tethering for Continuous-Flow Biocatalysis. *Angew. Chem.* **2017**, *129*, 2336–2341. [CrossRef]
140. Valikhani, D.; Bolivar, J.M.; Pfeiffer, M.; Nidetzky, B. Multivalency effects on the immobilization of sucrose phosphorylase in flow microchannels and their use in the development of a high-performance biocatalytic microreactor. *ChemCatChem* **2017**, *9*, 161–166. [CrossRef]
141. Dixon, M.; Webb, E. Enzyme techniques. In *Enzymes*, 3rd ed.; Academic Press: New York, NY, USA, 1979; pp. 7–11.

142. Yoon, S.K.; Choban, E.R.; Kane, C.; Tzedakis, T.; Kenis, P.J. Laminar flow-based electrochemical microreactor for efficient regeneration of nicotinamide cofactors for biocatalysis. *J. Am. Chem. Soc.* **2005**, *127*, 10466–10467. [CrossRef]
143. Cheikhou, K.; Tzédakis, T. Electrochemical microreactor for chiral syntheses using the cofactor NADH. *AIChE J.* **2008**, *54*, 1365–1376. [CrossRef]
144. Ruinatscha, R.; Buehler, K.; Schmid, A. Development of a high performance electrochemical cofactor regeneration module and its application to the continuous reduction of FAD. *J. Mol. Catal. B Enzym.* **2014**, *103*, 100–105. [CrossRef]
145. Rodríguez-Hinestroza, R.A.; López, C.; López-Santín, J.; Kane, C.; Benaiges, M.D.; Tzedakis, T. HLADH-catalyzed synthesis of β-amino acids, assisted by continuous electrochemical regeneration of NAD+ in a filter press microreactor. *Chem. Eng. Sci.* **2017**, *158*, 196–207. [CrossRef]
146. Fisher, K.; Mohr, S.; Mansell, D.; Goddard, N.J.; Fielden, P.R.; Scrutton, N.S. Electro-enzymatic viologen-mediated substrate reduction using pentaerythritol tetranitrate reductase and a parallel, segmented fluid flow system. *Catal. Sci. Technol.* **2013**, *3*, 1505–1511. [CrossRef]
147. Mano, N. Recent advances in high surface area electrodes for bioelectrochemical applications. *Curr. Opin. Electrochem.* **2020**, *19*, 8–13. [CrossRef]
148. Xiao, X.; Si, P.; Magner, E. An overview of dealloyed nanoporous gold in bioelectrochemistry. *Bioelectrochemistry* **2016**, *109*, 117–126. [CrossRef] [PubMed]
149. Salaj-Kosla, U.; Scanlon, M.D.; Baumeister, T.; Zahma, K.; Ludwig, R.; Conghaile, P.Ó.; MacAodha, D.; Leech, D.; Magner, E. Mediated electron transfer of cellobiose dehydrogenase and glucose oxidase at osmium polymer-modified nanoporous gold electrodes. *Anal. Bioanal. Chem.* **2013**, *405*, 3823–3830. [CrossRef] [PubMed]
150. Siepenkoetter, T.; Salaj-Kosla, U.; Magner, E. The immobilization of fructose dehydrogenase on nanoporous gold electrodes for the detection of fructose. *ChemElectroChem* **2017**, *4*, 905–912. [CrossRef]
151. Xiao, X.; Siepenkoetter, T.; Conghaile, P.O.; Leech, D.n.; Magner, E. Nanoporous gold-based biofuel cells on contact lenses. *ACS Appl. Mater. Interfaces* **2018**, *10*, 7107–7116. [CrossRef] [PubMed]
152. Bhattarai, J.K.; Neupane, D.; Nepal, B.; Mikhaylov, V.; Demchenko, A.V.; Stine, K.J. Preparation, modification, characterization, and biosensing application of nanoporous gold using electrochemical techniques. *Nanomaterials* **2018**, *8*, 171. [CrossRef]
153. Wang, X.; Falk, M.; Ortiz, R.; Matsumura, H.; Bobacka, J.; Ludwig, R.; Bergelin, M.; Gorton, L.; Shleev, S. Mediatorless sugar/oxygen enzymatic fuel cells based on gold nanoparticle-modified electrodes. *Biosens. Bioelectron.* **2012**, *31*, 219–225. [CrossRef]
154. Kizling, M.; Dzwonek, M.; Nowak, A.; Tymecki, Ł.; Stolarczyk, K.; Więckowska, A.; Bilewicz, R. Multi-Substrate Biofuel Cell Utilizing Glucose, Fructose and Sucrose as the Anode Fuels. *Nanomaterials* **2020**, *10*, 1534. [CrossRef]
155. Kumar, S.A.; Chen, S.-M. Electroanalysis of NADH using conducting and redox active polymer/carbon nanotubes modified electrodes—A review. *Sensors* **2008**, *8*, 739–766. [CrossRef]
156. Gooding, J.J. Nanostructuring electrodes with carbon nanotubes: A review on electrochemistry and applications for sensing. *Electrochim. Acta* **2005**, *50*, 3049–3060. [CrossRef]
157. Wang, J. Carbon-nanotube based electrochemical biosensors: A review. *Electroanal. Int. J. Devoted Fundam. Pract. Asp. Electroanal.* **2005**, *17*, 7–14. [CrossRef]
158. Zhu, Z.; Garcia-Gancedo, L.; Flewitt, A.J.; Xie, H.; Moussy, F.; Milne, W.I. A critical review of glucose biosensors based on carbon nanomaterials: Carbon nanotubes and graphene. *Sensors* **2012**, *12*, 5996–6022. [CrossRef] [PubMed]
159. Bollella, P.; Hibino, Y.; Kano, K.; Gorton, L.; Antiochia, R. Enhanced direct electron transfer of fructose dehydrogenase rationally immobilized on a 2-aminoanthracene diazonium cation grafted single-walled carbon nanotube based electrode. *ACS Catal.* **2018**, *8*, 10279–10289. [CrossRef]
160. Feizabadi, M.; Ajloo, D.; Soleymanpour, A.; Faridnouri, H. Study of electron transport in the functionalized nanotubes and their impact on the electron transfer in the active site of horseradish peroxidase. *J. Phys. Chem. Solids* **2018**, *116*, 313–323. [CrossRef]
161. Gross, A.J.; Chen, X.; Giroud, F.; Abreu, C.; Le Goff, A.; Holzinger, M.; Cosnier, S. A high power buckypaper biofuel cell: Exploiting 1, 10-phenanthroline-5, 6-dione with FAD-dependent dehydrogenase for catalytically-powerful glucose oxidation. *ACS Catal.* **2017**, *7*, 4408–4416. [CrossRef]

162. Sim, H.J.; Lee, D.Y.; Kim, H.; Choi, Y.-B.; Kim, H.-H.; Baughman, R.H.; Kim, S.J. Stretchable fiber biofuel cell by rewrapping multiwalled carbon nanotube sheets. *Nano Lett.* **2018**, *18*, 5272–5278. [CrossRef] [PubMed]
163. Jourdin, L.; Freguia, S.; Donose, B.C.; Chen, J.; Wallace, G.G.; Keller, J.; Flexer, V. A novel carbon nanotube modified scaffold as an efficient biocathode material for improved microbial electrosynthesis. *J. Mater. Chem. A* **2014**, *2*, 13093–13102. [CrossRef]
164. Barin, R.; Biria, D.; Rashid-Nadimi, S.; Asadollahi, M.A. Enzymatic CO_2 reduction to formate by formate dehydrogenase from Candida boidinii coupling with direct electrochemical regeneration of NADH. *J. CO_2 Util.* **2018**, *28*, 117–125. [CrossRef]
165. Bulutoglu, B.; Macazo, F.C.; Bale, J.; King, N.; Baker, D.; Minteer, S.D.; Banta, S. Multimerization of an Alcohol Dehydrogenase by Fusion to a Designed Self-Assembling Protein Results in Enhanced Bioelectrocatalytic Operational Stability. *ACS Appl. Mater. Interfaces* **2019**, *11*, 20022–20028. [CrossRef]
166. Gross, A.; Holzinger, M.; Cosnier, S. Buckypaper bioelectrodes: Emerging materials for implantable and wearable biofuel cells. *Energy Environ. Sci.* **2018**, *11*, 1670–1687. [CrossRef]
167. Choi, D.S.; Lee, H.; Tieves, F.; Lee, Y.W.; Son, E.J.; Zhang, W.; Shin, B.; Hollmann, F.; Park, C.B. Bias-Free In Situ H2O2 Generation in a Photovoltaic-Photoelectrochemical Tandem Cell for Biocatalytic Oxyfunctionalization. *ACS Catal.* **2019**, *9*, 10562–10566. [CrossRef]
168. Seelajaroen, H.; Bakandritsos, A.; Otyepka, M.; Zbořil, R.; Sariciftci, N.S. Immobilized Enzymes on Graphene as Nanobiocatalyst. *ACS Appl. Mater. Interfaces* **2019**, *12*, 250–259. [CrossRef]
169. Chen, K.; Arnold, F.H. Engineering new catalytic activities in enzymes. *Nat. Catal.* **2020**, *3*, 1–11. [CrossRef]
170. Wong, T.S.; Schwaneberg, U. Protein engineering in bioelectrocatalysis. *Curr. Opin. Biotechnol.* **2003**, *14*, 590–596. [CrossRef]
171. Yin, L.L.; Yuan, H.; Liu, C.; He, B.; Gao, S.-Q.; Wen, G.-B.; Tan, X.; Lin, Y.-W. A rationally designed myoglobin exhibits a catalytic dehalogenation efficiency more than 1000-fold that of a native dehaloperoxidase. *ACS Catal.* **2018**, *8*, 9619–9624. [CrossRef]
172. Chen, K.; Arnold, F.H. Engineering cytochrome P450s for enantioselective cyclopropenation of internal alkynes. *J. Am. Chem. Soc.* **2020**, *142*, 6891–6895. [CrossRef]
173. Brandenberg, O.F.; Miller, D.C.; Markel, U.; Ouald Chaib, A.; Arnold, F.H. Engineering Chemoselectivity in Hemoprotein-Catalyzed Indole Amidation. *ACS Catal.* **2019**, *9*, 8271–8275. [CrossRef] [PubMed]
174. Mateljak, I.; Monza, E.; Lucas, M.F.; Guallar, V.; Aleksejeva, O.; Ludwig, R.; Leech, D.; Shleev, S.; Alcalde, M. Increasing redox potential, redox mediator activity, and stability in a fungal laccase by computer-guided mutagenesis and directed evolution. *ACS Catal.* **2019**, *9*, 4561–4572. [CrossRef]
175. Aleksejeva, O.; Mateljak, I.; Ludwig, R.; Alcalde, M.; Shleev, S. Electrochemistry of a high redox potential laccase obtained by computer-guided mutagenesis combined with directed evolution. *Electrochem. Commun.* **2019**, *106*, 106511. [CrossRef]
176. Vidal, L.S.; Kelly, C.L.; Mordaka, P.M.; Heap, J.T. Review of NAD (P) H-dependent oxidoreductases: Properties, engineering and application. *Biochim. Biophys. Acta Proteins Proteom.* **2018**, *1866*, 327–347. [CrossRef]
177. Liu, X.; Bastian, S.; Snow, C.D.; Brustad, E.M.; Saleski, T.E.; Xu, J.-H.; Meinhold, P.; Arnold, F.H. Structure-guided engineering of Lactococcus lactis alcohol dehydrogenase LlAdhA for improved conversion of isobutyraldehyde to isobutanol. *J. Biotechnol.* **2013**, *164*, 188–195. [CrossRef]
178. Li, H.; Chen, J.; Li, Y. Enhanced activity of yqhD oxidoreductase in synthesis of 1, 3-propanediol by error-prone PCR. *Prog. Natl. Sci.* **2008**, *18*, 1519–1524. [CrossRef]
179. Guterl, J.K.; Sieber, V. Biosynthesis "debugged": Novel bioproduction strategies. *Eng. Life Sci.* **2013**, *13*, 4–18. [CrossRef]
180. Schrittwieser, J.H.; Velikogne, S.; Hall, M.l.; Kroutil, W. Artificial biocatalytic linear cascades for preparation of organic molecules. *Chem. Rev.* **2018**, *118*, 270–348. [CrossRef] [PubMed]
181. Dong, F.; Chen, H.; Malapit, C.A.; Prater, M.B.; Li, M.; Yuan, M.; Lim, K.; Minteer, S.D. Biphasic bioelectrocatalytic synthesis of chiral β-hydroxy nitriles. *J. Am. Chem. Soc.* **2020**, *142*, 8374–8382. [CrossRef]
182. Abdellaoui, S.; Macazo, F.C.; Cai, R.; De Lacey, A.L.; Pita, M.; Minteer, S.D. Enzymatic electrosynthesis of alkanes by bioelectrocatalytic decarbonylation of fatty aldehydes. *Angew. Chem. Int. Ed.* **2018**, *57*, 2404–2408. [CrossRef]

183. Kuk, S.K.; Singh, R.K.; Nam, D.H.; Singh, R.; Lee, J.K.; Park, C.B. Photoelectrochemical reduction of carbon dioxide to methanol through a highly efficient enzyme cascade. *Angew. Chem. Int. Ed.* **2017**, *56*, 3827–3832. [CrossRef]
184. Hickey, D.P.; Gaffney, E.M.; Minteer, S.D. Electrometabolic pathways: Recent developments in bioelectrocatalytic cascades. In *Electrocatalysis*; Springer: Cham, Switzerland, 2020; pp. 149–165.
185. Duca, M.; Weeks, J.R.; Fedor, J.G.; Weiner, J.H.; Vincent, K.A. Combining noble metals and enzymes for relay cascade electrocatalysis of nitrate reduction to ammonia at neutral pH. *ChemElectroChem* **2015**, *2*, 1086–1089. [CrossRef]
186. Chen, Q.; Hu, Y.; Zhao, W.; Zhu, C.; Zhu, B. Cloning, expression, and characterization of a novel (S)-specific alcohol dehydrogenase from Lactobacillus kefir. *Appl. Biochem. Biotechnol.* **2010**, *160*, 19. [CrossRef]
187. Vedha-Peters, K.; Gunawardana, M.; Rozzell, J.D.; Novick, S.J. Creation of a broad-range and highly stereoselective D-amino acid dehydrogenase for the one-step synthesis of D-amino acids. *J. Am. Chem. Soc.* **2006**, *128*, 10923–10929. [CrossRef]
188. Richter, N.; Neumann, M.; Liese, A.; Wohlgemuth, R.; Weckbecker, A.; Eggert, T.; Hummel, W. Characterization of a whole-cell catalyst co-expressing glycerol dehydrogenase and glucose dehydrogenase and its application in the synthesis of l-glyceraldehyde. *Biotechnol. Bioeng.* **2010**, *106*, 541–552. [CrossRef]
189. Wu, R.; Zhu, Z. Self-powered enzymatic electrosynthesis of l-3, 4-Dihydroxyphenylalanine in a hybrid bioelectrochemical system. *ACS Sustain. Chem. Eng.* **2018**, *6*, 12593–12597. [CrossRef]
190. Chen, H.; Prater, M.B.; Cai, R.; Dong, F.; Chen, H.; Minteer, S.D. Bioelectrocatalytic Conversion from N_2 to Chiral Amino Acids in a H_2/α-keto Acid Enzymatic Fuel Cell. *J. Am. Chem. Soc.* **2020**, *142*, 4028–4036. [CrossRef]
191. Cai, R.; Milton, R.D.; Abdellaoui, S.; Park, T.; Patel, J.; Alkotaini, B.; Minteer, S.D. Electroenzymatic C–C bond formation from CO_2. *J. Am. Chem. Soc.* **2018**, *140*, 5041–5044. [CrossRef] [PubMed]
192. Yuan, M.; Sahin, S.; Cai, R.; Abdellaoui, S.; Hickey, D.P.; Minteer, S.D.; Milton, R.D. Creating a Low-Potential Redox Polymer for Efficient Electroenzymatic CO2 Reduction. *Angew. Chem.* **2018**, *130*, 6692–6696. [CrossRef]
193. Schlager, S.; Dumitru, L.M.; Haberbauer, M.; Fuchsbauer, A.; Neugebauer, H.; Hiemetsberger, D.; Wagner, A.; Portenkirchner, E.; Sariciftci, N.S. Electrochemical reduction of carbon dioxide to methanol by direct injection of electrons into immobilized enzymes on a modified electrode. *ChemSusChem* **2016**, *9*, 631–635. [CrossRef]
194. Sciarria, T.P.; Batlle-Vilanova, P.; Colombo, B.; Scaglia, B.; Balaguer, M.; Colprim, J.; Puig, S.; Adani, F. Bio-electrorecycling of carbon dioxide into bioplastics. *Green Chem.* **2018**, *20*, 4058–4066. [CrossRef]
195. Cadoux, C.M.; Milton, R.D. Recent enzymatic electrochemistry for reductive reactions. *ChemElectroChem* **2020**. [CrossRef]
196. Lee, Y.S.; Ruff, A.; Cai, R.; Lim, K.; Schuhmann, W.; Minteer, S.D. Electroenzymatic Nitrogen Fixation Using an Organic Redox Polymer-Immobilized MoFe Protein System. *Angew. Chem.* **2020**. [CrossRef]
197. Milton, R.D.; Abdellaoui, S.; Khadka, N.; Dean, D.R.; Leech, D.; Seefeldt, L.C.; Minteer, S.D. Nitrogenase bioelectrocatalysis: Heterogeneous ammonia and hydrogen production by MoFe protein. *Energy Environ. Sci.* **2016**, *9*, 2550–2554. [CrossRef]
198. Lee, Y.S.; Yuan, M.; Cai, R.; Lim, K.; Minteer, S.D. Nitrogenase Bioelectrocatalysis: ATP-Independent Ammonia Production Using a Redox Polymer/MoFe Protein System. *ACS Catal.* **2020**, *10*, 6854–6861. [CrossRef]

Publisher's Note: MDPI stays neutral with regard to jurisdictional claims in published maps and institutional affiliations.

© 2020 by the authors. Licensee MDPI, Basel, Switzerland. This article is an open access article distributed under the terms and conditions of the Creative Commons Attribution (CC BY) license (http://creativecommons.org/licenses/by/4.0/).

MDPI
St. Alban-Anlage 66
4052 Basel
Switzerland
Tel. +41 61 683 77 34
Fax +41 61 302 89 18
www.mdpi.com

Catalysts Editorial Office
E-mail: catalysts@mdpi.com
www.mdpi.com/journal/catalysts

www.ingramcontent.com/pod-product-compliance
Lightning Source LLC
LaVergne TN
LVHW070643100526
838202LV00013B/866